Joseph P.S. Kung

A Source Book in Matroid Theory

Springer Science+Business Media, LLC

Author's address

Joseph P.S. Kung,
Department of Mathematics,
North Texas State University,
Denton, TX 76203-5116 (USA)

Library of Congress Cataloging in Publication Data

Kung, Joseph P.S.
A source book in matroid theory.

Bibliography: p.
Includes index.
1. Matroids. I. Title.
QA166.6.K86 511'.6 84-24334
ISBN 978-0-8176-3173-4 ISBN 978-1-4684-9199-9 (eBook)

DOI 10.1007/978-1-4684-9199-9

CIP Kurztitelaufnahme der Deutschen Bibliothek

Kung, Joseph P.S.:
A source book in matroid theory / Joseph P.S. Kung.
Boston; Basel; Stuttgart:
Birkhäuser, 1986

Originally published by Birhäuser Boston, Inc. 1986
Softcover reprint of the hardcover 1st edition 1986

Table of Contents

Chapter II: Linear representations of matroids

Commentary

Reprints

Chapter III. Enumeration in geometric lattices

Commentary

Reprints

Chapter IV: The Tutte decomposition

Commentary

Reprints

Chapter V: Recent advances

Commentary

Reprints

Foreword

by Gian-Carlo Rota

The subjects of mathematics, like the subjects of mankind, have finite lifespans, which the historian will record as he freezes history at one instant of time.

There are the old subjects, loaded with distinctions and honors. As their problems are solved away and the applications reaped by engineers and other moneymen, ponderous treatises gather dust in library basements, awaiting the day when a generation as yet unborn will rediscover the lost paradise in awe. Then there are the middle-aged subjects. You can tell which they are by roaming the halls of Ivy League universities or the Institute for Advanced Studies. Their high priests haughtily refuse fabulous offers from eager provincial universities while receiving special permission from the President of France to lecture in English at the Collège de France. Little do they know that the load of technicalities is already critical, about to crack and submerge their theorems in the dust of oblivion that once enveloped the dinosaurs.

Finally, there are the young subjects-combinatorics, for instance. Wild-eyed individuals gingerly pick from a mountain of intractable problems, childishly babbling the first words of what will soon be a new language. Childhood will end with the first *Séminaire Bourbaki.*

It could be impossible to find a more fitting example than matroid theory of a subject now in its infancy. The telltale signs, for an unfailing diagnosis, are the abundance of deep theorems, going together with a paucity of theories.

Like many another great idea, matroid theory was invented by one of the great American pioneers, Hassler Whitney. His paper, which is still today the best entry to the subject, flagrantly reveals the unique peculiarity of this field, namely, the exceptional variety of cryptomorphic definitions for a matroid, embarassingly unrelated to each other and exhibiting wholly different mathematical pedigrees. It is as if one were to condense all trends of present day mathematics onto a single finite structure, a feat that anyone would *a priori* deem impossible, were it not for the mere fact that matroids do exist.

The original motivation, both in Whitney and the later papers of Tutte, was graph theoretic. Matroids are objects that play the rôle of the dual graph when the graph is not planar. Almost every fact about graphs that can be formulated without using the term "vertex" has a matroidal analogue. The deepest insight obtained from this yoga is Tutte's homotopy theorem, truly a great combinatorial feat.

But one of the axioms for matroids, as was first recognized by Mac Lane, is Steinitz' axiom for linear independence of vectors. From this follows a second yoga: almost every fact about linear independence that can be stated without reference to the underlying field gives a theorem about matroids. For example, Higgs was able to develop a matroidal analogue of linear transformation.

The suspicion quickly followed that matroids could be represented by points in projective space, a suspicion quickly quashed by Mac Lane. In fact, the condition for representability of a matroid in a given projective space can be stated in terms of the absence of "obstructions" or forbidden configurations, as in the archetypal theorem of Kuratowski on planar graphs. Such theorems are difficult to come by, witness Seymour's theorem for GF(3).

Next, we find the lattice theorists – Birkhoff, Crapo, Rota – whose yoga is to see a matroid as its lattice of flats. Many deep enumerative properties, often associated with homological properties (Folkman), hold for matroids. The Whitney numbers of a matroid (a term introduced by Harper and Rota) exhibit some of the properties of binomial coefficients, as Dowling and Wilson proved, though not all, as the counterexample of Dilworth and Greene showed.

As if three yogas were not enough, there followed the Tutte-Grothendieck decomposition theory (developed most energetically by Brylawski), which displayed an astonishing analogy with K-theory. Zaslavsky's solution of Steiner's problem, which gives an explicit formula for the number of regions into which space is subdivided by a set of hyperplanes, is the finest application to date of this yoga.

What next? Kahn and Kung have just come along with a new yoga: a varietal theory of matroids, that brings universal algebra into an already crowded game. Similarly, the intricate decomposition theory of Seymour should lead to yet another yoga.

All these yogas lead to a deep suspicion. Anyone who has worked with matroids has come away with the conviction that the notion of a matroid is one of the richest and most useful concepts of our day. Yet, we long, as we always do, for *one* idea that will allow us to see through the plethora of disparate points of view. Whether this idea will ever come along will depend largely on who reads the essays collected in this fine volume.

Introduction

In every area of mathematics, there are certain crucial papers which anyone working in the area should be acquainted with. In this anthology, we have included a substantial portion of these papers in matroid theory. Each paper is provided with an extensive commentary. The first aim of the commentaries is to sketch the background and to explain the notation and terminology whenever they deviate from current usage. (This is particularly necessary for the older papers.) The second aim is to trace the influence of the paper in the subsequent development of matroid theory. A bibliography of related papers is included and the commentaries, as a whole, contain an annotated bibliography of those branches of matroid theory covered in this anthology. Thus, the commentaries should enable an uninitiated reader to get quickly into the heart of matroid theory – without getting bogged down in a morass of cryptomorphisms – by reading some of the seminal ideas in *urtext*.

An anthology, especially of such a flourishing subject as matroid theory, is by its very nature bound to be incomplete. The ideal choices of the editor are often overruled by the dictates of the real world and several papers had to be omitted for rather banal reasons. We had to forgo a planned section on the algorithmic aspects of matroids for just such reasons. However, we believe that the papers included give a balanced picture of the structural theory of matroids and, perhaps more importantly, indicate some of the most exciting connections of matroid theory with other areas of mathematics.

As a very small step towards compensating for the lack of encyclopedic coverage, we give here a list of books, lecture notes, and expository papers in matroid theory.

A. Books

BIRKHOFF, G.: *Lattice theory* (Third edition), Amer. Math. Soc. Colloq. Publ., Vol. 25, Amer. Math. Soc., Providence, R.I., 1967.

BRUTER, C.P.: *Les matroïdes*, Dunod, Paris, 1970.

BRYANT, V. and PERFECT, H.: *Independence theory in combinatorics*, Chapman and Hall, London and New York, 1980.

CRAPO, H.H. and ROTA, G.-C.: *On the foundations of combinatorial theory: Combinatorial geometries* (Preliminary edition), M.I.T. Press, Cambridge, Mass., 1970.

CRAWLEY, P. and DILWORTH, R. P.: *Algebraic theory of lattices,* Prentice-Hall, Englewood Cliffs, N.J., 1973.

LAWLER, E.L.: *Combinatorial optimization: Networks and matroids,* Holt, Rinehart and Winston, New York and London, 1976.

MAEDA, F. and MAEDA, S.: *Theory of symmetric lattices,* Springer-Verlag, Berlin and New York, 1970.

MIRSKY, L.: *Transversal theory,* Academic Press, New York and London, 1971.

TUTTE, W.T.: *Introduction to the theory of matroids*, American Elsevier, New York, 1971.

WELSH, D.J.A.: *Matroid theory*, Academic Press, London and New York, 1976.

WHITE (ed.), N.L.: *Theory of matroids,* Cambridge Univ. Press, Cambridge, 1986.

B. Lecture notes

BRUTER, C.P.: *Éléments de la théorie des matroïdes*, Lecture Notes in Math., Vol. 387, Springer-Verlag, Berlin and New York, 1974.

BRYLAWSKI, T. and KELLY, D.G.: *Matroids and combinatorial geometries*, Univ. North Carolina, Chapel Hill, N.C., 1980.

CRAPO, H.H.: *Constructions in combinatorial geometry*, Combinatorial Theory Advanced Science Seminar, Bowdoin College, Bowdoin, Maine, 1971.

FOURNIER, J.-C.: *Introduction à la notion de matroïde (Géométrie combinatoire)*, Univ. de Paris-Sud, Département de Math., Orsay, 1979.

GREENE, C.: *Lectures in combinatorial geometries* (Notes by D. Kennedy), Combinatorial Theory Advanced Science Seminar, Bowdoin College, Bowdoin, Maine, 1971.

HIGGS, D. A.: *Geometry*, Univ. Waterloo. Waterloo, Ont., 1967.
von RANDOW. R.: *Introduction to the theory of matroids*, Lecture Notes in Economics and Math. Systems, Vol. 109, Springer-Verlag, Berlin and New York, 1975.

C. Expository and other papers of general interest

BIRKHOFF, G.: Ordered sets in geometry, *Ordered sets* (I. Rival, ed.) pp. 407-443, Reidel, Dordrecht and Boston, 1982.
BLACKBURN, J.E., CRAPO, H. H. and HIGGS, D.A.: A catalogue of combinatorial geometries, *Math. Comp. 27* (1973), 155-166, *addendum, ibid. 27*(1973), loose microfiche suppl. A 12-G 12.
BRYLAWSKI, T. and KELLY, D. G.: Matroids and combinatorial geometries, *Studies in Combinatorics* (G.-C. Rota, ed.), pp. 179-217, Math. Assoc. Amer., Washington, D.C., 1978.
CHEUNG, A.L.C. and CRAPO, H. H.: A combinatorial perspective on algebraic geometry, *Advances in Math. 20*(1976), 388-414.
CRAPO, H.H.: The combinatorial theory of structures, *Matroid Theory and its Applications (Proc. Conf., Szeged, 1982)* (L. Lovász and A. Recski, eds.), North-Holland, Amsterdam, to appear.
GLAZEK, K.: Some old and new problems in the independence theory, *Colloq. Math.42*(1979), 127-189.
IRI, M. and FUJISHIGE, S.: Use of matroid theory in operations research, circuits and system theory, *Internat. J. Systems Sci. 12*(1981), 27-54.
MIRSKY, L. and PERFECT, H.: Applications of the notion of independence to problems of combinatorial analysis, *J. Combin. Theory 2*(1967), 327-357.
PERFECT, H.: Independence theory and matroids, *Math. Gaz. 65*(1981), 103-111.
TUTTE, W.T.: Lectures on matroids, *J. Res. Nat. Bur. Standards, Sect. B 69B*(1965), 1-47.
WEINBERG, L.: Matroids, oriented matroids, and polymatroids: theory and applications, *Circuit theory and design (Proc. Conf., The Hague, 1981)*, pp. 167-175, North-Holland, Amsterdam, 1981.
WELSH, D.J.A.: Combinatorial problems in matroid theory, *Combinatorial Mathematics and its Applications (Proc. Conf., Oxford, 1969)*, pp. 291-306, Academic Press, London, 1971.
WILSON, R.J.: An introduction to matroid theory, *Amer. Math. Monthly 80*(1973), 500-525.

Acknowledgements

We would like to thank the contributors for permitting us to reprint their papers. In addition, we acknowledge the generous response of the following institutions to our request for permission to reprint: the American Mathematical Society, Duke University Press, the Indiana University Mathematical Journal, the Johns Hopkins University Press, Springer-Verlag, and Studies in Applied Mathematics (M.I.T.). Finally, we express our special thanks to Gian-Carlo Rota for his help in getting this project off the ground.

Chapter I
Origins and basic concepts

1. The origins of matroid theory

There are many notions of dependence in algebra. Besides linear dependence of vectors, there are algebraic and p-dependence of elements in a field extension. (For a definition of p-dependence, see the commentary on Mac Lane [I. 4] in §1.4.) One of the historical forces behind the discovery of the concept of a matroid in the thirties was the recognition that these notions of dependence share many common properties, the most striking being the fact that the maximal independent sets all have the same cardinality. It was natural, in a decade when the axiomatic method was still a fresh idea, to attempt to find the fundamental properties of dependence common to these notions, postulate them as axioms, and derive their common properties from the axioms in a purely abstract manner. This was done by many. (See §1.5 for a complete survey; it was an early testimony to the naturalness and inevitability of the concept of a matroid that all these axiomatizations, discovered independently by very different mathematicians, are all equivalent.) However, except for Whitney's work, there was no attempt to go beyond the elementary facts and equivalences. This was perhaps due to the fact that a key example, independence of a set of edges in a graph, and hence a key concept, that of a dual graph, were not available to those approaching matroids from an algebraic point of view. Thus, while future historians of mathematics may debate when the *definition* of a matroid first appeared, there is no doubt that the *theory* of matroids began in Whitney's 1935 paper, "On the abstract properties of linear dependence". This paper is the first paper reprinted in our anthology.

There are three parts and an appendix in this paper. In the first part, four equivalent definitions – or, to use the terminology of Birkhoff [67], Chap. VI, §11, *crypto-isomorphisms* – of a matroid are given as indicated below.

Basic concept/Axioms	*Section*
I. Independent sets:	1 and 6
I_1: Incomparability	
I_2: Augmentation	

R. Rank function: 2

R_1: Normalization

R_2: Unit increase

R_3: Local upper semimodularity

B. Bases: 7

B_1: Incomparability

B_2: (Non-symmetric) basis exchange

C. Circuits: 8

C_1: Incomparability

C_2: (Strong) circuit elimination

(It may be helpful to remark that some of the axioms are referred to in the paper before they are stated.) These four axioms are proved to be equivalent in the following way:

$$C \underset{\S5}{\overset{\S8}{\rightleftarrows}} R \underset{\S6}{\overset{\S4}{\rightleftarrows}} I \overset{\S7}{\longleftrightarrow} B.$$

Here, $\longrightarrow$ means implication. In addition, two important theorems are proved in Part I. The first is *Theorem 3* (§3): if a rank function satisfies (R_2) and (R_3), then it is semimodular (Eqn. 3.3). The second is *Theorem 6* (§7): All bases contain the same number of elements. In §9, the final section, the notion of a fundamental set of circuits is introduced.

The second part is concerned with separability and duality. It may be regarded as sequel to [I. 3]. In §10, separability is defined in terms of the rank function and it is proved that every matroid has a unique decomposition into (connected) components. Next, a method for building up a matroid by starting with a circuit and adding elements inductively is described (Theorem 17). This result and Theorem 19 in [I. 3] are the precursors of similar theorems (see Tutte [66, Chap. 12] and Seymour [V. 1]) on how a 3-connected graph or matroid may be constructed inductively. Finally, components of matroids are characterized in terms of the rank function and circuits.

In §11, duality is defined in terms of the rank function. This definition is shown to be equivalent to one using basis complements. In contrast to graphs (see [I. 3]), the dual of a matroid is again a matroid. This fact is important historically in the emergence of matroid theory as a distinct area of mathematics. It is also proved that the decomposition of a matroid into components is preserved under duality.

The matroid structure of the column vectors of a matrix is studied in Part III. Given an $m \times n$ matrix over a field (here taken to be the reals, but this assumption is not essential), there are two associated structures. The first is the *column matroid,* the matroid with n elements defined by linear dependence of the column vectors. The second is the *row space,* the subspace of m-dimensional space spanned by the row vectors. (We remark that in this section, a hyperplane is simply a subspace, not necessarily of codimension one.) The connection between these structures is given in Theorem 27, which states that a subspace of dimension d determines a unique matroid of rank d. (Note however that different subspaces may define the same matroid.)

The next section, §13, is concerned with duality of column matroids. The main result (Theorem 28) states that if H' is the subspace orthogonal to the subspace H (relative to the usual inner product $<x, y> = \sum x_i y_i$), then the matroid determined by H' is dual to the matroid determined by H. This theorem contains the geometric essence of the relation between primal and dual programs in linear programming (Dantzig [63], Chap. 6), between a polytope and its Gale transform (see, for example, McMullen [79]), and between a linear error-correcting code and its dual code (see [IV. 2]).

Now let $C_1, \ldots, C_n$ be the columns of a matrix $\boldsymbol{M}$ and let P be a circuit in the column matroid. There exist numbers (unique up to a non-zero multiple) b_{Pi} such that

$$b_{Pi} \neq 0 \;\text{ if } C_i \in P \;\text{ and } b_{Pi} = 0 \;\text{ otherwise, and}$$

$$\sum_i b_{Pi} C_i = 0 .$$

The matrix (b_{Pi}), where P ranges over all the circuits of the column matroid, is called the *circuit matrix* of $\boldsymbol{M}$. The column matroid of the circuit matrix is dual to the column matroid of $\boldsymbol{M}$. Thus, if the rows are restricted to a fundamental set of circuits, the circuit matrix provides an effective way of constructing a matrix whose column matroid is dual to a given column matroid. Finally, a result (Theorem 32) relating the determinants of minors of the circuit matrix with rows restricted to a strict fundamental set of circuits is proved. We remark that the proof can be made easier by observing that if $P_1, \ldots, P_q$ is a strict fundamental set of circuits, then the circuit matrix restricted to the rows $P_1, \ldots, P_q$ is of the form

$$
\begin{array}{c|c|cccc|}
 & C_1 \cdots C_{n-q} & C_{n-q+1} & & \cdots & C_n \\
\hline
P_1 & & 1 & 0 & \cdots & 0 \\
P_2 & & 0 & 1 & & \\
\vdots & * & & & \ddots & \\
P_q & & 0 & & & 1 \\
\hline
\end{array}
$$

(The square matrix formed by the last q columns is the identity matrix.) Part III ends with an example (the Fano plane) of a matroid which can be represented as a column matroid only if the field is of characteristic 2.

In the appendix, *binary* matroids, i.e. matroids which can be represented as vectors over the finite field of two elements, are characterized by the following property (C^*): the symmetric difference of two circuits is a disjoint union of circuits. In particular, the circuits of a binary matroid are determined once a strict fundamental set is known.

As Whitney's paper initiated matroid theory, a list of related papers could, in principle, be a complete bibliography of matroid theory. We shall restricted ourselves to three related areas.

Equivalent axiomatizations of matroids

A curious feature of matroid theory not shared by other areas of mathematics is that there are many natural and quite different ways of defining a matroid. Many of these definitions are variations on notions in linear algebra. The more elementary ones can be classified using a scheme in Nicoletti [80] and Nicoletti and White [86]. Other definitions are less immediate and often surprising. We summarize several of these below. (The format is: *basic idea*; *main axiom*; *sources.*)

1. Closure or linear span; Mac Lane-Steinitz exchange axiom; Mac Lane [I. 4].
2. Marriage theorem; the Philip Hall conditions with the rank function instead of cardinality imply the existence of an independent transversal; Rado [42].

3. Arc coloring lemmas, duality in electrical networks; self-dual axioms relating circuits and cocircuits or bonds; Minty [66].
4. Greedy algorithms; the greedy algorithm finds a maximum weight (or lexicographically maximal) independent set for any non-negative weight function (or linear order) on the elements; Gale [68], Edmonds [71], Kruskal [56], Rado [57], Welsh [68].
5. Ports or collections of circuits containing a given element; excluded clutter minors; Seymour [76], Lehman [65].
6. Laplace expansion of a zero determinant; core extraction; Kung [77].
7. Contractions and deletions; existence of certain functions satisfying the Tutte decomposition; Oxley and Welsh [79].

Constructive aspects of matroid definitions

A matroid may be described in many ways according to the axiomatization chosen. For example, it may be described by listing its bases or by listing its circuits. The computational complexity of algorithms which construct one description given another is studied in Hausmann and Korte [81], and Robinson and Welsh [80]. See also Hull [75] and Dawson [80].

The language of matroid theory has proved useful in the recursion theory of vector spaces. Some of the papers on recursion theory in matroids and vector spaces are Baldwin [82], Crossley and Nerode [76], Downey [83], Metakides and Nerode [80], Nerode and Remmel [82, 83], and Remmel [80].

Variations on matroids

There are many combinatorial and geometric structures similar to matroids. Several are listed below (in the format: *name*; *motivating idea*; *source*).

1. Convexity spaces; convex closure; Bryant and Webster [72, 73, 77] and Prenowitz [61].
2. Oriented matroids; "signed" linear dependence over an ordered field; Bland [77], Bland and Las Vergnas [78], Folkman and Lawrence [78], Gutierrez Novoa [65], Las Vergnas [80], Rockefellar [69], Zaslavsky [75].
3. Polymatroids; combinatorial optimization, polyhedra, and submodular functions; Edmonds [70].

4. Combinatorial geometry on a category; Grothendieck topologies; Graves [71].
5. Supermatroids; polymatroids on arbitrary partially ordered sets; Dunstan, Ingleton and Welsh [72].
6. Primoids, Duoids, etc.; combinatorial generalizations of polytopes; Todd [76].
7. Bimatroids or linking systems; non-singular minors in a matrix, Laplace expansion of a bideterminant, matrix multiplication; Kung [78], Schrijver [78, 79].
8. Orthogonal matroids and Pfaffian structures; orthogonal and symplectic analogues of matroids, *Erlanger Programm* for matroids; Kung [78].
9. Permutation geometries; covering relations among subpermutations; Cameron and Deza [79].
10. Tabloids; rank of a submatrix, rank functions on pairs of subsets which are matroid rank functions when one subset is held constant, matrix multiplication; Hocquenghem [79].
11. Geometries on partially ordered sets; Faigle [79, 80], Pezzoli [81].
12. Anti-exchange closure; convex closure; Edelman [80], Edelman and Jamison [85].
13. Genoids; independent sets of vertices in a graph; Naddef, Sbihi and Tchuente [80].
14. Greedoids, Coxeteroids, selectors, etc.; greedy algorithms on strings, Bruhat order of Coxeter groups, etc.; Björner [pre], Crapo [84], Korte and Lovász [83, pre].

References

Baldwin, J.: Recursion theory and abstract dependence, *Patras Logic Symposium* (G. Metakides, ed.), pp. 67-76, North-Holland, Amsterdam, 1982.

Birkhoff, G.: *Lattice theory* (Third edition), Amer. Math. Soc. Colloq. Publ., Vol. 25, Amer. Math. Soc., Providence, R.I., 1967.

Björner, A.: On matroids, groups, and exchange languages, *Matroid Theory and its Applications (Proc. Conf., Szeged, 1982)* L. Lovász and A. Recski, eds.), North-Holland, Amsterdam, to appear.

Bland, R.G.: A combinatorial abstraction of linear programming, *J. Combin. Theory Ser. B 23*(1977), 33-57.

BLAND, R.G. and LAS VERGNAS, M.: Orientability of matroids, *J Combin. Theory Ser. B 24*(1978), 94-123.

BRYANT, V.W. and WEBSTER, R.J.: Convexity spaces. I. The basic properties; II. Separations; III. Dimension, *J. Math. Anal. Appl. 37*(1972), 206-213; *43*(1973), 321-327; *57*(1977), 382-392.

CAMERON, P.J. and DEZA, M.: On permutation geometries, *J. London Math. Soc. (2) 20*(1979), 373-386.

CRAPO, H.H.: Selectors, *Advances in Math., 54*(1984), 233-277.

CROSSLEY, J.N. and NERODE, A.: Effective dimension, *J. Algebra 41*(1976), 398-412.

DANTZIG, G.B.: *Linear programming and extensions,* Princeton Univ. Press, Princeton, N.J., 1963.

DAWSON, J.E.: A note on some algorithms for matroids, *J. Math. Anal. Appl. 75*(1980), 611-615.

DOWNEY, R.: Nowhere simplicity in matroids, *J. Austral. Math. Soc. Ser. A 35*(1983), 28-45.

DUNSTAN, F.D.J., INGLETON, A.W. and WELSH, D.J.A.: Supermatroids, *Combinatorics (Proc. Conf. Combinatorial Math., Math. Inst., Oxford, (1972)*, pp. 72-122, Inst. Math. Appl., Southend-on-Sea, 1972.

EDELMAN, P.H.: Meet-distributive lattices and the anti-exchange closure, *Algebra Universalis 10*(1980), 290-299.

EDELMAN, P.H. and JAMISON, R.E.: The theory of convex geometries, *Geometriae Dedicata 19*(1985), 247-270.

EDMONDS, J.: Submodular functions, matroids, and certain polyhedra, *Combinatorial Structures and their Applications,* pp. 69-87, Gordon and Breach, New York, 1970.

EDMONDS, J.: Matroids and the greedy algorithm, *Math. Programming 1*(1971), 127-136.

FAIGLE U.: The greedy algorithm for partially ordered sets, *Discrete Math. 28*(1979), 150-159.

FAIGLE, U.: Geometries on partially ordered sets, *J. Combin. Theory Ser. B 28*(1980), 26-51.

FOLKMAN, J. and LAWRENCE, J.: Oriented matroids, *J. Combin. Theory Ser. B 25*(1978), 199-236.

GALE, D.: Optimal assignments in an ordered set: An application of matroid theory, *J. Combin. Theory 4*(1968), 176-180.

GRAVES, W.H.: A categorical approach to combinatorial geometry, *J. Combin. Theory Ser. A 11*(1971), 222-232.

GUTIERREZ NOVOA, L.: On *n*-ordered sets and order completeness, *Pacific J. Math. 15*(1965) 1337-1345.

HAUSMANN, D. and KORTE, B.: Algorithmic versus axiomatic definitions of matroids, *Math. Programming Stud. No. 14*(1981), 99-111.

HULL, B.: Two algorithms for matroids, *Discrete Math. 13*(1975), 121-128.

HOCQUENGHEM, S.: Tabloïdes, *J. Combin. Theory Ser. B 26*(1979), 233-250.

KORTE, B. and LOVÁSZ, L.: Structural properties of greedoids, *Combinatorica 3*(1983), 359-374.

KORTE, B. and LOVÁSZ, L.: Greedoids – a structural framework for the greedy algorithm, *Progress in Combinatorial Optimization* (W.R. Pulleyblank, ed.), Academic Press, New York and London, 1984.

KORTE, B. and LOVÁSZ, L.: Posets, matroids and greedoids, *Matroid Theory and its Applications (Proc. Conf. Szeged, 1982)* (L. LOVÁSZ and A. RECSKI, eds.), North-Holland, Amsterdam, to appear.

KRUSKAL, J.B.: On the shortest spanning subtree of a graph and the travelling salesman problem, *Proc. Amer. Math. Soc. 7*(1956), 48-50.

KUNG, J.P.S.: The core extraction axiom for combinatorial geometries, *Discrete Math. 19*(1977), 167-175.

KUNG, J.P.S.: Bimatroids and invariants, *Advances in Math. 30*(1978), 263-275.

LAS VERGNAS, M.: Convexity in oriented matroids, *J. Combin. Theory Ser. B 29*(1980), 231-243.

LEHMAN, A.: Matroids and ports (Abstract), *Notices Amer. Math. Soc. 12*(1965), 342.

MCMULLEN, P.: Transforms, diagrams and representations, *Contributions to Geometry (Proc. Geometry-Symposium, Siegen, 1978)* (J. TOLKE and J. WILLS, eds.), pp. 192-130, Birkhäuser, Basel, 1979.

METAKIDES, G. and NERODE, A.: Recursion theory on fields and abstract dependence, *J. Algebra 65*(1980), 36-59.

MINTY, G.J.: On the axiomatic foundations of the theories of directed linear graphs, electrical networks and network-programming, *J. Math. Mech. 15*(1966), 485-520.

NADDEF, D., SBIHI, N. and TCHUENTE, M.: Independence systems with continuous cardinality of bases, *Discrete Math. 31*(1980), 107-109.

NERODE, A. and REMMEL, J.B.: Recursion theory on matroids, *Patras Logic Symposium* (G. Metakides, ed.), pp. 41-67, North-Holland, Amsterdam, 1982.

NERODE, A. and REMMEL, J.B.: Recursion theory on matroids II, *Southeast Asian Conference on Logic* (C.T. Chang and M.J. Wicks, eds.), pp. 133-183, Elsevier, North-Holland, New York, 1983.

NICOLETTI, G.: Generating cryptomorphic axiomatizations of matroids, *Geometry and Differential Geometry (Proc. Conf., Univ. Haifa, 1979)*, pp. 110-113, Lecture Notes in Math., Vol. 792, Springer-Verlag, New York and Berlin, 1980.

NICOLETTI, G. and WHITE, N.L.: Axiom systems, *Theory of matroids* (N.L. White, ed.) Chapter 2, Cambridge Univ. Press, Cambridge, 1986.

OXLEY, J.G. and WELSH, D.J.A.: The Tutte polynomial and percolation, *Graph Theory and Related Topics (Proc. Conf., Univ. Waterloo, Waterloo, Ont., 1977),* pp. 329-339, Academic Press, New York, 1979.

PEZZOLI, L.: Modular independence systems, *Boll. Un. Mat. Ital. B (5) 18*(1981), 575-590.

PRENOWITZ, W.: A contemporary approach to classical geometry, *Amer. Math. Monthly 68*(1961), no. 1, part II.

RADO, R.: A theorem on independence relations, *Quart. J. Math., Oxford Ser. 13*(1942), 83-89.

RADO, R.: Note on independence functions, *Proc. London Math. Soc. (3) 7*(1957), 300-320.

REMMEL, J.B.: Recursion theory on algebraic structures with an independent set, *Ann. Math. Logic 18*(1980), 153-191.

ROBINSON, G.C. and WELSH, D.J.A.: The computational complexity of matroid properties, *Math. Proc. Cambridge Philos. Soc. 87*(1980), 29-45.

ROCKEFELLAR, R.T.: The elementary vectors of a subspace of R^N, *Combinatorial Mathematics and its Applications (Proc. Conf., Univ. North Carolina, Chapel Hill, N.C., 1967)*, pp. 104-127, Univ. North Carolina Press, Chapel Hill, N.C., 1969.

SCHRIJVER, A.: *Matroids and linking systems,* Math. Centre Tracts 88, Mathematisch Centrum, Amsterdam, 1978.

SCHRIJVER, A.: Matroids and linking systems, *J. Combin. Theory Ser. B 26*(1979), 349-369.

SEYMOUR, P.D.: A forbidden minor characterization of matroid ports, *Quart. J. Math. Oxford Ser. (2) 27*(1976), 407-413.

TODD, M.J.: A combinatorial generalization of polytopes, *J. Combin. Theory Ser. B 20*(1976), 229-242.

TUTTE, W.T.: *Connectivity in graphs,* Mathematical Expositions, No. 15, Univ. Toronto Press, Toronto, Ont.; Oxford Univ. Press, London, 1966.

WELSH, D.J.A.: Kruskal's theorem for matroids, *Proc. Cambridge Philos. Soc. 64*(1968), 3-4.

ZASLAVSKY, T.: Combinatorial ordered geometry. I. Bilateral geometry, Unpublished manuscript, 1975.

2. Geometric lattices

Another motivation behind the discovery of matroids came from the axiomatic approach to geometry. This approach originated in this century in the classic work of Hilbert and Veblen. In the thirties, one of its goals was to formulate the axioms of geometry in such a way that the points do not play a special rôle. This required the study of the "algebraic" operations of intersection (or meet) and span (or join) of two subspaces and led naturally to lattice theory, which was then in its infancy. However, lattices were too general and extra axioms had to be imposed to recover the geometric properties. Most attempts opted for some variant of the axiom of modularity or Dedekind's axiom: if a, b and c are elements of the lattice with $c \leq a$, then

$$(a \wedge b) \vee c = a \wedge (b \vee c).$$

As modular atomic lattices are direct sums of points and lattices of flats of a projective geometry (see Birkhoff [35]), this axiom is too strong to use in most combinatorial situations. The "right" axiom for doing combinatorial geometry, that of upper semimodularity, was singled out only after matroids were discovered. This was accomplished in Birkhoff's 1935 paper "Abstract linear dependence in lattices" which established the fundamental connection between simple matroids (or combinatorial geometries) and geometric lattices.

Briefly, the content of this paper is as follows. In §2, a linearly complete (i.e. closed) set in a matroid is defined. It is then proved that the product (i.e. the intersection) of two closed sets is again closed. In §3, the join or supremum $a \cup b$ of two closed sets a and b is defined in two ways: by using the dependence relation in the matroid and more generally, by showing that in any collection C of subsets closed under intersection, there exists a unique minimum subset in C containing any pair of subsets in C. From this, it is concluded that the closed sets of a matroid form a lattice. In the next section, the lattices of closed sets of matroids are characterized lattice-theoretically by the properties of *semimodularity, viz.* if b and c cover a, then $b \cup c$ cover b, and *atomicity, viz.* every element in the lattice is a join of points (or elements of rank one). Lattices satisfying these properties are called *geometric lattices.* A geometric lattice defines a simple matroid (i.e. a matroid with no circuits of size one or two) or combinatorial geometry on its points and conversely, a simple matroid defines a geometric lattice. The rank $r(a)$ of a closed set in the matroid equals the length of a maximal chain stretched from the minimum to a. This observation implies that the rank function of a matroid is semimodular. In §5, an application to connectivity is given. Finally, it is noted in §6 that a simple matroid is the direct sum of a finite number of single points and projective geometries if and only if the order dual of its lattice of closed sets is also geometric lattice.

Related work on the connection between a matroid and its lattice of closed sets are: (a) Generalizing the dependence relation to elements of the lattice other than points, Dilworth [44], Finkbeiner [51, 60]; (b) Generalizing the set of points to a partially ordered set and the geometric lattice to a semimodular lattice, Faigle [80].

References

BIRKHOFF, G.: Combinatorial relations in projective geometries, *Annals of Math. (2)* 36(1935), 743-748.

DILWORTH, R.P.: Dependence relations in a semimodular lattice, *Duke Math. J. 11*(1944), 575-587.

FAIGLE, U.: Geometries on partially ordered sets, *J. Combin. Theory Ser. B 28*(1980), 26-51.

FINKBEINER, D.T.: A general dependence relation for lattices, *Proc. Amer. Math. Soc. 2*(1951), 756-759.

FINKBEINER, D.T.: A semimodular imbedding of lattices, *Canad. J. Math. 12*(1960), 582-591.

3. Orthogonal duality

The concept of duality has a threefold importance in matroid theory. It is of historical importance because it indicated that matroid theory goes beyond being merely a convenient language for working with notions of dependence. It is also one of the basic constructions of matroid theory; indeed, most of the basic constructions can be derived from duality, deletion and extension. Finally, it serves as a common generalization of many "dual" constructions in combinatorics and geometry (see §1 of this commentary). Duality was first discovered for graphs in the paper "Non-separable and planar graphs" by Whitney. Although this paper predates the discovery of matroids and some of its sections are superceded, it remains an unrivalled introduction to the cycle matroid of a graph, duality and the matroidal aspects of planar graphs.

The first part of this paper deals with separability in graphs. In §2, the rank and nullity of a graph are defined combinatorially. (Equivalent definitions are: the rank [nullity] of a graph is the rank [nullity] of the vertex-edge incidence matrix, considered as a matrix over the integers modulo 2.) In §3, two equivalent definitions of separability are given. A graph is said to be *non-separable* if it is connected and has no cut-vertex, or, equivalently, if every pair of vertices is on a circuit. §4 is concerned with the decomposition of a graph into non-separable components. This section can be replaced by §10 in [I. 1]. In §5, a circuit of graphs (i.e. a graph obtained by identifying at least two vertices in a pair of non-separable graphs or by replacing the edges of a circuit by non-separable graphs) is shown to be non-separable (Theorem 16). It is also shown that the graph formed by taking the union of the non-separable graphs $G_1, \ldots, G_m$ (which are not necessarily disjoint) has non-separable components $G_1, \ldots, G_m$ if and only if no subset of the graphs $G_1, \ldots, G_m$ forms a circuit of graphs. This result is used in §6 to prove Theorem 19, which states that a non-separable graph can be built up by taking first a circuit and adding edges or "suspended chains" so that at each stage, the partially constructed graph is non-separable. This theorem is a useful tool in proofs by induction (cf. Theorem 17 in [I. 1]).

Part II is about duality in planar graphs. In §7, congruence (which is the same as isomorphism) of graphs is defined. In the next section, a dual graph G' is defined using rank. (If the dual graph G' does not exist, this definition yields the dual matroid of the cycle matroid.) The extent to which a dual graph G' is determined by G is investigated. (In Theorem 22, "equivalence" can be replaced by "2-isomorphism" to obtain a necessary and sufficient condition for two graphs G' and G'' to be duals of the same graph G. See Whitney [33].) It is also shown that non-separable components are preserved under duality. (This portion may be replaced by §11 in [I. 1].) §9 begins with a preliminary result. This states that if G is dual to G' and α is an edge in G whose corresponding edge in G' is α', then $G \setminus \alpha$, the graph obtained from G by deleting the edge α, is dual to G'/α', the graph obtained from G' by contracting α'. (In fact, the condition that duality interchanges deletion and contraction and a non-degeneracy condition characterize duality as an operation on matroids. See Kung [84].) The main theorem, asserting that a graph is planar if and only if it has a dual graph, is now proved. Using this theorem, the graphs K_5 and $K_{3,3}$, and their subdivisions, are shown to be non-planar. (A complete discussion of the relation of the main theorem to Kuratowski's theorem can be found in Whitney [33]. See also Tutte [63] and Parsons [71].) The paper concludes with an elegant proof of Euler's formula using duality.

The concept of duality for planar graphs has been generalized to "Trinity" in Tutte [75]. Duality for infinite planar graphs is treated in Thomassen [80, 82] and duality for infinte matroids in Brualdi and Scrimger

[68], Higgs [69, 69′], Klee [71], Las Vergnas [71], Oxley [78], and Sierpinski [52, Chap. 1]. The notion of duality extends also to *clutters* (i.e. collections of pairwise incomparable subsets of a set). In this very general context, duality provides a framework for many algorithms and minimax theorems in combinatorial optimization. A good introduction can be found in the seminal paper [70] of Edmonds and Fulkerson. Other related papers are Crapo [67], Matthews [79], and McKee [81].

References

BRUALDI, R.A. and SCRIMGER, E.B.: Exchange systems, matchings, and transversals, *J. Combin. Theory 5*(1968), 244-257.

CRAPO, H.H.: Geometric duality, *Rend. Sem. Mat. Univ. Padova 38*(1967), 23-26.

EDMONDS, J. and FULKERSON, D.R.: Bottleneck extrema, *J. Combin. Theory 8*(1970), 299-306.

HIGGS, D.A.: Matroids and duality, *Colloq. Math. 20*(1969), 215-220.

HIGGS, D.A.: Infinite graphs and matroids, *Recent Progress in Combinatorics (Proc. Third Waterloo Conf. on Combinatorics, 1968),* pp. 245-253, Academic Press, New York, 1969.

KLEE, V.: The greedy algorithm for finitary and cofinitary matroids, *Combinatorics (Proc. Sympos. Pure Math., Vol. 19, Univ. California, Los Angeles, Calif., 1968),* pp. 137-152, Amer. Math. Soc., Providence, R. I., 1971.

KUNG, J.P.S.: A characterization of orthogonal duality in matroid theory, *Geometriae Dedicata 15*(1983), 69-72.

LAS VERGNAS, M.: Sur la dualité en théorie des matroïdes, *Théorie des matroïdes (Rencontre Franco-Britannique, Brest, 1970),* pp. 67-85, Lecture Notes in Math., Vol. 211, Springer-Verlag, Berlin, 1971.

MATTHEWS, L.: Duality in independence structures, *Third Symposium on Operations Research (Univ. Mannheim, Mannheim, 1978) Section 2,* pp. 173-179, Operations Research Verfahren, 32, Hain, Meisenheim, 1979.

MCKEE, T.A.: A quantifier for duality, *Discrete Math. 34*(1981), 315-318.

OXLEY, J.G.: Infinite matroids, *Proc. London Math. Soc. (3) 37*(1978), 259-272.

PARSONS, T.D.: On planar graphs, *Amer. Math. Monthly 78*(1971), 76-78.

SIERPINSKI, W.: *General topology,* Univ. Toronto Press, Toronto, 1952.

THOMASSEN, C.: Planarity and duality of finite and infinite graphs, *J. Combin. Theory Ser. B 29*(1980), 244-271.

THOMASSEN, C.: Duality of infinite graphs, *J. Combin. Theory Ser. B* *33*(1982), 137-160.
TUTTE, W.T.: How to draw a graph, *Proc. London Math. Soc. (3) 13*(1963), 743-767.
TUTTE, W.T.: Duality and trinity, *Infinite and finite sets (Colloq., Keszthely, 1973), Vol. III*, pp. 1459-1472, Colloq. Math. Soc. János Bolyai, Vol. 10, North-Holland, Amsterdam, 1975.
WHITNEY, H.: Planar graphs, *Fundamenta Math.* *21*(1933), 73-84.

4. Exchange lattices

Apart from providing the right level of generality for the dual graph construction, the first application of matroid theory is in the study of field extensions. The groundwork for this application was laid in Mac Lane's paper, "A lattice formulation for transcendence degrees and p-bases," the fourth paper reprinted in this chapter.

A major portion of Mac Lane's paper is devoted to exchange axioms in lattices, of which this paper gives the definitive account. The most useful of these is Axiom E_1 in §2:

> *Let a be an element of a lattice L and p and q be points (i.e. elements covering the minimum) of L. Then $a < a \vee p \leqslant a \vee q$ implies $q \leqslant a \vee p$.*

This axiom is now called the Mac Lane-Steinitz exchange axiom. It is often reformulated in terms of closure operations. Let S be a set. A function $A \mapsto \overline{A}$, taking a subset A of S to a subset $\overline{A}$ of S, is said to be a closure operation if $A \subseteq \overline{A}$, $\overline{A} = \overline{\overline{A}}$, and, $A \subseteq B$ implies $\overline{A} \subseteq \overline{B}$. A closure operation is said to *have finite character* if for every subset A, there exists a finite subset $\{a_1, \ldots, a_r\}$ of A such that $\overline{A} = \overline{\{a_1, \ldots, a_r\}}$. In this context, the Mac Lane-Steinitz exchange axiom becomes

> *Let $A \subseteq S$ and $p, q \in S$. Then $p, q \notin \overline{A}$ and $p \in \overline{A \cup \{q\}}$ imply $q \in \overline{A \cup \{p\}}$.*

Another, more pictorial, reformulation of this axiom is the following "partition" axiom.

> *Let X be a closed set and let $\{Y_i\}$ be the collection of closed sets covering X in the lattice of closed sets. Then the collection $\{Y_i - X\}$ of difference sets partitions $S - X$ into disjoint blocks.*

A closure operation on S having finite character and satisfying the Mac Lane-Steinitz exchange axiom defines a matroid on S. This axiomatization is especially useful for studying the geometric aspects of matroid theory. Several other exchange axioms (some of them "pointless") are studied. The introduction (§1) contains a useful guide and summary.

The applications to field extensions are found in §§5 and 6. In §5, it is shown that the lattice of relatively algebraically closed subfields between two given fields is a exchange (or geometric) lattice. These lattices are non-modular if the rank is at least 3 – indeed, the only modular elements are the maximum, the minimum, and the points – and show the need to go to a more general axiom than the modular axiom. (It is an intriguing open problem to find a characterization of these "algebraic" lattices similar to the characterization of lattices of subspaces as connected modular geometric lattices.) The notion of p-dependence is considered in §6. A subset $z_1, \ldots, z_n$ of field elements is said to be *p-dependent* over a subfield R if the elements $z_1^{1/p}, \ldots, z_n^{1/p}$ are algebraically dependent over R. The notion of p-dependence defines a matroid and the basis theorem (§3) is applied to show that the "relative degree of imperfection" is an invariant. A very accessible account of the applications of matroid theory to the study of field extensions can be found in §1 of Mac Lane [76]. Another survey is Mac Lane [40].

References

Mac Lane, S.: Modular fields, *Amer. Math. Monthly 47*(1940), 259-274.

Mac Lane, S.: Topology and logic as a source of algebra, *Bull. Amer. Math. Soc. 82*(1976), 1-40.

5. Survey of the early literature

Their historical interest aside, the early papers in matroid theory are sources of deep ideas which are now and then in danger of being forgotten. In this section, we give a brief but complete survey of the matroid literature up to 1945, ten years after Whitney's founding paper. This survey complements the existing reviews in two ways. First, *Mathematical Reviews* began in 1940 and so many of the pioneering papers were not reviewed there. Second, we are able to indicate the influence of the early papers in the later development of matroid theory.

The summaries are arranged here first by the year of publication of the paper and then by alphabetical order of the author. Although the boundaries are not precise, papers in graph theory (mostly due to Whitney) which contributed concepts to matroid theory are included. However, because they would take us too far afield, the important papers of Mac Lane using the theory of geometric lattices in the study of field extensions in characteristic p are not summarized here. A clear and informative overview of these papers can be found in §1 of Mac Lane [76].

Summaries of early papers in matroid theory

1910/1930

E. Steinitz, Algebraische Theorie der Körper, *J. Reine Angew. Math. 137* (1910), 167-309; reprinted as a book, R. Baer and H. Hasse, eds., de Gruyter, Berlin, 1930.

This is perhaps the earliest paper in which the ideas of matroid theory appear in a recognizable form. These ideas occur in §22 on the transcendental extensions of fields. There, the fundamental properties of algebraic dependence are isolated and elementary theorems (such as the fact that two maximal algebraically independent sets have the same cardinality) are proved using just those properties.

1930

C. Kuratowski, Sur les problèmes des courbes gauches en Topologie, *Fundamenta Math. 15*(1930), 271-283.

In this famous paper is the Kuratowski condition for planarity: A graph is embeddable in the plane if and only if it contains no subgraph isomorphic to a subdivision of K_5 or $K_{3,3}$. This condition provides the model for all the "forbidden minors" theorems in matroid theory.

1932

H. Whitney, A logical expansion in mathematics, *Bull. Amer. Math. Soc. 38*(1932), 572-579.

H. Whitney, The coloring of graphs, *Annals of Math. (2) 33*(1932), 688-718.

In the first paper, a new formulation of the principle of inclusion and exclusion using characteristic functions of subsets is given. Among other applications, this is used to proved that the number $M(G; \lambda)$ of ways to (properly) color the vertices of a graph G using λ colors is a polynomial in λ with zero constant term of degree V, where V is the number of vertices in G. The polynomial $M(G; \lambda)$ is called the *chromatic polynomial* of G and first appeared in G.D. Birkhoff [12]. If we write

$$M(G; \lambda) = \sum_{i=0}^{V-1} m_i(G) \lambda^{V-i},$$

then the coefficients $m_i(G)$ are given by

$$m_i(G) = \sum_j (-1)^{i+j} m_{ij}(G),$$

where $m_{ij}(G)$ is the number of subgraphs in G of rank i and nullity j. In addition, $m_i(G)$ is interpreted in terms of broken circuits. Linearly order the edges of G in some way. A *broken circuit* is a subset of edges of the form $C \setminus \{e\}$, wher C is a circuit of G and e is the largest edge (in the linear order) in C. The number $(-1)^i m_i(G)$ is the number of subsets of edges in G of size i containing no broken circuits.

In the second paper, the study of the numbers $m_i(G)$ and $m_{ij}(G)$, considered as functions on the set of graphs, is continued. The first result is that $(-1)^i m_i(G) > 0$. (This result is generalized to matroids in Rota [III. 1].) Next the numbers α_k and β_k, are defined by

$$\alpha_{kl} = \text{number of subsets of broken circuits } \{C_1, \ldots, C_l\} \text{ such that}$$
$$|C_1 \cup \ldots \cup C_l| = k,$$
$$\alpha_k = \sum_l (-1)^l \alpha_{kl},$$
$$(-1)^i \beta_i = \sum_k \binom{k}{i} \alpha_k$$

are studied. The numbers m_i, α_i and β_i can be expressed in terms of each other and in certain cases, α_i and β_i have simpler expressions than m_i.

In Part II, the "graphcopy" functions N_T are introduced. Let T be a graph. If G is a graph, let $N_T(G)$ be the number of subgraphs of G isomorphic to T. The collections $\{N_T(G)$: T is a graph$\}$ and $\{N_T(G)$: T is a non-separable graph$\}$ are shown to be algebraically independent as functions defined on graphs. (Two remarks. This result generalizes easily to "matroidcopy" functions. For graphs, a strengthening can be found in Erdös, Lovász and Spencer [79].) The graphcopy functions $N_T(G)$ are important in that the functions $m_{ij}(G)$ can be expressed uniquely as polynomials with no constant term in $N_T(G)$. Finally, criteria for a given graphcopy function N_T to appear non-trivially in the polynomial expression of a given m_{ij} are given.

Part III is concerned with sequences $f_i(a)$ and $m_i(a)$ of functions related by

$$m_0(a)+m_1(a)x+m_2(a)x^2+\cdots=\exp(f_1(a)+f_2(a)x+f_3(a)x^3+\cdots)$$

The functions $m_i(a)$ are expressed as polynomials in $f_i(a)$. These polynomials are (apart from factors of $1/i!$) the Bell polynomials discovered later by Bell. (See Bell [34] or Riordan [58], p. 35.) For use in Part IV, a variant of this theory involving sequences f_{ij} and m_{ij} is developed.

The preceding theory is applied to the functions $m_{ij}(G)$ in Part IV. The functions $f_{ij}(G)$ related to $m_{ij}(G)$ are easier to work with for two reasons. First, $f_{ij}(G)$ is additive under disjoint union of graphs. Second, $f_{ij}(G)$ is a linear homogeneous polynomial in the graphcopy functions $N_T(G)$; indeed, $f_{ij}(G)$ is the linear part in the polynomial expression of $m_{ij}(G)$ in terms of $N_T(G)$. To prove these facts, a theory of "generalized" graph is worked out. (A *generalized* graph is simply a sequence (g_T: T a graph) of non-negative integers and a "real" graph is a sequence (g_T) for which there exists a graph G such that $N_T(G)=g_T$. Every generalized graph can be written as an integral linear combination of "real" graphs.) Next, it is proved that $\{m_{10}\}\cup\{m_{ij}\colon j>0\}$ and $\{f_{10}\}\cup\{f_{ij}\colon j>0\}$ are algebraically independent sets of functions on graphs. Finally, the difficult problem of finding all the inequalities between the functions $m_{ij}(G)$ (or $f_{ij}(G)$) is posed.

In Part V, $M(G;\lambda)$, $m_{ij}(G)$, and other functions are computed for graphs with a small number of vertices.

Last, but not least, is the note. In this note is the first appearance (due to Foster) of the Tutte decomposition

$$m_{i,j}(G)=m_{i,j}(G\setminus e)+m_{i,j-1}(G/e)$$

which is equivalent to

$$M(G;\lambda)=M(G\setminus e;\lambda)-M(G/e;\lambda).$$

§IV of this anthology is devoted to further developments of this idea.

References to related work on the chromatic polynomial can be found in §§III and IV. The notion of broken circuits is further studied in Hoggar [75], Wilf [76, 77], Brylawski [77], and Brylawski and Oxley [81].

1933

H. Whitney, On the classification of graphs, *Amer. J. Math.* 55(1933), 236-244.
H. Whitney, 2-isomorphic graphs, *Amer. J. Math.* 55(1933), 245-254.

For matroid theorists, the interest of the first paper lies in the concept of 2-isomorphism of graphs. Consider the following operations on graphs.
(V) Join two disjoint graphs G and H by identifying a vertex in G with a vertex in H, or, reversing this, break up a graph with a cut vertex into two disjoint graphs.
(T) This operation can be applied only to graphs of the form $H_1 \cup H_2$, where H_1 and H_2 are graphs having exactly two vertices a and b in common and $\{a, b\}$ is an edge in both H_1 and H_2. Delete all the edges $\{a, c_1\}, \ldots, \{a, c_m\}$ and $\{b, d_1\}, \ldots, \{b, d_n\}$ of H_1 incident on a and b and replace them by the edges $\{b, c_1\}, \ldots, \{b, c_m\}$ and $\{a, d_1\}, \ldots, \{a, d_n\}$. That is, turn the graph H_1 around at a and b.
Two graphs G and H are said to be 2-*isomorphic* if G can be obtained from H by applying a sequence of these two operations.

In the second paper, it is proved that two graphs G and H are 2-isomorphic if and only if there is a one-to-one correspondence between their edges so that circuits correspond to circuits – that is, if and only if their cycle matroids are isomorphic. The proof is by induction and uses Theorem 18 in [I. 3]. It is also observed that 2-isomorphism corresponds to "electrical equivalence."

A related paper is Truemper [80].

H. Whitney, A set of topological invariants for graphs, *Amer. J. Math.* 55(1933), 231-235.

A *topological invariant* for graphs is a function defined on the set of graphs which remains unchanged when an edge is subdivided (i. e. which is invariant under series extensions). If G is a graph of rank $R(G)$ and nullity $N(G)$, then

$$m_i'(G) = \sum_j (-1)^{i+j} m_{R(G)-j,\, N(G)-i}(G),$$

where the numbers $m_{i,j}(G)$ are defined as in Whitney [32], are topological invariants. Further, $m'_i(G)$ can be interpreted in terms of broken cut sets. (In the context of matroid theory, $m'_i(G)$ are the coefficients of the characteristic polynomial of the dual or cut set matroid of G. The coefficients $m'_i(G)$ are topologically invariant because series extension in G corresponds to parallel extension in the dual matroid and the characteristic polynomial is unchanged under parallel extensions.) An example is given of two graphs which are not 2-isomorphic (see preceding summary for a definition) but have the same invariants m'_i.

Related papers are Tutte [IV. 1], Crapo [67] and Tutte [74].

H. Whitney, Planar graphs, *Fundamenta Math. 21*(1933), 73-84.

Kuratowski [30] (see above) proved that a graph is planar if and only if (K) it does not contain as a subgraph a subdivision of K_5 or $K_{3,3}$. Whitney [I. 3] proved that a graph is planar if and only if (W) the orthogonal dual of its cycle matroid is also graphic. In this paper, a direct proof that (K) is equivalent to (W) is given.

The proof that (W) implies (K) consists of a counting argument that the duals of the cycle matroids of K_5 and $K_{3,3}$ cannot be graphic. The proof that (K) implies (W) is more difficult and proceeds by induction. Let G be a graph not containing subdivisions of K_5 or $K_{3,3}$. The cases when G is separable or has two edges in series are easily disposed of. Thus, one may assume that G is non-separable and every vertex has degree at least 3. Remove an edge $\{a, b\}$ from G to obtain a smaller graph G_0. Let A and B be the set of edges in G_0 incident on a and b respectively. Then, as A and B are cut sets of G, the sets A' and B' corresponding to A and B in the dual of G_0 are circuits. There are now two cases, depending on whether the circuits A' and B' have a vertex in common. These cases are handled by rather intricate constructions.

Related papers are Tutte [63] and Thomassen [80, 82].

1935

G. Birkhoff, Combinatorial relations in projective geometries, *Annals of Math. (2) 36*(1935), 743-748.

The main result is the following geometric representation of complemented modular lattices: A complemented modular lattice of finite rank is the direct product of a finite Boolean algebra and lattices of subspaces of a finite-dimensional vector space. This is proved by showing that in a connected complemented modular lattice, the points satisfy a set of axioms of projective geometry similar to those of Veblen.

Related papers are Baer [42], Faigle and Herrmann [81], and Kahn and Kung [pre].

1935/36

T. NAKASAWA, Zur Axiomatik der linearen Abhängigkeit, I, II, and III, *Sci. Rep. Toyko Bunrika Daigaku,* *2*(1935), 235-255, *3*(1936), 123-136, and *3*(1936), 45-69.

The first paper in this series contains an axiomatization of matroids (called here B_1-spaces) contemporaneous with but independent of Whitney's. Thus, Nakasawa and Whitney were the first to write down explicitly an axiom system for matroids.

Throughout these papers, Nakasawa uses a proof-theoretic notation derived from the *cycle calculus* of G. Thomsen (see Thomsen [33]). To give an idea of this notation, here are Nakasawa's axioms for a matroid in his notation. A matroid (in Nakasawa's terminology, a "B_1-space") on a set B_1 of elements is defined by specifying which finite sequences ("cycles") $a_1 a_2 \ldots a_s$ of elements from B_1 are dependent ("valid"). The sentence

$$a_1 a_2 \ldots a_s = 0$$

or, more briefly,

$$a_1 a_2 \ldots a_s,$$

means $a_1 a_2 \ldots a_s$ is dependent. The sentence

$$a_1 a_2 \ldots a_s \neq 0$$

means $a_1 a_2 \ldots a_s$ is not dependent. The dependent sequences satisfy the following axioms:

1. (Reflexivity) $\quad a a.$
2. (Sucession) $\quad a_1 a_2 \ldots a_s \rightarrow a_1 a_2 \ldots a_s x.$

3. (Interchangability) $a_1 \ldots a_i \ldots a_s \rightarrow a_i \ldots a_1 \ldots a_s$.

4. (Transitivity) $a_1 \ldots a_s \neq 0$, $xa_1 \ldots a_s$, and $a_1 \ldots a_s y$ $\rightarrow x a_1 \ldots a_{s-1} y$.

Here, → means "implies." The rest of Part I is devoted to deriving elementary consequences of the axioms. Some of the topics treated are closure, rank, meets and joins of closed sets, and the submodular inequality. In the final section, several "reduction rules" for dependent sets are derived, a typical example being

$$a_1 \ldots a_n x_1 \ldots x_k \text{ and } x_1 \ldots x_k y \rightarrow a_1 \ldots a_n x_1 \ldots x_{k-1} y \text{ or } x_1 \ldots x_k.$$

In Part II, the notion of a B_2-space is introduced. A *B_2-space* on a set B_2 is given by a dependence relation on finite sequences of elements from B_2. This dependence relation satisfies Axioms 2, 3, and 4 as well as

1*. $a \neq 0$ and $a\,a$

5. (Intersection) $a_1 \ldots a_s x y \rightarrow$ there exists z, $a_1 \ldots a_s z$ and $x y z$.

It is proved that Axiom 5 implies the modular equality for the rank function. The remainder of Part II is concerned with circuits ("prime cycles"), connectivity, and direct sums.

Part III begins by relating the results and definitions in Parts I and II to the work of Whitney and others. In particular, it is proved that the axioms for a B_1-space are equivalent to Whitney's axioms for a matroid. It is also observed that a connected B_2-space is a projective geometry. Part III concludes with a discussion of the logical dependence and independence of the various axioms.

1936

K. Menger (with F. Alt and O. Schreiber), New foundations of projective and affine geometry, *Annals of Math. (2) 37*(1936), 456-481.

The aim of this paper is to find a lattice-theoretic (or "algebraic") formulation of the axioms of affine and projective geometry and using it, to determine the theorems which are true in both affine and projective geometry.

Along the way, several elementary theorems (such as upper semimodularity of the rank function) which are also true in matroid theory are derived.

1937

F. Klein-Barmen, Birkhoffsche und harmonische Verbände, *Math. Zeitschrift 42*(1937), 58-81

A Birkhoff lattice is a (upper or lower) semimodular lattice. A harmonic lattice is a lattice which is both upper and lower semimodular. In this paper, several elementary results (which have also appeared elsewhere) are proved. Examples: A Birkhoff lattice satisfies the Jordan-Dedekind chain condition. A lattice is upper (or lower) semimodular if and only if its rank function satisfies the upper (or lower) semimodular inequality. A harmonic lattice is modular. There is a bibliography of the author's earlier papers on general lattice theory.

S. Mac Lane, A combinatorial condition for planar graphs, *Fundamenta Math. 28*(1937), 22-32.

A set $\mathscr{C}$ of circuits in a graph G is said to be *complete* if every circuit in G can be expressed uniquely as a sum modulo 2 of a subset of circuits in $\mathscr{C}$. The main theorem states: A graph is planar if and only if the graph contains a complete set of circuits such that no edge appears in more than two of these circuits. In addition, the latter condition is shown (without using the main theorem) to be equivalent to Whitney's condition for planarity (see [I. 3]).

Related papers are Dworakowski and Kaczmarski [74], Lefschetz [65], O'Neil [73], Thomassen [80, 82], and Tutte [63].

B.L. Van Der Waerden, *Moderne Algebra* (2. Aufl.), Springer-Verlag, Berlin, 1937.

We restrict our attention to §64. In order to prove that the transcendence degree of an extension of a field is well-defined, three fundamental properties of algebraic dependence are proved:

1. u_i is algebraically dependent on $u_1, \ldots, u_n$;
2. If v is algebraically dependent on $u_1, \ldots, u_n$ but not on $u_1, \ldots, u_{n-1}$, then u_n is algebraically dependent on $u_1, \ldots, u_{n-1}, v$;

3. If w is algebraically dependent on $v_1, \ldots, v_s$ and every v_j is in turn algebraically dependent on $u_1, \ldots, u_n$, then w is algebraically dependent on $u_1, \ldots, u_n$.

It is observed that these properties are shared by linear dependence of vectors and that these are just the properties needed in the proofs of the elementary theorems.

L.R. Wilcox, Modularity in the theory of lattices, *Annals of Math. (2) 40* (1939), 490-505.

For matroid theorists, the most important part of this paper is the definition of the key notion of modular pair. An ordered pair (a, b) of elements in a lattice is said to be a *modular pair* – in symbols, $(a, b)M$ – if for all elements $c \leqslant a$,

$$(a \wedge b) \vee c = a \wedge (b \vee c).$$

However, the motivation of this paper is more in infinite or continuous lattice theory. The aim is to develop a lattice version of affine geometry and in particular, to generalize the construction of a continuous geometry (in the sense of von Neumann [62]) to lattices of affine subspaces of a vector space. Much of the paper is concerned with lattices satisfying the axioms:

A. (*Symmetry*) $(a, b)M$ implies $(b, a)M$.

B. (*Local modularity*) If $a \wedge b \neq 0$, then $(a, b)M$.

Lattices satisfying Axiom A are called *symmetric* lattices. Lattices satisfying Axiom B are called by the author "*semi-modular*" lattices. (This is a somewhat non-standard terminology.) "Semi-modular" lattices generalize lattices of affine subspaces and a notion of parallelism can be defined. In the final section is the following useful technical result. In a symmetric lattice, the Jordan-Dedekind chain condition holds, a rank function r can be defined, and, $(a, b)M$ if and only if

$$r(a \vee b) + r(a \wedge b) = r(a) + r(b).$$

This paper is a major influence in the theory of infinite lattices. A definitive account of the work it initiated can be found in Maeda and Maeda [70].

Related papers dealing with modularity in geometric lattices are Wilcox [44], Stanley [III. 3], Brylawski [75], and Rao and Rao [76]. Lattice-theoretic aspects of parallelism are studied in Hsu [49] and locally modular geometric lattices in Kantor [74, 76].

1940

R.P. DILWORTH, Note on complemented modular lattices, *Bull. Amer. Math. Soc.* *46*(1940), 74-76. (MR 1, 197)

An element a in a lattice L is said to be "neutral" or, to use a more standard terminology, *distributive* if $a \vee (b \wedge c) = (a \vee b) \wedge (a \vee c)$ for all b and c in L. In this paper, the following conditions are shown to be equivalent for an element a in a complemented modular lattice L: (i) a is distributive; (ii) a has a unique complement a^c; (iii) $L \cong [0, a] \times [0, a^c]$. This theorem is generalized to relatively complemented lattices in Dilworth [50].

O. HAUPT, G. NÖBELING and C. PAUC, Über Abhängigkeitsräume, *J. Reine Angew. Math.* *181*(1940), 193-217. (MR 1, 169)
O. HAUPT, G. NÖBELING and C. PAUC, Sekanten und Paratingenten in topologischen Abhängigkeitsräumen, *J. Reine Angew. Math.* *182*(1940), 105-121. (MR 2, 130)

These papers were motivated by attempts (see, for example, Torrance [37]) to axiomatize the notion from differential geometry of tangent space at a point. A "dependence space" (which is equivalent to a matroid) is defined using dependence relations. The key axiom is the exchange axiom: If $\{x_1, \ldots, x_n\}$ is independent and $\{x_1, \ldots, x_n, y\}$, $\{x_1, \ldots, x_n, z\}$ are both dependent, then $\{x_2, \ldots, x_n, y, z\}$ is dependent. Various notions – including that of basis, closure, rank, and connectivity – are developed in much the usual manner. There are interesting examples of matroids arising from differential geometry.

1941

R.P. DILWORTH, The arithmetic theory of Birkhoff lattices, *Duke Math. J.* *8*(1941), 286-299. (MR 3, 100)
R.P. DILWORTH, Ideals in Birkhoff lattices, *Trans. Amer. Math. Soc.* *49*(1941), 325-353. (MR 2, 343)

Let a be an element in a lattice. A set of meet-irreducible elements $x_1, \ldots, x_r$ is said to be a *reduced decomposition* if

$$a = x_1 \wedge x_2 \wedge \ldots \wedge x_r$$

and for all i

$$x_i \not\geq x_1 \wedge \ldots \wedge x_{i-1} \wedge x_{i+1} \wedge \ldots x_r .$$

The meet-irreducibles x_i are said to be components of a in the reduced decomposition. The Kurosh-Ore theorem (see, for example, Birkhoff [67]) states that in a modular lattice, the number of components in a reduced decomposition of an element a is a fixed number depending only on a. In the first paper, the Kurosh-Ore theorem is generalized as follows. In a semi-modular lattice satisfying both the ascending and descending chain conditions, the number of components in a reduced decomposition of an element a is a fixed number depending only on a if and only if the upper interval $[a, 1]$ (here, 1 is the maximum of the lattice) is a modular lattice. This result is extended to a wider class of semi-modular lattices in the second paper. A related paper is Dilworth [40].

L.R. Wilcox, A topology for semi-modular lattices, *Duke Math. J.* 8(1941), 273-285. (MR 3, 56)

This paper is concerned with the question of constructing an extension of a topology given on the set of points of a geometric lattice to the entire geometric lattice. Let L be a geometric lattice of finite rank with rank function r and let P be the set of points in L. Suppose that P is a metric space with metric δ. If $S, T \subseteq P$, define the distance $\delta(S, T)$ between S and T by

$$\delta(S, T) = \inf\{\delta(p, q) : p \in S, q \in T\} .$$

The metric δ is *compatible* with the lattice structure of L if the following axioms are satisfied:

W1: If $p_1, \ldots, p_n$ is an independent set and p is any point, then for every $\varepsilon > 0$ there exists an $\eta > 0$ such that for all q_i with $\delta(p_i, q_i) < \eta$,

$$|\delta(p, q_1 \vee q_2 \vee \ldots \vee q_n) - \delta(p, p_1 \vee p_2 \vee \ldots \vee p_n)| < \varepsilon .$$

(That is, $\delta(p, p_1 \vee \ldots \vee p_i \vee \ldots \vee p_n)$ is continuous as a function of p_i whenever $p_1, \ldots, p_n$ is independent.)

W2: For every element a in L, the set $\{p: p \text{ is a point and } p \leqslant a\}$ is closed in the metric topology.

If the metric δ is compatible, then the topology given by the metric δ can be extended to a topology (perhaps not necessarily metric) on all of L. This topology is described in terms of convergence. Let (a^i) be a sequence of elements in L. Define $L(a^i)$ to be the set of points in L satisfying: there exist a positive integer m and a sequence (p^i) of points such that for all $i>m$, $p^i \leqslant a^i$ and p^i converges to p in the metric topology. The topology on L is now defined by specifying that the sequence (a^i) converges to the join $\bigvee\{p: p \in L(a^i)\}$ whenever the sequence $(r(a^i))$ of ranks eventually becomes constant and

$$\lim_{i \to \infty} r(a^i) = r(\bigvee\{p:\ p \in L(a^i)\}).$$

This topology is Hausdorff and the set L_k of all elements in L of rank k is an open and closed set. No examples are given in this paper. However, it is not hard to see that when L is the lattice of all subspaces of a vector space of dimension d over a metric field, the topology on L_k given here is the same as the topology on L_k given by identifying each subspace with the $\binom{d}{k}$-tuple of its Plücker coordinates and imposing a suitable product topology.

1942

R. Rado, A theorem on independence relations, *Quart. J. Math., Oxford Ser. 13*(1942), 83-89. (MR 4, 269c)

An *independence function* I on a set S is a function $I(x_1, \ldots, x_m)$ defined on finite sequences of elements of S to $\{0, 1\}$ satisfying

(i) $I(x_1, \ldots, x_m) \geqslant I(x_1, \ldots, x_m, x_{m+1})$;

(ii) $I(x_1, \ldots, x_m) = I(x_{\sigma 1}, \ldots, x_{\sigma m})$
for every permutation σ of $\{1, \ldots, m\}$;

(iii) $I(x, x) = 0$;

(iv) $I(x_1, \ldots, x_m) I(y_1, \ldots, y_{m+1}) \leqslant \sum_{j=1}^{m+1} I(x_1, \ldots, x_m, y_j)$.

The sets $\{x_1, \ldots, x_m\}$ for which $I(x_1, \ldots, x_m\} = 1$ form the independent sets of a matroid on S and conversely. A generalization of Philip Hall's marriage theorem (now known as Rado's theorem) is proved. This states:

Let M be a matroid on S with rank function r and let $A_1, \ldots, A_n$ be a list of subsets of S. There exists a system of distinct representatives which is independent in the matroid M if and only if for all $J \subseteq \{1, \ldots, n\}$,

$$r\left(\bigcup_{j \in J} A_j\right) \geqslant |J| .$$

The proof of this theorem proceeds by induction on n. In addition, it is observed that the "generalized Hall conditions" is equivalent to the existence of an independent system of distinct representatives if and only if the function r is the rank function of a matroid on S.

Rado's theorem is an important tool in matching theory; for example, the common transversal theorem (Ford and Fulkerson [62], p. 76) is a simple consequence. It is also a basic result in the study of transversal matroids and their generalizations. Papers which are directly related to Rado's theorem include Aigner and Dowling [71], Brualdi [71[3]], Györi and Milner [78], Kundu and Lawler [73], Las Vergnas [70], McDiarmid [73, 75, 75′], Perfect [69, 69′, 72], Rado [49] and Welsh [68, 69, 70, 71, 71′].

1944

R.P. Dilworth, Dependence relations in a semi-modular lattice, *Duke Math. J. 11*(1944), 575-587.(MR 6, 143g)

In Birkhoff [I. 2], it is proved that a geometric lattice induces the structure of a simple matroid on its points. The aim of this paper is to extend this result to elements of higher rank in a geometric lattice. This is accomplished by showing how a matroid structure can be defined on the points of a quasi-modular lattice. A *quasi-modular* lattice L is a lattice in which a covers $a \cap b$ and $a \cap b \neq 0$ (the minimum of L) imply $a \cup b$ covers b. In a quasi-modular lattice, the rank function $\varrho(a)$ (which equals the length of a maximal chain stretched between 0 and a) is well-defined and satisfies the properties of normalization, unit increase, and quasi-modularity (i.e. $\varrho(a) + \varrho(b) \geqslant \varrho(a \cup b) + \varrho(a \cap b)$ if $a \cap b \neq 0$). Let P be the set of points of a quasi-modular lattice L. Then, the subsets $T \subseteq P$ such that

$$\varrho\left(\bigvee_{p\in T} p\right) \geqslant |T|$$

form the independent sets of a matroid on P. If the inequality is in fact an equality, then T is said to be *normal*. The normal subsets play a key rôle in the proof that the independent sets as specified above satisfy the matroid axioms. In addition, it is shown that the sets $\{p: p \leqslant t\}$, where t is an element of the lattice L, are closed subsets of the matroid. Thus, the quasi-modular lattice L embeds in a rank preserving way into the geometric lattice of closed sets of the matroid. (In fact, this result generalizes to: every finite lattice can be embedded into a geometric lattice. See Crawley and Dilworth [73], Chap. 14 and also Finkbeiner [60].) An important example is when the quasi-modular lattice is obtained from a geometric lattice L by removing or "truncating" all the elements of rank i, where $1 \leqslant i \leqslant n-1$. The geometric lattice $D_n(L)$ obtained as above is called a *completed Dilworth truncation*. Specializing further, when L is the Boolean algebra of all subsets of a set S, $D_2(L)$ is the lattice of partitions of S.

This seminal paper founded the combinatorial theory of submodular functions which is useful in combinatorial optimization and the theory of packing and covering. Further work on the relation between matroids and submodular functions can be found in Crapo and Rota [70], Chap. 7, Dunstan [76], Dunstan, Ingleton and Welsh [72], Edmonds [70], Edmonds and Rota [66], Ingleton [59], McDiarmid [73], Nguyen [77, 77/78], Pym and Perfect [70], and Welsh [71, 71′].

Simplicial geometries are the completed Dilworth truncations of Boolean algebras. Thus, they are generalizations of partition lattices. In addition, they have intimate connections with algebraic topology. The main reference is Crapo and Rota [71]. Other results, mainly on coordinatization, can be found in Cordovil [80, 82], Cordovil and Las Vergnas [79], Lindström [79], and Todd [76].

Finally, geometric aspects of the completed Dilworth truncation are studied in Crapo [67/68, 70], Kung [83], and Mason [77, 81].

L.R. Wilcox, Modularity in Birkhoff lattices, *Bull. Amer. Math. Soc.* *50*(1944), 135-138. (MR 5, 170)

Let L be a lattice with no infinite ascending or descending chains. Then L is upper semimodular if and only if the relation of being a modular pair is symmetric. (See Wilcox [39].) An example is given to show that this result is in general false if the finiteness assumption is dropped.

For related work, see Maeda [65] and the references in there.

References

AIGNER, M. and DOWLING, T.A.: Matching theory for combinatorial geometries, *Trans. Amer. Math. Soc. 158*(1971), 231-245

BAER, R.: A unified theory of projective spaces and finite abelian groups, *Trans. Amer. Math. Soc. 52*(1942), 283-343.

BELL, E.T.: Exponential polynomials, *Annals of Math. 35*(1934), 258-277.

BIRKHOFF, G.: *Lattice theory* (Third edition), Amer. Math. Soc. Colloq. Publ., Vol. 25, Amer. Math. Soc., Providence, R.I., 1967.

BIRKHOFF, G.D.: A determinantal formula for the number of ways of coloring a map, *Annals of Math. (2) 14*(1912), 42-46.

BRUALDI, R.A.: Generalized transversal theory, *Théorie des matroïdes (Rencontre Franco-Brittanique, Brest, 1970),* pp. 5-30, Lecture Notes in Math., Vol. 211, Springer-Verlag, Berlin, 1971.

BRUALDI, R.A.: Menger's theorem and matroids, *J. London Math. Soc. (2) 4*(1971), 46-50.

BRUALDI, R.A.: On families of finite independence structures, *Proc. London Math. Soc. (3) 22*(1971), 265-293.

BRYLAWSKI, T.: Modular constructions for combinatorial geometries, *Trans. Amer. Math. Soc. 203*(1975), 1-44.

BRYLAWSKI, T.: The broken-circuit complex, *Trans. Amer. Math. Soc. 234*(1977), 417-433.

BRYLAWSKI, T. and OXLEY, J.: The broken-circuit complex: its structure and factorization, *Europ. J. Combin. 2*(1981), 107-121.

CORDOVIL, R.: Representation over a field of full simplicial matroids, *Europ. J. Combin. 1*(1980), 201-205.

CORDOVIL, R.: On Reid's 3-simplicial matroid theorem, *Combinatorica 2*(1982), 135-141.

CORDOVIL, R. and LAS VERGNAS, M.: Géométries simpliciales unimodulaires, *Discrete Math. 26*(1979), 213-217.

CRAPO, H.H.: A higher invariant for matroids, *J. Combin. Theory 2*(1967), 406-417.

CRAPO, H.H.: The joining of exchange geometries, *J. Math. Mech. 17*(1967/68), 837-852.

CRAPO, H.H.: Geometric duality and the Dilworth completion, *Combinatorial Structures and their Applications,* pp. 37-46, Gordon and Breach, New York, 1970.

CRAPO, H.H. and ROTA, G.-C.: *On the foundations of combinatorial theory: Combinatorial geometries* (Preliminary edition), M.I.T. Press, Cambridge, Mass., 1970.

CRAPO, H.H. and ROTA, G.-C.: Simplicial geometries, *Combinatorics (Proc. Sympos. Pure Math., Vol.* 19), pp. 71-75, Amer. Math. Soc., Providence, R.I., 1971.

CRAWLEY, P. and DILWORTH, R.P.: *Algebraic theory of lattices,* Prentice-Hall, Englewood Cliffs, N.J., 1973.

DILWORTH, R.P.: Lattices with unique irreducible decompositions, *Annals of Math. (2) 41*(1940), 771-777.

DILWORTH, R.P.: The structure of relatively complemented lattices, *Annals of Math. (2) 51*(1950), 348-359.

DUNSTAN, F.D.J.: Matroids and submodular functions, *Quart. J. Math. Oxford Ser. (2) 27*(1976), 339-348.

DUNSTAN, F.D.J., INGLETON, A.W. and WELSH, D.J.A.: Supermatroids, *Combinatorics (Proc. Conf. Combinatorial Math., Math. Inst., Oxford, 1972),* pp. 72-122, Inst. Math. Appl., Southend-on-Sea, 1972.

DWORAKOWSKI, K. and KACZMARSKI, J.: On some properties of flat graphs, *Demonstratio Math. 7*(1974), 5-12.

EDMONDS, J.: Submodular functions, matroids, and certain polyhdedra, *Combinatorial Structures and their Applications,* pp. 69-87, Gordon and Breach, New York, 1970.

EDMONDS, J. and ROTA, G.-C.: Submodular set functions, Abstract, *Waterloo Combinatorics Conf.,* Univ. of Waterloo, Ontario, 1966.

ERDÖS, P., LOVÁSZ, L. and SPENCER, J.: Strong independence of graphcopy functions, *Graph Theory and Related Topics,* pp. 165-172, Academic Press, New York, 1979.

FAIGLE, U. and HERRMANN, C.: Projective geometry on partially ordered sets, *Trans. Amer. Math. Soc. 266*(1981), 319-332.

FINKBEINER, D.T.: A semimodular imbedding of lattices, *Canad. J. Math. 12*(1960), 582-591.

FORD, L.R. and FULKERSON, D.R.: *Flows in Networks,* Princeton Univ. Press, Princeton, N.J., 1962.

GYÖRI, E. and MILNER, E.C.: A theorem of transversal theory for matroids of finite character, *Discrete Math. 23*(1978), 235-240.

HOGGAR, S.G.: Chromatic polynomials and broken cycles, *J. Combin. Theory Ser. B. 19*(1975), 77-86.

HSU, C.: On lattice-theoretic characterization of the parallelism in affine geometry, *Annals of Math. (2) 50*(1949), 1-7.

INGLETON, A.W.: A note on independence functions and rank, *J. London Math. Soc. 34*(1959), 49-56.

KAHN, J. and KUNG, J.P.S.: A classification of modularly complemented geometric lattices, *Europ. J. Combin., to appear.*

KANTOR, W.M.: Dimension and embedding theorems for geometric lattices, *J. Combin. Theory Ser. A 17*(1974), 173-195.

KANTOR, W.M.: Some highly geometric lattices, *Colloquio Internazionale sulle Teorie Combinatorie (Roma, 1973), Tomo I,* pp. 183-191, Accad. Naz. Lincei, Rome, 1976.

KUNDU, S. and LAWLER, E.L.: A matroid generalization of a theorem of Mendelsohn and Dulmage, *Discrete Math. 4*(1973), 159-163.

KUNG, J.P.S.: A factorization theorem for comaps of geometric lattices, *J. Combin. Theory Ser. B 34*(1983), 40-47.

LAS VERGNAS, M.: Sur un théorème de Rado, *C. R. Acad. Sci. Paris Sér. A-B 270*(1970), A804-A806.

LEFSCHETZ, S.: Planar graphs and related topics, *Proc. Nat. Acad. Sci. U. S. A. 54*(1965), 1763-1765.

LINDSTRÖM, B.: Nonregular simplicial matroids, *Discrete Math. 28*(1979), 101-102.

MAC LANE, S.: Topology and logic as a source of algebra, *Bull. Amer. Math. Soc. 82*(1976), 1-40.

MAEDA, F. and MAEDA, S.: *Theory of symmetric lattices,* Springer-Verlag, New York-Berlin, 1970.

MAEDA, S.: On the symmetry of the modular relation in atomic lattices, *J. Sci. Hiroshima Univ. Ser. A-I 29*(1965), 165-170.

MASON, J.H.: Matroids as the study of geometrical configurations, *Higher Combinatorics* (M. Aigner, Ed.), pp. 133-176, Reidel, Dordrecht, 1977.

MASON, J.H.: Glueing matroids together: a study of Dilworth truncations and matroid analogues of exterior and symmetric powers, *Algebraic Methods in Graph Theory (Proc. Conf., Szeged, 1978),* pp. 519-561, Colloq. Math. Soc. János Bolyai 25, North-Holland, Amsterdam, 1981.

MCDIARMID, C.J.H.: Independence structures and submodular functions, *Bull. London Math. Soc. 4*(1972), 196-198.

MCDIARMID, C.J.H.: A note on a theorem of R. Rado on independent transversals, *Bull. London Math. Soc. 5*(1973), 315-316.

MCDIARMID, C.J.H.: Rado's theorem for polymatroids, *Math. Proc. Cambridge Philos. Soc. 78*(1975), 513-514.

MCDIARMID, C.J.H.: Extensions of Menger's theorem, *Quart. J. Math. Oxford Ser. (2) 26*(1975), 141-157.

VON NEUMANN, J.: *Collected works, Vol. 4: Continuous geometries and other topics,* Pergamon Press, London, 1962.

NGUYEN, H.Q.: Functors of the category of combinatorial geometries and strong maps, *Discrete Math. 20*(1977/78), 143-158.

NGUYEN, H.Q.: Semimodular functions and combinatorial geometries, *Trans. Amer. Math. Soc. 238*(1978), 355-383.

O'NEIL, P.V.: A short proof of Mac Lane's planarity theorem, *Proc. Amer. Math. Soc. 37*(1973), 617-618.

PERFECT, H.: A generalization of Rado's theorem on independent transversals, *Proc. Cambridge Philos. Soc. 66*(1969), 513-515.

PERFECT, H.: Independence spaces and combinatorial problems, *Proc. London Math. Soc. (3) 19*(1969), 17-30.

PERFECT, H.: Independent transversals with constraints, *J. London Math. Soc.* (2) 5(1972), 385-386.

PERFECT, H. and PYM, J.S.: Submodular functions and independence structures, *J. Math. Anal. Appl. 30*(1970), 1-31.

RADO, R.: Axiomatic treatment of rank in infinite sets, *Canad. J. Math. 1*(1949), 337-343.

RAO, A. R. and RAO, K.P.S.B.: Modular pairs in combinatorial geometries, *J. London Math. Soc. (2) 13*(1976), 385-386.

RIORDAN, J.: *An introduction to combinatorial analysis,* Wiley, New York, 1958.

THOMASSEN, C.: Planarity and duality of finite and infinite graphs, *J. Combin. Theory Ser. B 29*(1980), 244-271.

THOMASSEN, C.: Duality of infinite graphs, *J. Combin. Theory Ser. B. 33*(1982), 137-160.

THOMSEN, G.: *Grundlagen der Elementargeometrie,* Hamburger mathematische Einzelschriften 15, Teubner, Leipzig, 1933.

TODD, M.J.: Characterizing binary simplicial matroids, *Discrete Math. 16*(1976), 61-70.

TORRANCE, C.C.: Tangent lines and planes in topological spaces, *Trans. Amer. Math. Soc. 41*(1937), 193-207.

TRUEMPER, K.: On Whitney's 2-isomorphism theorem for graphs, *J. Graph Theory 4*(1980), 39-43.

TUTTE, W.T.: How to draw a graph, *Proc. London Math. Soc. (3) 13*(1963), 743-767.

TUTTE, W.T.: Codichromatic graphs, *J. Combin. Theory Ser. B 16*(1974), 168-174.

WELSH, D.J.A.: Applications of a theorem by Rado, *Mathematika 15*(1968), 199-203.

WELSH, D.J.A.: Transversal theory and matroids, *Canad. J. Math. 21*(1969), 1323-1330.

WELSH, D.J.A.: On matroid theorems of Edmonds and Rado, *J. London Math. Soc. (2)*(1970), 251-256.

WELSH, D.J.A.: Related classes of set functions, *Studies in Pure Mathematics (Presented to Richard Rado)*, pp. 261-269, Academic Press, London, 1971.
WELSH, D.J.A.: Generalized versions of Hall's theorem, *J. Combin. Theory Ser. B 10*(1971), 95-101.
WILF, H.S.: Which polynomials are chromatic?, *Colloquio Internazionale sulle Teorie Combinatorie (Roma 1973), Tomo I*, pp. 247-256, Accad. Naz. Lincei, Rome, 1976.
WILF, H.S.: A note on $P(-\lambda; G)$, *J. Combin. Theory Ser. B 22*(1977), 296.

6. Strong maps

The final portion of this chapter is concerned with maps or morphisms between matroids or geometric lattices. Since there are many natural geometric and algebraic methods which construct from a given set of vectors another set of vectors "related" to it, there are correspondingly many different notions of morphisms between matroids. Among these, the most extensively studied are the notions of strong maps, weak maps, and comaps.

Chronologically, the first of these notions to appear is the notion of a weak map. A function $\tau: G \to H$ from a geometric lattice G to a geometric lattice H is said to be a *weak map* if it preserves joins of points (i.e. $\tau(p_1 \vee \dots \vee p_m) = \tau(p_1) \vee \dots \vee \tau(p_m)$ for points $p_1, \dots, p_m$) and decreases rank (i.e. for any element x of G, $r_H(\tau(x)) \leq r_G(x)$). Equivalently, τ is a weak map if: for any subset I of points of G, τ restricted to I is one-to-one and $\tau(I)$ is an independent set of points in H imply I is an independent set in G. Thus, the points of H may be thought of as being "in more special position" than the points of G. The intuition behind this observation is supported by the following example. Let $M(R)$ be a set of vectors with coordinates in an integral domain R and let P be a prime ideal in R. The set $M(R/P)$ of vectors obtained from $M(R)$ by regarding the coordinates as elements in R/P is said to be obtained from $M(R)$ by *specialization of coordinates*. The function mapping a vector in $M(R)$ to its specialization in $M(R/P)$ induces a weak map of the lattice of closed sets of $M(R)$ to the lattice of closed sets of $M(R/P)$.

Weak maps (or maps, as they were then called) were first defined in Higgs [66]. A survey can be found in Kung and Nguyen [86].

A more restricted notion of morphisms is the notion of strong maps. A weak map $\sigma: G \to H$ is said to be a *strong map* if it preserves joins of (arbitrary) elements in G (i.e. $\sigma(x \vee y) = \sigma(x) \vee \sigma(y)$ for elements x and y in G). Equivalently, σ is a strong map if the inverse image of a closed set of points is also a closed set of points in G. Thus, strong maps are analogues for matroids of continuous maps in topology.

The foundations of the theory of strong maps were laid in the paper "Strong maps of geometries" by Higgs, the last paper reprinted in this chapter. (See also Crapo [67] for an independent development.) This paper has three main parts. In the first part, the notions of strong map, subgeometry and quotient are defined. Essentially, a *subgeometry* S of a geometric lattice G is a join-closed subset of G such that the injection $i: S \to G$ is a strong map. A *quotient* Q is a meet-closed subset of G such that the *canonical surjection* $G \to Q$, $y \to \bigwedge \{x \in Q: x \geq y\}$ is a strong map. (A quotient may be thought of as a matroid on a geometric lattice.) If a is an element of G, then the set $G/a = [a, 1] = \{x: x \geq a\}$ is a quotient of G. Now let $\sigma: G \to H$ be a strong map. An element x in G is said to be *σ-closed* if it is the maximum element in G mapped by σ to $\sigma(x)$, i.e. if $x = \bigvee \{y \in G: \sigma(y) = \sigma(x)\}$. The σ-closed elements form a quotient Q of G. If σ is surjective, then $Q \cong H$. Every strong map can be factored into a surjection followed by an injection.

The lift construction is described in the second part. Let Q be a quotient of G and $\sigma: G \to Q$ the canonical surjection. An element x in G is said to be *σ-independent* if the rank of its image in Q equals its rank in G. If x is σ-independent and $y \leq x$ then y is also σ-independent. It turns out that there is at most a "single layer" of σ-independent elements in G but not in Q lying between the σ-independent elements in Q and the σ-dependent elements in Q. The lift construction consists of adding this layer of elements to Q, thus obtaining a quotient $L = L(G \to Q)$ of G between G and Q. If $G \neq Q$, the rank of L is one greater than the rank of Q.

The factorization theorem is proved in the third part. To state this, the notion of contraction is needed. The *kernel* $\ker \sigma$ of the strong map $\sigma: G \to H$ is the unique σ-closed element of G mapped to the minimum of H. If σ is surjective and $H \cong G/\ker \sigma$, then σ is said to be a *contraction*. A contraction is the analogue of a projection from a subspace. The factorization theorem states that every strong map can be factored into an injection followed by a contraction. This is proved in the following way. Consider first a surjective strong map $\sigma: G \to H$. By relabelling elements, it may be assumed that H is a quotient of G. Let n be the difference $\operatorname{rank}(G) - \operatorname{rank}(H)$ and let E be a geometric lattice of rank n. The function $G \oplus E \to H$ obtained by taking the direct product of σ and the function sending every element of E to the minimum of H is a strong map. Let $L = L^n(G \oplus E \to H)$ be the quotient of $G \oplus E$ obtained by applying the lift construction n times. Both G and E are subgeometries of L. Moreover, if a is the image of 1_E, the maximum of E, under the injection $E \to L$, then $H = L/a$ and the strong map $L \to H$ is a contraction. Thus, σ is factored into an injection $G \to L$ followed by a contraction $L \to H$. The general case now follows easily. The geometric lattice L is now called the *Higgs major* of the strong map σ.

The factorization theorem offers much evidence that strong maps are the right choice of morphisms for many purposes. In particular, it asserts that the category of strong maps is the smallest category (in the sense of containing the fewest morphisms) of geometric lattices containing injections and contractions as morphisms. In this connection, we remark that strong maps are often the right morphisms to make many algebraic constructions functorial. See, for example, Graves [71] and Orlik and Solomon [80] (p. 171).

The third notion of a morphism is the notion of a comap. A function γ: $G \to H$ from a geometric lattice G to a geometric lattice H is a said to be a *comap* if it preserves the relation of covers or equals and meets of modular pairs of elements. The comap γ is said to be *normalized* if it sends the minimum of G to the minimum of H. The geometric model behind a normalized comap is a retraction or "section". Let S be a set of vectors in the vector space V and let U be a subspace of V. The function defined from the lattice of closed sets of S to the lattice of subspaces contained in U which sends an element y to the intersection of U with the subspace spanned by the vectors in y is a normalized comap. It is called the retraction of S onto U. More generally, let G be a geometric lattice and u a modular element in G. The map γ: $G \to [0, u]$, $y \mapsto y \wedge u$ is a normalized comap and is called the *retraction of G onto the modular element u*. As for strong maps, there is a factorization theorem for normalized comaps. This states that every normalized comap can be factored into an injection followed by a retraction onto a modular flat.

Comaps were first studied in Crapo [67/68]. The factorization theorem was stated in Crapo [71] and proved in Kung [83].

Much work has been done on strong maps since Higgs' paper. The following is a list of papers on strong maps.

Surveys: Brylawski [86], Crapo [71], Kung [86], Mason [77].

Other proofs of the factorization theorem: Kung [77], Las Vergnas [80], Welsh [76], p. 315.

Extension of the factorization theorem: Wilde [77].

Elementary strong maps, modular filters, and single element extensions: Cheung [74], Cheung and Crapo [73], Cordovil [83], Crapo [65, 71], Dowling and Kelly [74, 76].

Elementary factorizations and majors: Cheung and Crapo [73], Kelly and Kennedy [78], Kennedy [75].
Principal maps and transversal matroids: Brown [74], Dowling and Kelly [74], Kung [78], Oxley, Prendergast and Row [82].
Bimatroid multiplication and strong maps: Kung [78], Schrijver [79].
Strong maps of geometries on partially ordered sets: Faigle [80].
Enumerative aspects: Crapo [68], Las Vergnas [77/78, 80].
Geometric maps: Sachs [71, 72].

References

Brown, T.J.: Transversal theory and F-products, *J. Combin. Theory Ser. A* *17*(1974),290- 298.

Brylawski, T.: Constructions, in *Theory of matroids* (N.L. White, ed.), Cambridge Univ. Press, Cambridge, 1986.

Cheung, A.L.C.: Compatibility of extensions of a combinatorial geometry, Thesis, Univ. Waterloo, Ontario, Canada, 1974.

Cheung, A.L.C. and Crapo, H.H.: On relative position in extensions of combinatorial geometries, preprint, 1973.

Cordovil, R.: Sur la compatibilité des extensions ponctuelles d'un matroïde, *J. Combin. Theory Ser. B 34*(1983), 209-223.

Crapo, H.H.: Single-element extensions of matroids, *J. Res. Nat. Bur. Standards Sect. B 69B*(1965), 55-65.

Crapo, H.H.: Structure theory for geometric lattices, *Rend. Sem. Mat. Univ. Padova 38*(1967), 14-22.

Crapo, H.H.: The joining of exchange geometries, *J. Math. Mech. 17*(1967/68), 837-852.

Crapo, H.H.: Möbius inversion in lattices, *Archiv der Math. (Basel)* *19*(1968), 595-607.

Crapo, H.H.: Orthogonal representations of combinatorial geometries, *Atti del Convegno di Geometria Combinatoria e sue Applicazioni (Univ. Perugia, Perugia, 1970),* pp. 175-186, Inst. Math., Univ. Perugia, Perugia, 1971.

Crapo, H.H.: *Constructions in combinatorial geometry,* Notes, N.S.F. Combinatorial Theory Advanced Science Seminar, Bowdoin College, Bowdoin, Maine, 1971.

Dowling, T.A. and Kelly, D.G.: Elementary strong maps and transversal geometries, *Discrete Math. 7*(1974), 209-224.

DOWLING, T.A. and KELLY, D.G.: Elementary strong maps between combinatorial geometries, *Colloquio Internazionale sulle Teorie Combinatorie (Roma, 1973), Tomo II,* pp. 121-152, Accad. Naz. Lincei, Rome, 1976.

FAIGLE, U.: Über Morphismen halbmodularer Verbände, *Aequationes Math. 21*(1980), 53-67.

GRAVES, W.H.: An algebra associated to a combinatorial geometry, *Bull. Amer. Math. Soc. 77*(1971), 757-761.

HIGGS, D.A.: Maps of geometries, *J. London Math. Soc. 41*(1966), 612-618.

KELLY, D.G. and KENNEDY, D.: The Higgs factorization of a geometric strong map, *Discrete Math. 22*(1978), 139-146.

KENNEDY, D.: Majors of geometric strong maps, *Discrete Math. 12*(1975), 309-340.

KUNG, J.P.S.: The core extraction axiom for combinatorial geometries, *Discrete Math. 19*(1977), 167-175.

KUNG, J.P.S.: The alpha function of a matroid-I: Transversal matroids, *Stud. Appl. Math. 58*(1978), 263-275.

KUNG, J.P.S.: Bimatroids and invariants, *Advances in Math. 30*(1978), 238-249.

KUNG, J.P.S.: A factorization theorem for comaps of geometric lattices, *J. Combin. Theory Ser. B 34*(1983), 40-47.

KUNG, J.P.S.: Strong maps, in *Theory of matroids* (N.L. White, ed.), Cambridge Univ. Press, Cambridge, 1986.

KUNG, J.P.S. and NGUYEN, H.Q.: Weak maps, in *Theory of matroids* (N.L. White, ed.), Cambridge Univ. Press, Cambridge, 1986.

LAS VERGNAS, M.: Acyclic and totally cyclic orientations of combinatorial geometries, *Discrete Math. 20*(1977/78), 51-61.

LAS VERGNAS, M.: On the Tutte polynomial of a morphism of matroids, *Combinatorics 79, Part I, Ann. Discrete Math. 8*(1980), 7-20.

MASON, J.H.: Matroids as the study of geometrical configurations, *Higher Combinatorics* (M. Aigner, ed.), pp. 133-176, Reidel, Dordrecht, 1977.

ORLIK, P. and SOLOMON, L.: Combinatorics and topology of complements of hyperplanes, *Invent. Math. 56*(1980), 167-189.

OXLEY, J.G., PRENDERGAST, K. and ROW, D.H.: Matroids whose ground sets are domains of functions, *J. Austral. Math. Soc. Ser. A 32*(1982), 380-387.

SACHS, D.: Geometric mappings on geometric lattices, *Canad. J. Math. 23*(1971), 22-35.

SACHS, D.: A note on geometric mappings, *Rend. Sem. Mat. Padova 47*(1972), 23-28.

Schrijver, A.: Matroids and linking systems, *J. Combin. Theory Ser. B* *26*(1979), 349-369.
Welsh, D.J.A.: *Matroid theory,* Academic Press, London and New York, 1976.
Wilde, P.J.: Matroids with given restrictions and contractions, *J. Combin. Theory Ser. B 22*(1977), 122-130.

Reprinted from *American Journal of Mathematics*,
volume 57, pages 800 through 804,
by permission of the Johns Hopkins University Press.

ON THE ABSTRACT PROPERTIES OF LINEAR DEPENDENCE.[1]

By HASSLER WHITNEY.

1. Introduction. Let $C_1, C_2, \cdots, C_n$ be the columns of a matrix $\boldsymbol{M}$. Any subset of these columns is either linearly independent or linearly dependent; the subsets thus fall into two classes. These classes are not arbitrary; for instance, the two following theorems must hold:

(a) Any subset of an independent set is independent.

(b) If $\boldsymbol{N}_p$ and $\boldsymbol{N}_{p+1}$ are independent sets of p and $p+1$ columns respectively, then $\boldsymbol{N}_p$ together with some column of $\boldsymbol{N}_{p+1}$ forms an independent set of $p+1$ columns.

There are other theorems not deducible from these; for in § 16 we give an example of a system satisfying these two theorems but not representing any matrix. Further theorems seem, however, to be quite difficult to find. Let us call a system obeying (a) and (b) a "matroid." The present paper is devoted to a study of the elementary properties of matroids. The fundamental question of completely characterizing systems which represent matrices is left unsolved. In place of the columns of a matrix we may equally well consider points or vectors in a Euclidean space, or polynomials, etc.

This paper has a close connection with a paper by the author on linear graphs;[2] we say a subgraph of a graph is independent if it contains no circuit. Although graphs are, abstractly, a very small subclass of the class of matroids, (see the appendix), many of the simpler theorems on graphs, especially on non-separable and dual graphs, apply also to matroids. For this reason, we carry over various terms in the theory of graphs to the present theory. Remarkably enough, for matroids representing matrices, dual matroids have a simple geometrical interpretation quite different from that in the case of graphs (see § 13).

The contents of the paper are as follows: In Part I, definitions of matroids in terms of the concepts rank, independence, bases, and circuits are considered, and their equivalence shown. Some common theorems are deduced (for instance Theorem 8). Non-separable and dual matroids are studied in

[1] Presented to the American Mathematical Society, September, 1934.

[2] "Non-separable and planar graphs," *Transactions of the American Mathematical Society*, vol. 34 (1932), pp. 339-362. We refer to this paper as G.

Part II; this section might replace much of the author's paper G. The subject of Part III is the relation between matroids and matrices. In the appendix, we completely solve the problem of characterizing matrices of integers modulo 2, of interest in topology.

I. Matroids.

2. Definitions in terms of rank. Let a set M of elements $e_1, e_2, \cdots, e_n$ be given. Corresponding to each subset N of these elements let there be a number $r(N)$, the *rank* of N. If the three following postulates are satisfied, we shall call this system a *matroid.*

(R_1) *The rank of the null subset is zero.*

(R_2) *For any subset N and any element e not in N,*

$$r(N+e)=r(N)+k, \qquad (k=0 \text{ or } 1).$$

(R_3) *For any subset N and elements e_1, e_2 not in N, if $r(N+e_1) = r(N+e_2) = r(N)$, then $r(N+e_1+e_2)=r(N)$.*

Evidently *any subset of a matroid is a matroid.* In what follows, M is a fixed matroid. We make the following definitions:

$$\rho(N) = \text{number of elements in } N.$$

$$n(N) = \rho(N) - r(N) = \textit{nullity} \text{ of } N.$$

N is *independent,* or, the elements of N are independent, if $n(N)=0$; otherwise, N, and its set of elements, are *dependent.*

Lemma 1. *For any N, $r(N) \geqq 0$ and $n(N) \geqq 0$. If $N \subset M$, then $r(N) \leqq r(M)$, $n(N) \leqq n(M)$.*

Lemma 2. *Any subset of an independent set is independent.*

e is *dependent on* N if $r(N+e)=r(N)$; otherwise e is *independent of* N.

A *base* is a maximal independent submatroid of M, i. e. a matroid B in M such that $n(B)=0$, while $B \subset N$, $B \neq N$ implies $n(N)>0$. See also Theorem 7. A *base complement* $A = M - B$ is the complement in M of a base B. A *circuit* is a minimal dependent matroid, i. e. a matroid P such that $n(P)>0$, while $N \subset P$, $N \neq P$ implies $n(N)=0$.[3]

Theorem 1. *N is independent if and only if it is contained in a base, or, if and only if it contains no circuit.*

---

[3] Compare G, Theorem 9.

THEOREM 2. *A circuit is a minimal submatroid contained in no base, i. e. containing at least one element from each base complement. A base is a maximal submatroid containing no circuit. A base complement is a minimal submatroid containing at least one element from each circuit.*

The above facts follow at once from the definitions. Note the reciprocal relationship between circuits and base complements. Note also that the definitions of independence and of being a circuit depend only on the given subset, while the property of being a base depends on the relationship of the subset to M.

3. Properties of rank. Our object here is to prove Theorem 3. The following definition will be useful:

$$\Delta(M, N) = r(M + N) - r(M). \tag{3.1}$$

LEMMA 3. $\Delta(M + e_2, e_1) \leqq \Delta(M, e_1)$.

Suppose first $r(M + e_1) = r(M) + 1$; then $r(M + e_1 + e_2) = r(M) + k$, $k = 1$ or 2. If $k = 2$, then $r(M + e_2) = r(M) + 1$, on account of (R_2), and the inequality holds; if $k = 1$, $r(M + e_2) = r(M) + l$, $l = 0$ or 1, and it holds again. If $r(M + e_2) = r(M) + 1$, the same reasoning applies. If finally $r(M + e_1) = r(M + e_2) = r(M)$, the inequality follows from (R_3).

LEMMA 4. $\Delta(M + N, e) \leqq \Delta(M, e)$.

If $N = e_1 + \cdots + e_p$, the last lemma gives

$$\Delta(M + N, e) \leqq \Delta(M + e_1 + \cdots + e_{p-1}, e) \leqq \cdots \leqq \Delta(M, e).$$

THEOREM 3. $\Delta(M + N_2, N_1) \leqq \Delta(M, N_1)$, *or,*

$$r(M + N_1 + N_2) \leqq r(M + N_1) + r(M + N_2) - r(M). \tag{3.2}$$

This is true if N_1 contains but a single element. For the general case, we apply the last lemma and induction, setting $N_1 = N' + e$:

$$\begin{aligned} \Delta(M + N_2, N_1) &= \Delta(M + N_2 + e, N') + \Delta(M + N_2, e) \\ &\leqq \Delta(M + e, N') + \Delta(M, e) = \Delta(M, N_1). \end{aligned}$$

(3.2) is evidently equivalent to:

$$r(M_1 + M_2) \leqq r(M_1) + r(M_2) - r(M_1 M_2). \tag{3.3}$$

4. Deduction of (I_1), (I_2) from (R_1), (R_2), (R_3). The first postulate

4

on independent sets below obviously holds if (R_1) and (R_2) hold. To prove (I_2), take N, N' as given there; then

$$r(N) = p, \qquad r(N') = p + 1.$$

We must show that for some i, $\Delta(N, e'_i) = 1$. (Then e'_i does not lie in N.) If this is not so, then on using Lemma 4 we find

$$\begin{aligned} 1 = r(N') - r(N) &\leqq \Delta(N, N') \\ &= \Delta(N, e'_1) + \Delta(N + e'_1, e'_2) + \cdots + \Delta(N + e'_1 + \cdots + e'_p, e'_{p+1}) \\ &\leqq \Delta(N, e'_1) + \Delta(N, e'_2) + \cdots + \Delta(N, e'_{p+1}) = 0, \end{aligned}$$

a contradiction.

5. Deduction of (C_1), (C_2) from (R_1), (R_2), (R_3). We shall need here a theorem showing how the nullity (or rank) of a matroid may be determined when we know what circuits it contains.

LEMMA 5. *Each element of a circuit is dependent on the rest of the circuit.*

If e is an element of the circuit P, then $n(P) = 1$, $n(P - e) = 0$; hence $r(P) = \rho(P) - 1 = \rho(P - e) = r(P - e)$.

LEMMA 6. *If e is dependent on P_1 but on no proper subset of P_1, then $P = P_1 + e$ is a circuit.*

As $\Delta(P_1, e) = 0$, $r(P) = r(P_1) \leqq \rho(P_1) < \rho(P)$, $n(P) > 0$, and P contains a circuit P'. If P' does not contain e, take e' in P'; then

$$\Delta(P_1 - e', e') \leqq \Delta(P' - e', e') = 0,$$

hence $r(P_1 - e') = r(P_1)$, and

$$\begin{aligned} \Delta(P_1 - e', e) &= r(P_1 - e' + e) - r(P_1 - e') \\ &\leqq r(P_1 + e) - r(P_1) = \Delta(P_1, e) = 0, \end{aligned}$$

and e is dependent on the proper subset $P_1 - e'$ of P_1, a contradiction. Therefore P' contains e. As P' is a circuit, e is dependent on the rest of P'; hence $P' = P$.

THEOREM 4. *If e is not in N, there is a circuit in $N + e$ which contains e if and only if e is dependent on N.*

Suppose $P_1 + e = P$ is a circuit, $P_1 \subset N$. Then

$$\Delta(N, e) \leqq \Delta(P_1, e) = 0,$$

and e is dependent on N. Suppose, conversely, $\Delta(N, e) = 0$. Let P_1 be a smallest subset of N on which e is dependent; then by the last lemma, $P = P_1 + e$ is a circuit. (It may be that $P = e$.)

THEOREM 5. *If N is formed element by element, then $n(N)$ is just the number of times that adding an element increases the number of circuits present.*

Say $N = e_1 + \cdots + e_p$. Then if O is the null set,

$$r(N) = \Delta(O, e_1) + \Delta(e_1, e_2) + \cdots + \Delta(e_1 + \cdots + e_{p-1}, e_p).$$

Each $\Delta(e_1 + \cdots + e_{i-1}, e_i) = 0$ or 1, and $= 0$ if and only if e_i is dependent on $e_1 + \cdots + e_{i-1}$, i. e. if and only if there is a circuit in $e_1 + \cdots + e_i$ containing e_i. The number of terms is $p = \rho(N)$, and the theorem follows.

We turn now to the proof of (C_1) and (C_2). The first is obvious. To prove the second, take P_1, P_2, e_1, e_2 as given. As

$$\Delta(P_1 - e_2, e_2) = \Delta(P_2 - e_1, e_1) = 0,$$

we have

$$\Delta(P_1 + P_2 - e_2, e_2) = \Delta(P_1 + P_2 - e_1 - e_2, e_1) = 0.$$

These equations give

$$r(P_1 + P_2 - e_1 - e_2) = r(P_1 + P_2 - e_2) = r(P_1 + P_2).$$

Using (R_2) gives

$$r(P_1 + P_2 - e_1) = r(P_1 + P_2 - e_1 - e_2);$$

hence the required circuit P_3 exists, by Theorem 4.

6. Postulates for independent sets. Let M be a set of elements. Let any subset N of M be either "independent" or "dependent." Let the two following postulates be satisfied:

(I_1) *Any subset of an independent set is independent.*

(I_2) *If $N = e_1 + \cdots + e_p$ and $N' = e'_1 + \cdots + e'_{p+1}$ are independent, then for some i such that e'_i is not in N, $N + e'_i$ is independent.*

The resulting system is equivalent to a matroid, as we now show. Given any subset N of M, we let $r(N)$ be the number of elements in a largest independent subset of N. Obviously Postulates (R_1) and (R_2) are satisfied; we must prove (R_3). Say

$$r(N+e_1)=r(N+e_2)=r(N)=r.$$

Then $r(N+e_1+e_2)=r$ or $r+1$. If it equals $r+1$, there is an independent set $N'=e'_1+\cdots+e'_{r+1}$ in $N+e_1+e_2$. Let $N''=e_1''+\cdots+e_r''$ be an independent set in N. By (I_2) there is an i such that $N''+e'_i$ is an independent set of $r+1$ elements. But $N''+e'_i$ lies in $N+e_1$ or in $N+e_2$, and hence $r(N+e_1)$ or $r(N+e_2)\geqq r+1$, a contradiction. Therefore $r(N+e_1+e_2)=r$, as required.

We have shown how to deduce either set of postulates (R) or (I) from the other. Moreover the definitions of the rank and the independence or dependence of any subset of M agree under the two systems, and hence they are equivalent.

7. Postulates for bases. Let M be a set of elements, and let each subset either be or not be a "base." We assume

(B_1) *No proper subset of a base is a base.*

(B_2) *If B and B' are bases and e is an element of B, then for some element e' in B', $B-e+e'$ is a base.*

We shall prove the equivalence of this system with the preceding one. We write here $e_1e_2\cdots$ instead of $e_1+e_2+\cdots$ for short.

THEOREM 6. *All bases contain the same number of elements.*

For suppose

$$B=e_1\cdots e_pe_{p+1}\cdots e_qe_{q+1}\cdots e_r,$$
$$B'=e_1\cdots e_pe'_{p+1}\cdots e'_q$$

are bases, with exactly $e_1,\cdots,e_p$ in common, and $r>q$. We might have $p=0$. $q>p$, on account of (B_1). By (B_2), we can replace e_{p+1} in B by an element e' of B', giving a base B_1. $e'=e'_{i_1}$ is one of the elements $e'_{p+1},\cdots,e'_q$, for otherwise B_1 would be a proper subset of B. Hence

$$B_1=e_1\cdots e_pe'_{i_1}e_{p+2}\cdots e_qe_{q+1}\cdots e_r.$$

If $q>p+1$, we replace e_{p+2} in B_1 by an element e'_{i_2} of B', giving a base B_2. Continuing in this manner, we obtain finally the base

$$B_{q-p} = e_1 \cdots e_p e'_{p+1} \cdots e'_q e_{q+1} \cdots e_r.$$

But this contains B' as a proper subset, contradicting (B_1).

We shall say a subset of M is independent if it is contained in a base. (I_1) obviously holds; we shall prove (I_2). Let N, N' be independent sets in the bases B, B'. Say

$$B = e_1 \cdots e_p e_{p+1} \cdots e_q e_{q+1} \cdots e_r e_{r+1} \cdots e_s,$$
$$B' = e_1 \cdots e_p e'_{p+1} \cdots e'_q e'_{q+1} \cdots e'_r e_{r+1} \cdots e_s,$$
$$N = e_1 \cdots e_p e_{p+1} \cdots e_q, \qquad N' = e_1 \cdots e_p e'_{p+1} \cdots e'_q e'_{q+1}.$$

Then N and N' have just $e_1, \cdots, e_p$ in common, and B and B' have just these elements and $e_{r+1}, \cdots, e_s$ in common. By (B_2), there is an element e'_{i_1} of B' such that

$$B_1 = B - e_{q+1} + e'_{i_1}$$

is a base. (This element cannot be any of $e_1, \cdots, e_p, e_{r+1}, \cdots, e_s$, by (B_1).) If i_1 is one of the numbers $p + 1, p + 2, \cdots, q + 1$, then $N + e'_{i_1}$ is in a base B_1, as required. Suppose not; then there is a base

$$B_2 = B_1 - e_{q+2} + e'_{i_2},$$

with $i_2 \neq i_1$. If $p + 1 \leqq i_2 \leqq q + 1$, $N + e'_{i_2}$ is in a base B_2. If not, we find a base B_3, etc. We can drop out each of the $r - q$ elements $e_{q+1}, \cdots, e_r$ in turn; as there are only $r - q - 1$ elements e'_i with $i > q + 1$, we find at some point a base containing $e_1, \cdots, e_q, e'_j$ with $p + 1 \leqq j \leqq q + 1$. Then e'_j is in N', and $N + e'_j$ is in a base and is thus independent, as required.

The definitions of base and independent sets in the two systems (I) and (B) are easily seen to agree. Suppose (I_1) and (I_2) hold. (B_1) obviously holds; using (I_2), we prove that all bases contain the same number of elements; (B_2) now follows at once from (I_2). Hence the two systems are equivalent.

THEOREM 7. *B is a base in M if and only if*

$$r(B) = r(M), \qquad n(B) = 0.$$

Evidently B is a base under the given conditions. To prove the converse, we note first that there exists a base with $r(M)$ elements, as $r(M)$ is the maximum number of independent elements in M (see § 6). By Theorem 6, all bases have this many elements, and the equations follow.

THEOREM 8. *If B is a base and N is independent, then for some N' in B, $N + N'$ is a base.*

This follows from repeated application of Postulate (I_2) and the last theorem.

8. Postulates for circuits. Let M be a set of elements, and let each subset either be or not be a "circuit." We assume:

(C_1) *No proper subset of a circuit is a circuit.*

(C_2) *If P_1 and P_2 are circuits, e_1 is in both P_1 and P_2, and e_2 is in P_1 but not in P_2, then there is a circuit P_3 in $P_1 + P_2$ containing e_2 but not e_1.*

(C_2) may be phrased as follows: If the circuits P_1 and P_2 have the common element e, then $P_1 + P_2 - e$ is the union of a set of circuits.

We shall define the rank of any subset of M, and shall then show that the postulates for rank are satisfied. Let $e_1, \cdots, e_p$ be any ordered set of elements of M. Set $\Gamma_i = 0$ if there is a circuit in $e_1 + \cdots + e_i$ containing e_i, and set $\Gamma_i = 1$ otherwise (compare Theorem 5). Let the "rank" of $(e_1, \cdots, e_p)$ be

$$r(e_1, \cdots, e_p) = \sum_{i=1}^{p} \Gamma_i.$$

LEMMA 7. $r(e_1, \cdots, e_{q-2}, e_{q-1}, e_q) = r(e_1, \cdots, e_{q-2}, e_q, e_{q-1})$.

To prove this, let N be the ordered set $e_1, \cdots, e_{q-2}$, and set

$$r(N) = r, \qquad r(N, e_{q-1}) = r_1, \qquad r(N, e_q) = r_2,$$
$$r(N, e_{q-1}, e_q) = r_{12}, \qquad r(N, e_q, e_{q-1}) = r_{21}.$$

CASE 1. There is no circuit in $N + e_{q-1}$ containing e_{q-1}, and none in $N + e_q$ containing e_q. Then

$$r_1 = r_2 = r + 1.$$

If there is a circuit in $N + e_{q-1} + e_q$ containing e_{q-1} and e_q, then

$$r_{12} = r_1 = r_2 = r_{21};$$

otherwise,

$$r_{12} = r_1 + 1 = r_2 + 1 = r_{21}.$$

CASE 2. There is a circuit P_2 in $N + e_{q-1}$ containing e_{q-1}, and a circuit P_1 in $N + e_{q-1} + e_q$ containing e_{q-1} and e_q. Then, by (C_2), there is a circuit P_3 in $N + e_q$ containing e_q. Hence

$$r_{12} = r_1 = r = r_2 = r_{21}.$$

CASE 3. There is a circuit P_2 as above, but no circuit P_1 as above. If there is a circuit P_3 as above, the last set of equations hold. Otherwise,

$$r_{12} = r_1 + 1 = r + 1 = r_2 = r_{21}.$$

CASE 4. There is a circuit in $N + e_q$ containing e_q. This case overlaps the two preceding ones; the proof above applies here also.

LEMMA 8. *The rank of any subset N is independent of the ordering of the elements of N.*

We saw above that interchanging the last two elements of any subset does not alter the rank; hence, evidently, interchanging any two adjacent elements leaves the rank unchanged. Any ordering of M may be obtained from any other by a number of interchanges of adjacent elements; the rank remains unchanged at each step, proving the lemma.

Postulates (R_1) and (R_2) are obviously satisfied. To prove (R_3), suppose $r(N + e_1) = r(N + e_2) = r(N)$. Then there is a circuit in $N + e_1$ containing e_1 and one in $N + e_2$ containing e_2; hence $r(N + e_1 + e_2) = r(N)$.

The definitions of rank and of circuits under the two systems (R), (C) agree, and hence the systems are equivalent.

9. Fundamental sets of circuits. The circuits $P_1, \cdots, P_q$ of a matroid M form a *fundamental set of circuits* if $q = n(M)$ and the elements $e_1, \cdots, e_n$ of M can be ordered so that P_i contains e_{n-q+i} but no e_{n-q+j} $(j > i)$. The set is *strict* if P_i contains e_{n-q+i} but no e_{n-q+j} $(0 < j < i$ or $j > i)$. These sets may be called sets *with respect to* $e_{n-q+1}, \cdots, e_n$.

THEOREM 9. *If $B = e_1 + \cdots + e_{n-q}$ is a base in $M = e_1 + \cdots + e_n$, then there is a strict fundamental set of circuits with respect to $e_{n-q+1}, \cdots, e_n$; these circuits are uniquely determined.*

As $r(B) = r(M)$, $\Delta(B, e_i) = 0$ $(i = n - q + 1, \cdots, n)$. Hence, by Theorem 4, there is a circuit P_i containing e_i and elements (possibly) of B. $P_{n-q+1}, \cdots, P_n$ is the required set. Suppose, for a given i, there were also a circuit $P'_i \neq P_i$. Then Postulate (C_2) applied to P_i and P'_i would give us a circuit P in B, which is impossible.

This theorem corresponds to the theorem that if a square submatrix N of a matrix M is non-singular, then N can be turned into the unit matrix by a linear transformation on the rows of M.

THEOREM 10. *If $P_1, \cdots, P_q$ form a fundamental set of circuits with* *respect to* $e_{n-q+1}, \cdots, e_n$, *then there is a unique strict set* $P'_1, \cdots, P'_q$ *with respect to* $e_{n-q+1}, \cdots, e_n$.

Set $B = M - (e_{n-q+1} + \cdots + e_n)$. The existence of $P_1, \cdots, P_q$ shows that $r(M) = r(M - e_n) = \cdots = r(B)$. Hence $\rho(B) = n - q = r(M) = r(B)$, and B is a base, by Theorem 7. Theorem 9 now applies.

Note that a matroid is not uniquely determined by a fundamental set of circuits (but see the appendix). This is shown by the following two matroids, in each of which the first two circuits form a strict fundamental set:

M, with circuits 1234, 1256, 3456;
M', with circuits 1234, 1256, 13456, 23456.

II. Separability, Dual Matroids.

10. Separable matroids. If $M = M_1 + M_2$, then $r(M) \leqq r(M_1) + r(M_2)$, on account of (3.3). If it is possible to divide the elements of M into two groups, M_1 and M_2, each containing at least one element, such that

$$r(M) = r(M_1) + r(M_2), \tag{10.1}$$

or, which is equivalent (as M_1 and M_2 have no common elements),

$$n(M) = n(M_1) + n(M_2), \tag{10.2}$$

we shall say M is *separable*; otherwise, M is *non-separable.*[4] Any single element forms a non-separable matroid. Any maximal non-separable part of M is a *component* of M.[5]

Theorem 11. *If*

$$M = M_1 + M_2, \quad r(M) = r(M_1) + r(M_2),$$
$$M'_1 \subset M_1, \quad M'_2 \subset M_2, \quad M' = M'_1 + M'_2,$$

then

$$r(M') = r(M'_1) + r(M'_2).$$

Set $M_1'' = M_1 - M_1'$, $M_2'' = M_2 - M_2'$. The relations (see Theorem 3)

$$\begin{aligned} r(M) &= \Delta(M_1 + M_2', M_2'') + \Delta(M', M_1'') + r(M') \\ &\leqq \Delta(M_2', M_2'') + \Delta(M_1', M_1'') + r(M') \\ &= r(M_2) - r(M_2') + r(M_1) - r(M_1') + r(M') \end{aligned}$$

[4] Compare G, Theorem 15.
[5] See G, § 4.

together with the fact that $r(M) = r(M_1) + r(M_2)$ show that $r(M') \geqq r(M'_1) + r(M'_2)$ and hence $r(M') = r(M'_1) + r(M'_2)$.

THEOREM 12.[6] *If $M = M_1 + M_2$, $r(M) = r(M_1) + r(M_2)$, M' is non-separable, and $M' \subset M$, then either $M' \subset M_1$ or $M' \subset M_2$.*

For suppose $M' = M'_1 + M'_2$, $M'_1 \subset M_1$, $M'_2 \subset M_2$, and M'_1 and M'_2 each contain an element. By the last theorem, $r(M') = r(M'_1) + r(M'_2)$, which cannot be.

THEOREM 13. *If M_1 and M_2 are non-separable matroids with a common element e, then $M = M_1 + M_2$ is non-separable.*

For suppose $M = M'_1 + M'_2$, $r(M) = r(M'_1) + r(M'_2)$. By the last theorem, $M_1 \subset M'_1$ or $M_1 \subset M'_2$, and $M_2 \subset M'_1$ or $M_2 \subset M'_1$; this shows that either M'_1 or M'_2 is void.

THEOREM 14. *No two distinct components of M have common elements.*

This is a consequence of the last theorem. From this follows:

THEOREM 15.[7] *Any matroid may be expressed as a sum of components in a unique manner.*

THEOREM 16.[8] *A non-separable matroid M of nullity 1 is a circuit, and conversely.*

If M_1 is a proper non-null subset of the non-separable matroid M, and $M_2 = M - M_1$, then $r(M) < r(M_1) + r(M_2)$. Hence

$$1 = n(M) > n(M_1) + n(M_2),$$

and $n(M_1) = 0$, proving that M is a circuit.

Conversely, if $M = M_1 + M_2$ is a circuit, and M_1 and M_2 each contain elements, then

$$\begin{aligned} r(M_1) + r(M_2) &= \rho(M_1) + \rho(M_2) - n(M_1) - n(M_2) \\ &= \rho(M) > r(M), \end{aligned}$$

showing that M is non-separable.

---

[6] Compare G, Lemma, p. 344.

[7] Compare G, Theorem 12.

[8] Compare G, Theorem 10.

LEMMA 9. *Let $M = M_1 + M_2$ be non-separable, and let M_1 and M_2 each contain elements but have no common elements. Then there is a circuit P in M containing elements of both M_1 and M_2.*

Suppose there were no such circuit. Say $M_2 = e_1 + \cdots + e_s$. Using Theorem 4, we see that

$$\Delta(M_1 + e_1 + \cdots + e_{i-1}, e_i) = \Delta(e_1 + \cdots + e_{i-1}, e_i) \quad (i = 1, \cdots, s),$$

and hence $r(M) = r(M_1) + r(M_2)$, a contradiction.

THEOREM 17.[9] *Any non-separable matroid M of nullity $n > 0$ can be built up in the following manner: Take a circuit M_1; add a set of elements which forms a circuit with one or more elements of M_1, forming a non-separable matroid M_2 of nullity 2 (if $n(M) > 1$); repeat this process till we have $M_n = M$.*

As $n > 0$, M contains a circuit M_1. If $n > 1$, we use the preceding lemma $n - 1$ times. The matroid at each step is non-separable, by Theorems 16 and 13.

THEOREM 18.[10] *Let $M = M_1 + \cdots + M_p$, and let $M_1, \cdots, M_p$ be non-separable. Then the following statements are equivalent:*

(1) *$M_1, \cdots, M_p$ are the components of M.*

(2) *No two of the matroids $M_1, \cdots, M_p$ have common elements, and there is no circuit in M containing elements of more than one of them.*

(3) $r(M) = r(M_1) + \cdots + r(M_p)$.

We cannot replace rank by nullity in (3); see G, p. 347.

(2) follows from (1) on application of Theorems 13 and 16.

To prove (1) from (2), take any M_i. If it is not a component of M, there is a larger non-separable submatroid M'_i of M containing it. By Lemma 9, there is a circuit P in M'_i containing elements of M_i and elements not in M_i; P must contain elements of some other M_j, a contradiction.

Next we prove (3) from (1). If $p > 1$, M is separable; say $M = M'_1 + M'_2$, $r(M) = r(M'_1) + r(M'_2)$. By Theorem 12, each M_i is in either M'_1 or M'_2; hence M'_1 and M'_2 are each a sum of components of M. If one of these

[9] See G, Theorem 19; also Whitney, "2-isomorphic graphs," *American Journal of Mathematics*, vol. 55 (1933), p. 247, footnote.

[10] Compare G, Theorem 17.

contains more than one component, we separate it similarly, etc. (3) now follows easily.

Finally we prove (1) from (3). Let M' be a component of M, and suppose it has an element in M_i. As

$$r(M) = r(M_i) + \sum_{j \neq i} r(M_j),$$

M' is contained in M_i, by Theorem 12; as M_i is non-separable, $M' = M_i$.

THEOREM 19.[11] *The elements e_1 and e_2 are in the same component of M if and only if they are contained in a circuit P.*

If e_1 and e_2 are both in P, they are part of a non-separable matroid, which lies in a single component of M. Suppose now e_1 and e_2 are in the same component M_0 of M, and suppose there is no circuit containing them both. Let M_1 be e_1 plus all elements which are contained in a circuit containing e_1. By Lemma 9, there is a subset M^* of $M_0 - M_1$ which forms with part of M_1 a circuit P_3. P_3 does not contain e_1. If e'_4 is an element of P_3 in M_1, there is a circuit P_1 in M_1 containing e_1 and e'_4. Let e_3 be an element of M^*. Then in $M_1 + M^*$ there are circuits P_1 and P_3 which contain e_1 and e_3 respectively, and have a common element.

Let M' be a smallest subset of M_0 which contains circuits P'_1 and P'_3 such that one contains e_1, the other contains e_3, and they have common elements. Then P'_1 and P'_3 are distinct, and $M' = P'_1 + P'_3$. Let e_4 be a common element. By Postulate (C_2), there is a circuit P_1 in $M' - e_4$ containing e_1, and a circuit P_3 in $M' - e_4$ containing e_3. By the definition of M', P_1 and P_3 have no common elements. By Postulate (C_1), P_1 is not contained in P'_1; hence it contains an element e_5 of $M' - P'_1$. P_3 does not contain e_5. As P_3 is not contained in P'_3, it contains an element e_6 of P'_1. But now P'_1 contains e_1, P_3 contains e_3, $P'_1 + P_3$ have a common element e_6, and $P'_1 + P_3$ does not contain e_5 and is thus a proper subset of M', a contradiction. This proves the theorem.

11. Dual matroids. Suppose there is a $1 - 1$ correspondence between the elements of the matroids M and M', such that if N is any submatroid of M and N' is the complement of the corresponding matroid of M', then

$$r(N') = r(M') - n(N). \tag{11.1}$$

---

[11] Compare D. König, *Acta Litterarum ac Scientiarum Szeged*, vol. 6, pp. 155-179, 4. (p. 159). The present theorem shows that a "glied" is the same as a component.

We say then that M' is a *dual* of M.[12]

THEOREM 20. *If M' is a dual of M, then*

$$r(M') = n(M), \qquad n(M') = r(M).$$

Set $N = M$; then $n(N) = n(M)$. In this case N' is the null matroid, and $r(N') = 0$. (11.1) now gives $r(M') = n(M)$. Also

$$n(M') = \rho(M') - r(M') = \rho(M) - n(M) = r(M).$$

THEOREM 21. *If M' is a dual of M, then M is a dual of M'.*

Take any N and corresponding N' as before. The equations

$$r(N') = r(M') - n(N), \qquad r(M') = n(M),$$
$$\rho(N) + \rho(N') = \rho(M)$$

give

$$\begin{aligned} r(N) &= \rho(N) - n(N) = \rho(N) - [r(M') - r(N')] \\ &= \rho(N) - n(M) + [\rho(N') - n(N')] \\ &= \rho(M) - n(M) - n(N') = r(M) - n(N'), \end{aligned}$$

as required.

THEOREM 22. *Every matroid has a dual.*

This is in marked contrast to the case of graphs, for only a planar graph has a dual graph (see G, Theorem 29).

Let M' be a set of elements in 1 — correspondence with elements of M. If N' is any subset of M', let N be the complement of the corresponding subset of M, and set $r(N') = n(M) - n(N)$. (R_1), (R_2), (R_3) are easily seen to hold in M', as they hold in M; hence M' is a matroid. Obviously $r(M') = n(M)$, and M' is a dual of M.

THEOREM 23. *M and M' are duals if and only if there is a $1 - 1$ correspondence between their elements such that bases in one correspond to base complements in the other.*

Suppose first M and M' are duals. Let B be a base in either matroid, say in M, and let B' be the complement of the corresponding submatroid of the other matroid, M'. Then

---

[12] Compare G, § 8. Theorems 20, 21, 24, 25 correspond to Theorems 20, 21, 23, 25 in G. Note that *two duals of the same matroid are isomorphic*, that is, there is a $1 - 1$ correspondence between their elements such that corresponding subsets have the same rank. Such a statement cannot be made about graphs. Compare H. Whitney, "2-isomorphic graphs," *American Journal of Mathematics*, vol. 55 (1933), pp. 245-254.

$$r(B') = r(M') - n(B) = r(M'),$$
$$n(B') = r(M) - r(B) = 0,$$

and B' is a base in M', by Theorem 7.

Suppose, conversely, that bases in one correspond to base complements in the other. Let N be a submatroid of M and let N' be the complement of the corresponding submatroid of M'. There is a base B' in M' with $r(N')$ elements in N', by Theorem 8. The complement in M of the submatroid corresponding to B' in M' is a base B in M with $\rho(N') - r(N') = n(N')$ elements in $M - N$, and hence with $r(M) - n(N')$ elements in N. This shows that

$$r(N) = r(M) - n(N') + k, \qquad k \geqq 0.$$

In a similar fashion we see that

$$r(N') = r(M') - n(N) + k', \qquad k' \geqq 0.$$

As B contains $r(M)$ elements and B' contains $r(M')$ elements, $r(M) + r(M') = \rho(M)$. Hence, adding the above equations,

$$k + k' = r(N) + r(N') + n(N) + n(N') - r(M) - r(M')$$
$$= \rho(N) + \rho(N') - \rho(M) = 0.$$

Hence $k = 0$, and the first equation above shows that M and M' are duals.

There are various other ways of stating conditions on certain submatroids of M and M' which will ensure these matroids being duals.[13]

THEOREM 24. *Let $M_1, \cdots, M_p$ and $M'_1, \cdots, M'_p$ be the components of M and M' respectively, and let M'_i be a dual of M_i $(i = 1, \cdots, p)$. Then M' is a dual of M.*

Let N be any submatroid of M, and let the parts of N in $M_1, \cdots, M_p$ be $N_1, \cdots, N_p$. Let N'_i be the complement in M'_i of the submatroid corresponding to N_i; then $N' = N'_i + \cdots + N'_p$ is the complement in M' of the submatroid corresponding to N. By Theorems 18 and 11 we have

$$r(N') = r(N'_1) + \cdots + r(N'_p), \qquad n(N) = n(N_1) + \cdots + n(N_p).$$

Also

$$r(M') = r(M'_1) + \cdots + r(M'_p), \qquad r(N'_i) = r(M'_i) - n(N_i);$$

adding the last set of equations gives $r(N') = r(M') - n(N)$, as required.

[13] See for instance a paper by the author "Planar graphs," *Fundamenta Mathematicae*, vol. 21 (1933), pp. 73-84, Theorem 2. Cut sets may of course be defined in terms of rank.

THEOREM 25. *Let M and M' be duals, and let $M_1, \cdots, M_p$ be the components of M. Let $M'_1, \cdots, M'_p$ be the corresponding submatroids of M'. Then $M'_1, \cdots, M'_p$ are the components of M', and M'_i is a dual of M_i $(i = 1, \cdots, p)$.*

The complement in M of the submatroid corresponding to M'_i in M' is $\sum_{j \neq i} M_j$. Hence, as M and M' are duals and the M_j $(j \neq i)$ are the components of $\sum_{j \neq i} M_j$ (see Theorem 18),

$$r(M'_i) = r(M') - n(\sum_{j \neq i} M_j) = r(M') - \sum_{j \neq i} n(M_j).$$

Adding gives

$$\sum_i r(M'_i) = pr(M') - (p-1) \sum_j n(M_j) = pr(M') - (p-1)n(M)$$
$$= pr(M') - (p-1)r(M') = r(M').$$

Therefore, by Theorem 12, each component of M' is contained in some M'_i. In the same way we see that each component of M is contained in a matroid corresponding to a component of M'; hence the components of one matroid correspond exactly to the components of the other.

Let N_i be any submatroid of M_i, and let N' and N'_i be the complements in M' and M'_i of the submatroid corresponding to N_i. The equations

$$r(M') = \sum_j r(M'_j), \qquad r(N') = r(N'_i) + \sum_{j \neq i} r(M'_j),$$
$$r(N') = r(M') - n(N_i),$$

give

$$r(N'_i) = r(M'_i) - n(N_i),$$

which shows that M'_i is a dual of M_i.

THEOREM 26. *A dual of a non-separable matroid is non-separable.*

This is a consequence of the last theorem.

III. MATRICES AND MATROIDS.

12. Matrices, matroids, and hyperplanes. Consider the matrix

$$M = \left\| \begin{matrix} a_{11} & \cdots & a_{1n} \\ \cdot & \cdot & \cdot \\ a_{m1} & \cdots & a_{mn} \end{matrix} \right\|;$$

let its columns be $C_1, \cdots, C_n$. Any subset $\boldsymbol{N}$ of these columns forms a matrix, and this matrix has a rank, $r(\boldsymbol{N})$. If we consider the columns as abstract elements, we have a matroid M. The proof of this is simple if we consider the rank of a matrix as the number of linearly independent columns in it. (R_1) and (R_2) are then obvious. To prove (R_3), suppose $r(\boldsymbol{N}+C_1) = r(\boldsymbol{N}+C_2) = r(\boldsymbol{N})$; then C_1 and C_2 can each be expressed as a linear combination of the other columns of $\boldsymbol{N}$, and hence $r(\boldsymbol{N}+C_1+C_2) = r(\boldsymbol{N})$. The terms independent and base carry over to matrices and agree with the ordinary definitions; a base in $\boldsymbol{M}$ is a minimal set of columns in terms of which all remaining columns of $\boldsymbol{M}$ may be expressed.

We may interpret $\boldsymbol{M}$ geometrically in two different ways; the second is the more interesting for our purposes:

(a) Let E_m be Euclidean space of m dimensions. Corresponding to each column C_i of $\boldsymbol{M}$ there is a point X_i in E_m with coördinates $a_{1i}, \cdots, a_{mi}$. The subset $C_{i_1}, \cdots, C_{i_p}$ of $\boldsymbol{M}$ is linearly independent if and only if the points $O = (0, \cdots, 0)$, $X_{i_1}, \cdots, X_{i_p}$ are linearly independent in E_m, i.e. if and only if these $p+1$ points determine a hyperplane in E_m of dimension p. A base in $\boldsymbol{M}$ corresponds to a minimal set of points $X_{i_1}, \cdots, X_{i_p}$ in E_m such that each X_j of $\boldsymbol{M}$ lies in the hyperplane determined by $O, X_{i_1}, \cdots, X_{i_p}$. Then p is the rank of $\boldsymbol{M}$.

(b) Let E_n be Euclidean space of n dimensions. Let $R_1, \cdots, R_m$ be the rows of $\boldsymbol{M}$. If $Y_1, \cdots, Y_m$ are the corresponding points of E_n: $Y_i = (a_{i1}, \cdots, a_{in})$, then the points $O, Y_1, \cdots, Y_m$ determine a hyperplane $H = H(\boldsymbol{M})$, which we shall call the *hyperplane associated with* $\boldsymbol{M}$. The dimension $d(H)$ of H is $r(\boldsymbol{M})$. Let $\boldsymbol{N} = C_{i_1} + \cdots + C_{i_p}$ be a subset of $\boldsymbol{M}$, and let E' be the p-dimensional coördinate subspace of E_n containing the x_{i_1} and ... and the x_{i_p} axes. The j-th row of $\boldsymbol{N}$ corresponds to the point Y'_j in E' with coördinates $(a_{ji_1}, \cdots, a_{ji_p})$; this is just the projection of Y_j onto E'. If H' is the hyperplane in E' determined by the points $O, Y'_1, \cdots, Y'_m$, then H' is exactly the projection of H onto E', and

$$d(H') = r(\boldsymbol{N}). \tag{12.1}$$

Let $\boldsymbol{N} = (C_{i_1}, \cdots, C_{i_p})$ be any subset of $\boldsymbol{M}$, and let E', H' correspond to $\boldsymbol{N}$. Then $\boldsymbol{N}$ *is independent if and only if*

$$d(H') = p,$$

and is a base if and only if

$$d(H') = d(H) = p.$$

THEOREM 27. *There is a unique matroid M associated with any hyperplane H through the origin in E_n.*

Let M contain the elements $e_1, \cdots, e_n$, one corresponding to each coördinate of E_n. Given any subset $e_{i_1}, \cdots, e_{i_p}$, we let its rank be the dimension of the projection of H onto the corresponding coördinate hyperplane E' of E_n. It was seen above that if $\boldsymbol{M}$ is any matrix determining H, then M is the matroid associated with $\boldsymbol{M}$.

13. Orthogonal hyperplanes and dual matroids. We prove the following theorem:

THEOREM 28. *Let H be a hyperplane through the origin in E_n, of dimension r, and let H' be the orthogonal hyperplane through the origin, of dimension $n - r$. Let M and M' be the associated matroids. Then M and M' are duals.*

We shall show that bases in one matroid correspond to base complements in the other; Theorem 23 then applies. Let

$$\boldsymbol{M} = \begin{Vmatrix} a_{11} \cdots a_{1n} \\ \cdot \quad \cdot \quad \cdot \quad \cdot \\ a_{r1} \cdots a_{rn} \end{Vmatrix}, \qquad \boldsymbol{M}' = \begin{Vmatrix} b_{11} \quad \cdots b_{1n} \\ \cdot \quad \cdot \quad \cdot \quad \cdot \quad \cdot \\ b_{n-r,1} \cdots b_{n-r,n} \end{Vmatrix}$$

be matrices determining H and H' respectively. Say the first r columns of $\boldsymbol{M}$ form a base in $\boldsymbol{M}$, i. e. the corresponding determinant A is $\neq 0$. As H and H' are orthogonal, we have for each i and j

$$a_{i1}b_{j1} + a_{i2}b_{j2} + \cdots + a_{in}b_{jn} = 0.$$

Keeping j fixed, we have a set of r linear equations in the b_{jk}. Transpose the last $n - r$ terms in each equation to the other side, and solve for b_{jk}. We find

$$b_{jk} = \frac{-1}{A} \sum_{l=r+1}^{n} b_{jl} \begin{vmatrix} a_{11} \cdots a_{1l} \cdots a_{1r} \\ \cdot \quad \cdot \quad \cdot \quad \cdot \quad \cdot \quad \cdot \quad \cdot \\ a_{r1} \cdots a_{rl} \cdots a_{rr} \end{vmatrix} = \sum_{l=r+1}^{n} c_{kl}b_{jl} \qquad (k = 1, \cdots, r).$$

This is true for each $j = 1, \cdots, n - r$, and the c_{kl} are independent of j. Thus the k-th column of $\boldsymbol{M}'$ is expressed in terms of the last $n - r$ columns. As this is true for $k = 1, \cdots, r$, the last $n - r$ columns form a base in $\boldsymbol{M}'$, as required.

14. The circuit matrix of a given matrix. Consider the matrix $\boldsymbol{M}$ of § 12. Suppose the columns $C_{i_1}, \cdots, C_{i_p}$ form a circuit, i. e. the corresponding elements of the corresponding matroid form a circuit. Then these columns are linearly dependent, and there are numbers $b_1, \cdots, b_n$ such that

$$(14.1) \qquad \begin{array}{ll} a_{i1}b_1 + \cdots + a_{in}b_n = 0 & (i = 1, \cdots, m), \\ b_j = 0 \quad (j \neq i_1, \cdots, i_p), & b_j \neq 0 \quad (j = i_1, \cdots, i_p). \end{array}$$

The b_j are all $\neq 0$ $(j = i_1, \cdots, i_p)$, for otherwise a proper subset of the columns would be dependent, contrary to the definition of a circuit. (They are uniquely determined except for a constant factor; see Lemma 11.) Suppose the circuits of $\boldsymbol{M}$ are $P_1, \cdots, P_s$. Then there are corresponding sets of numbers $b_{i1}, \cdots, b_{in}$ $(i = 1, \cdots, s)$, forming a matrix

$$\boldsymbol{M}' = \left\| \begin{array}{ccc} b_{11} & \cdots & b_{1n} \\ \cdot & \cdot & \cdot \\ b_{s1} & \cdots & b_{sn} \end{array} \right\|,$$

the *circuit matrix* of the matrix $\boldsymbol{M}$.

THEOREM 29. *Let $P_1, \cdots, P_q$ be a fundamental set of circuits in $\boldsymbol{M}$* (see § 9). *Then the corresponding rows of the circuit matrix $\boldsymbol{M}'$ form a base for the rows of $\boldsymbol{M}'$. Hence $r(\boldsymbol{M}') = q = n(\boldsymbol{M})$.*

Suppose the columns of $\boldsymbol{M}$ are ordered so that P_i contains C_{n-q+i} but no column C_{n-q+j} $(j > i)$. Then if the corresponding row of $\boldsymbol{M}'$ is $R'_i = (b_{i1}, \cdots, b_{in})$, we have $b_{i,n-q+i} \neq 0$ and $b_{i,n-q+j} = 0$ $(j > i)$. Hence the rows $R'_1, \cdots, R'_q$ of $\boldsymbol{M}'$ are linearly independent, and $r(\boldsymbol{M}') \geqq q$. Hence $r(\boldsymbol{M}') = n(\boldsymbol{M}) = q$, and each row of $\boldsymbol{M}'$ may be expressed in terms of $R'_1, \cdots, R'_q$.

THEOREM 30. *If $\boldsymbol{M}'$ is the circuit matrix of $\boldsymbol{M}$ and H', H are the corresponding hyperplanes, then H' is the hyperplane of maximum dimension orthogonal to H.*

This is a consequence of (14.1) and the last theorem.

THEOREM 31. *The matroids corresponding to a matrix and its circuit matrix are duals.*

This follows from the last theorem and Theorem 28.

15. On the structure of a circuit matrix. Let M be any matroid, and M', its dual. If there exists a matrix $\boldsymbol{M}$ corresponding to M, it is perhaps most easily constructed by considering it as the circuit matrix of a matrix $\boldsymbol{M}'$

5

corresponding to M'. Let H and H' be the hyperplanes corresponding to $\boldsymbol{M}$ and $\boldsymbol{M}'$. We shall say the set of numbers $(a_1, \cdots, a_n)$ is in $Z_{i_1 \ldots i_p}$ if

$$a_j \neq 0 \quad (j = i_1, \cdots, i_p), \qquad a_j = 0 \quad (j \neq i_1, \cdots, i_p).$$

If $(a_1, \cdots, a_n)$ is in H and in $Z_{i_1 \ldots i_p}$, then the columns $C_{i_1}, \cdots, C_{i_p}$ of $\boldsymbol{M}'$ are dependent, evidently.

Lemma 10. *Let $(b_1, \cdots, b_n)$ be a point of H. If it is in $Z_{i_1 \ldots i_p}$, then the matroid $N' = e_{i_1} + \cdots + e_{i_p}$ is the union of a set of circuits in M'.*

Here e_i in M' corresponds to C_i in $\boldsymbol{M}$. We need merely show that for each i_s there is a circuit P in N' containing e_{i_s}. Let $k_1 = i_s, k_2, \cdots, k_q$ be a minimal set of numbers from $(i_1, \cdots, i_p)$ containing i_s such that there is a point $(c_1, \cdots, c_n)$ of H in $Z_{k_1 \ldots k_q}$; then $e_{k_1} + \cdots + e_{k_q}$ is the required circuit. For if it were not a circuit, there would be a proper subset $(l_1, \cdots, l_r)$ of $(k_1, \cdots, k_q)$ and a point $(d_1, \cdots, d_n)$ of H in $Z_{l_1 \ldots l_r}$. No $l_i = k_1$, on account of the minimal property of $(k_1, \cdots, k_q)$. Say $l_1 = k_t$, and set

$$a_i = d_{k_t} c_i - c_{k_t} d_i \qquad (i = 1, \cdots, n).$$

Then $(a_1, \cdots, a_n)$ is in H and in $Z_{m_1 \ldots m_u}$ with $(m_1, \cdots, m_u)$ a proper subset of $(k_1, \cdots, k_q)$ containing k_1, again a contradiction.

Lemma 11. *If $P = e_{i_1} + \cdots + e_{i_p}$ is a circuit of M' and $(b_1, \cdots, b_n)$ and $(b'_1, \cdots, b'_n)$ are in H and in $Z_{i_1 \ldots i_p}$, then these two sets are proportional.*

For otherwise, $(c_1, \cdots, c_n)$ with $c_i = b'_{i_1} b_i - b_{i_1} b'_i$ would be a point of H in some $Z_{k_1 \ldots k_q}$ with $(k_1, \cdots, k_q)$ a proper subset of $(i_1, \cdots, i_p)$, and P would not be a circuit.

It is instructive to show directly that Postulate (C_2) holds for matrices: P_1 and P_2 are represented by rows $(b_1, \cdots, b_n)$ and $(b'_1, \cdots, b'_n)$ of $\boldsymbol{M}$, lying in $Z_{12 i_1 \ldots i_p}$ and $Z_{1 k_1 \ldots k_q}$ respectively, where $k_1, \cdots, k_q \neq 2$. Set $c_i = b'_1 b_i - b_1 b'_i$; then $(c_1, \cdots, c_n)$ is in H and in $Z_{2 l_1 \ldots l_r}$, with $(l_1, \cdots, l_r)$ a subset of $(i_1, \cdots, i_p, k_1, \cdots, k_q)$; the existence of P_3 now follows from Lemma 10.

Theorem 32. *Let $\boldsymbol{M}$ be the circuit matrix of $\boldsymbol{M}'$. Let $P_1, \cdots, P_q$ form a strict fundamental set of circuits in M' with respect to $e_{n-q+1}, \cdots, e_n$, and let the first q rows in $\boldsymbol{M}$ correspond to $P_1, \cdots, P_q$. Let $(i_1, \cdots, i_s)$ be any set of numbers from $(1, \cdots, q)$, let $(j_1, \cdots, j_s)$ be any set from $(1, \cdots, n-q)$, and let $(i'_1, \cdots, i'_{q-s})$ be the set complementary to $(i_1, \cdots, i_s)$* in $(1, \cdots, q)$.

Then the determinant D in $\mathbf{M}$ with rows $i_1, \cdots, i_s$ and columns $j_1, \cdots, j_s$ equals zero if and only if the determinant D' with rows $1, \cdots, q$ and columns $j_1, \cdots, j_s$, $n-q+i'_1, \cdots, n-q+i'_{q-s}$ equals zero, or, if and only if there exists a circuit P in $\mathbf{M}'$ containing none of the columns $e_{j_1}, \cdots, e_{j_s}$, $e_{n-q+i'_1}, \cdots, e_{n-q+i'_{q-s}}$.

In the matrix of the last $q = r(\mathbf{M})$ columns of $\mathbf{M}$, the terms along the main diagonal and only those are $\neq 0$. If we expand D' by Laplace's expansion in terms of the columns $n-q+i'_1, \cdots, n-q+i'_{q-s}$, we see at once that $D' = 0$ if and only if $D = 0$.

Suppose $D = 0$. Then there is a set of numbers $(\alpha_1, \cdots, \alpha_q)$, not all zero, with $\alpha_i = 0$ $(i \neq i_1, \cdots, i_s)$, such that

$$b_k = \alpha_1 b_{1k} + \cdots + \alpha_q b_{qk} = 0 \qquad (k = j_1, \cdots, j_s),$$

$(b_{i1}, \cdots, b_{in})$ being the i-th row of $\mathbf{M}$., $b_k = 0$ also for $k = n-q+i'_1, \cdots, n-q+i'_{q-s}$, as each term is zero for such k. The point $(b_1, \cdots, b_n)$ is in H. Any circuit given by Lemma 10 is the required circuit P.

Suppose the circuit P exists. Then it is represented by a row $(b_1, \cdots, b_n)$ in $\mathbf{M}$. As the first q rows of $\mathbf{M}$ are of rank $q = r(\mathbf{M})$, $(b_1, \cdots, b_n)$ can be expressed in terms of them; say $b_k = \Sigma \alpha_i b_{ik}$. As $b_k = 0$ $(k = n-q+i'_1, \cdots, n-q+i'_{q-s})$, certainly $\alpha_k = 0$ $(k = i'_1, \cdots, i'_{q-s})$. $D = 0$ now follows from the fact that $b_k = 0$ $(k = j_1, \cdots, j_s)$.

16. A matroid with no corresponding matrix.[14] The matroid M' has seven elements, which we name $1, \cdots, 7$. The bases consist of all sets of three elements except

(16.1) 124, 135, 167, 236, 257, 347, 456.

Defining rank in terms of bases, we have: Each set of k elements is of rank k if $k \leqq 2$ and of rank 3 if $k \geqq 4$; a set of three elements is of rank 2 if the set is in (16.1) and is of rank 3 otherwise. It is easy to see that the postulates for rank are satisfied. (R_3) in the case that N contains two elements is satisfied vacuously. For suppose $r(N+e_1) = r(N+e_2) = r(N) = 2$. Then $N + e_1$ and $N + e_2$ are both in (16.1); but any two of these sets have but a single element in common.

[14] After the author had noted that M' satisfies (C^*) and corresponds to no linear graph, and had discovered a matroid with nine elements corresponding to no matrix, Saunders MacLane found that M' corresponds to no matrix, and is a well known example of a finite projective geometry (see O. Veblen and J. W. Young, *Projective Geometry*, pp. 3-5).

If there exists a matrix $\boldsymbol{M}'$, corresponding to M', then let $\boldsymbol{M}$ be its circuit matrix. 123 is a base in M', and hence

$$(16.2) \qquad 124, \quad 135, \quad 236, \quad 1237$$

form a fundamental set of circuits in M'. Let R_1, R_2, R_3, R_4 be the corresponding rows of $\boldsymbol{M}$. By multiplying in succession row 1, column 2, rows 2, 3, 4, and columns 4, 5, 6, 7 by suitable constants $\neq 0$, we bring $\boldsymbol{M}$ into the following form:

$$(16.3) \qquad \boldsymbol{M} = \left\| \begin{array}{ccc|cccc} 1 & 1 & 0 & 1 & 0 & 0 & 0 \\ 1 & 0 & a & 0 & 1 & 0 & 0 \\ 0 & 1 & b & 0 & 0 & 1 & 0 \\ 1 & c & d & 0 & 0 & 0 & 1 \\ \cdot & \cdot & \cdot & \cdot & \cdot & \cdot & \cdot \end{array} \right\|;$$

a, b, c and d are $\neq 0$. We now apply Theorem 32 with

$$(i_1, \cdots, i_s;\ j_1, \cdots, j_s) = (1, 4;\ 1, 2), \quad (2, 4;\ 1, 3), \quad (3, 4;\ 2, 3),$$

i. e. using the circuits 347, 257, 167. This gives

$$\begin{vmatrix} 1 & 1 \\ 1 & c \end{vmatrix} = \begin{vmatrix} 1 & a \\ 1 & d \end{vmatrix} = \begin{vmatrix} 1 & b \\ c & d \end{vmatrix} = 0,$$

and hence $c = 1$, $a = d = b$. Using the circuit 456, with sets $(1, 2, 3;\ 1, 2, 3)$ gives $2a = 0$, $a = 0$, a contradiction.

In regard to this example, see the end of the paper.

APPENDIX.

Matrices of Integers mod 2.

We wish to characterize those matroids M corresponding to matrices $\boldsymbol{M}$ of integers mod 2,[15] i. e. matrices whose elements are all 0 or 1, where rank etc. is defined mod 2. We shall consider linear combinations, *chains:*

$$(\text{A. }1) \qquad \alpha_1 e_1 + \cdots + \alpha_n e_n \qquad (\alpha\text{'s integers mod } 2)$$

in the elements of M. The α's may be taken as 0 or 1; (A. 1) may then be interpreted as the submatroid N whose elements have the coefficient 1. Conversely, any $N \subset M$ may be written as a chain. Submatroids are added

[15] See O. Veblen, "Analysis situs," 2nd ed., *American Mathematical Society Colloquium Publications*, Ch. I and Appendix 2.

(mod 2) by adding the corresponding chains (mod 2). For instance, $(e_1 + e_2) + (e_2 + e_3) \equiv e_1 + e_3 \pmod 2$.

Any sum (mod 2) of circuits in M we shall call a *cycle* in M. N is the *true sum* of $N_1, \cdots, N_s$ if these latter have no common elements and $N = N_1 + \cdots + N_s$. We consider matroids which satisfy the following postulate:

(C^*) *Each cycle is a true sum of circuits.*

Postulate (C_2) is a consequence of (C^*). For the cycle $P_1 + P_2$ is a submatroid containing e_2 but not e_1; The existence of P_3 now follows from (C^*).

A simple example of a matroid not satisfying (C^*) is given by the matroid M' at the end of § 9.

THEOREM 33. *A circuit is a minimal non-null cycle, and conversely.*

This is proved with the aid of Postulates (C_1) and (C^*).

THEOREM 34. *Let $P_1, \cdots, P_q$ be a strict fundamental set of circuits in M with respect to $e_{n-q+1}, \cdots, e_n$. Then there are exactly 2^q cycles in M, formed by taking all sums* (mod 2) *of $P_1, \cdots, P_q$.*

First, each sum $P_{i_1} + \cdots + P_{i_s}$ (mod 2) is a cycle, containing $e_{n-q+i_1}, \cdots, e_{n-q+i_s}$ and elements (perhaps) from $B = e_1, \cdots, e_{n-q}$; obviously distinct sums give distinct cycles. Now let Q be any cycle in M; say Q contains $e_{n-q+k_1}, \cdots, e_{n-q+k_r}$ and elements (perhaps) from B. Set $Q' = P_{k_1} + \cdots + P_{k_r}$; then $Q + Q'$ is a cycle containing elements from B alone. But B is a base (see the proof of Theorem 10), and hence contains no circuits. Consequently $Q + Q'$ is the null cycle, and $Q = Q'$.

THEOREM 35. *As soon as the circuits of a strict fundamental set are known, all the circuits may be determined.*

This is a consequence of the last two theorems. It is to be contrasted with the final remark of § 9.

Remark. The word "strict" may be omitted in the last two theorems.

THEOREM 36. *Let $e_1, \cdots, e_n$ be a set of elements, and let $P_1, \cdots, P_q$ be any subsets such that P_i contains e_{n-q+i} and possibly elements from $e_1, \cdots, e_{n-q}$ alone. Then there is a unique matroid M satisfying* (C^*), *with $P_1, \cdots, P_q$ as a strict fundamental set of circuits.*

We form the 2^q cycles of Theorem 34. Those cycles which contain no other non-null cycle as a proper subset we call circuits; in particular, $P_1, \cdots, P_q$ are circuits. To prove (C*), let Q be a non-null cycle. If it is not a circuit, it contains a circuit P as a proper subset. Q and P are sums (mod 2) from $P_1, \cdots, P_q$, hence the same is true of $Q - P$, and $Q - P$ is one of the 2^q cycles. If it is not a circuit, we again extract a circuit, etc.

This theorem furnishes a simple method of constructing all matroids satisfying (C*).

We turn now to the study of matrices of integers (mod 2)

$$\boldsymbol{M} = \left\| \begin{matrix} a_{11} & \cdots & a_{1n} \\ \cdot & \cdot & \cdot \\ a_{m1} & \cdots & a_{mn} \end{matrix} \right\| \qquad \text{(each } a_{ij} = 0 \text{ or } 1\text{)}.$$

Any linear combination (mod 2) of the columns

$$\text{(A. 2)} \qquad \alpha_1 C_1 + \cdots + \alpha_n C_n \qquad (\alpha\text{'s integers mod } 2)$$

is a set of numbers $(\Sigma\alpha_i a_{1i}, \cdots, \Sigma\alpha_i a_{mi})$, which we call a *chain* (mod 2) in $\boldsymbol{M}$. As before, we may take each coefficient as 0 or 1, and we may consider any chain merely as a submatrix of $\boldsymbol{M}$. The chain is a *cycle* if each of the corresponding numbers is $\equiv 0$ (mod 2). The columns $C_{i_1}, \cdots, C_{i_p}$ are *independent* (mod 2) if there exists no set of integers $\alpha_1, \cdots, \alpha_n$ not all $\equiv 0$ (mod 2), with $\alpha_i = 0$ $(i \neq i_1, \cdots, i_p)$, such that $\Sigma\alpha_i C_i$ is a cycle, i. e. if no non-null subset of $C_{i_1}, \cdots, C_{i_p}$ is a cycle. Using this definition, the terms base, circuit, rank, nullity etc. (mod 2) can be defined as in Part I.

Let M be a set of elements $e_1, \cdots, e_n$ corresponding to $C_1, \cdots, C_n$ in $\boldsymbol{M}$, and let $e_{i_1} + \cdots + e_{i_p}$ be a circuit in M if and only if $C_{i_1}, \cdots, C_{i_p}$ is a circuit in $\boldsymbol{M}$. We shall show that *M is a matroid satisfying* (C*) *and the definitions of cycle in M and $\boldsymbol{M}$ agree.*

We show first that each circuit is a cycle in $\boldsymbol{M}$. If $C_{i_1}, \cdots, C_{i_p}$ is a circuit, then these columns are dependent; hence $\Sigma\alpha_i C_i$ is a cycle, with $\alpha_i = 0$ $(i \neq i_1, \cdots, i_p)$. Moreover $\alpha_i = 1$ $(i = i_1, \cdots, i_p)$, for otherwise a proper subset of $C_{i_1}, \cdots, C_{i_p}$ would be dependent. Hence $C_{i_1} + \cdots + C_{i_p}$ is a cycle. Next, any sum (mod 2) of circuits is a cycle, evidently. Next we prove (C*). Suppose $Q = C_{i_1} + \cdots + C_{i_p}$ is a cycle. Let $(k_1, \cdots, k_q)$ be a minimal subset of $(i_1, \cdots, i_p)$ such that $P = C_{k_1} + \cdots + C_{k_q}$ is a cycle; then P is a circuit. $Q - P$ is a cycle; from it we extract a circuit, just as above, etc. It follows from (C*) that the definitions of cycle in M and $\boldsymbol{M}$ agree. Theorems 33, 34 and 35 now apply to $\boldsymbol{M}$ also.

We are now ready to prove the final theorem:

THEOREM 37. *Let M be any matroid satisfying* (C*). *Suppose* $\rho(M) = n$, *and* $e_1 + \cdots + e_{n-q}$ *is a base. Then if* $\boldsymbol{M}_1$ *is any matrix of integers* (*mod* 2) *with* $n - q$ *columns which are independent* (*mod* 2), *columns* $C_{n-q+1}, \cdots, C_n$ *can be adjoined in a unique manner to* $\boldsymbol{M}_1$, *forming a matrix* $\boldsymbol{M}$ *of which the corresponding matroid is* M.

Let $P_1, \cdots, P_q$ be a strict fundamental set of circuits in M with respect to $e_{n-q+1}, \cdots, e_n$ (Theorem 9). Say $P_1 = e_{i_1} + \cdots + e_{i_p} + e_{n-q+1}$. Set $C_{n-q+1} \equiv C_{i_1} + \cdots + C_{i_p}$ (mod 2); this determines C_{n-q+1} as a column of 0's and 1's so that $P'_1 = C_{i_1} + \cdots + C_{i_p} + C_{n-q+1}$ is a circuit. (P'_1 is a cycle; (C*) shows that it is a single circuit, as $C_1 + \cdots + C_{n-q}$ contains no circuit.) C_{n-q+1} evidently must be chosen in this manner. We choose the remaining columns of $\boldsymbol{M}$ similarly. Let M' be the matroid corresponding to $\boldsymbol{M}$. Then $P'_1, \cdots, P'_q$ is a strict set of circuits in M'. These same sets form a strict set in M; hence, by Theorem 35, the circuits in M' correspond to those in M. Consequently $M' = M$, completing the proof.

We end by noting that the matroid M' of § 16 satisfies Postulate (C*) but corresponds to no linear graph. For letting 123 be a base and (16.2) a fundamental set of circuits and determining the matroid as in Theorem 36, we come out with exactly M'. A corresponding matrix of integers mod 2 is constructed from (16.3) with $a = b = c = d = 1$; we interchange rows and columns in the left-hand portion, leave out the last row and column of the right-hand portion, and interchange these two parts. (The relation $2a = 0$ is of course true mod 2.)

On the other hand, it is easily seen that if the element 7 is left out, there is a corresponding graph, which must be of the following sort: It has four vertices a, b, c, d, and the arcs corresponding to the elements $1, \cdots, 6$ are

$$ab, \quad ac, \quad ad, \quad bc, \quad bd, \quad cd.$$

There is no way of adding the required seventh arc.

The problem of characterizing linear graphs from this point of view is the same as that of characterizing matroids which correspond to matrices (mod 2) with exactly two ones in each column.

HARVARD UNIVERSITY.

Reprinted from *American Journal of Mathematics*,
volume 57, pages 800 through 804,
by permission of the Johns Hopkins University Press.

ABSTRACT LINEAR DEPENDENCE AND LATTICES.

By Garrett Birkhoff.

1. *Introduction.* In a preceding paper,[1] Hassler Whitney has shown that it is difficult to distinguish theoretically between the properties of linear dependence of ordinary vectors, and those of elements of a considerably wider class of systems, which he has called "matroids."

Now it is obviously impossible to incorporate all of the heterogeneous abstract systems which are constantly being invented, into a body of systematic theory, until they have been classified into two or three main species. The purpose of this note is to correlate matroids with abstract systems of a very common type,[2] which I have called "lattices."

As this correlation is purely formal, the discussion will be descriptive rather than detailed.

2. *The fundamental construction.* Let us refer to Whitney's definition of a matroid as a set M whose subsets have a numerical rank function with certain properties (R_1)-(R_3). It is easy to define mutual dependence by abstraction from the theory of vectors, and to see that if we suppress elements of rank zero, and regard mutually dependent elements as merely repeated occurrences of the same element, then we get a matroid M^* with the same structural properties as M. Moreover M^* has the additional property that no element is dependent on any other element.[3]

Now let us call a subset of M^* "linearly complete" if and only if it contains all elements dependent on it. By what we have just shown, distinct elements of M^* are distinct linearly complete subsets. Moreover

Lemma 1. *The product $N \cdot N'$ of any two linearly complete subsets N and N' of M^*, is linearly complete.*

For suppose e any element of M^* not in $N \cdot N'$; by symmetry, we can suppose e not in N. Therefore

[1] H. Whitney, "On the abstract properties of linear dependence," *American Journal of Mathematics*, vol. 57 (1935), pp. 509-533. His main definition is in his section two.

[2] Cf. the author's "On the combination of subalgebras," *Proceedings of the Cambridge Philological Society*, vol. 39 (1933), pp. 441-464. This article will be referred to subsequently as "Subalgebras."

[3] This assertion corresponds to S. MacLane's assumption "without loss of generality" of his conditions (R_4)-(R_5) in his note on "Some interpretations of abstract linear dependence, etc.," *supra*.

$$1 = r(N+e) - r(N) \qquad \text{[by definition of } N\text{]}$$
$$\leqq r(N \cdot N' + e) - r(N \cdot N') \qquad \text{[by Whitney's Lemma 4].}$$

Hence $r(N \cdot N' + e) \geqq r(N \cdot N') + 1$, and e is not dependent on $N \cdot N'$, proving the lemma.

Observe now that M^* is itself linearly complete, and has the property $M^* \cdot N = N$ for any subset N. Finally, the operation of intersection is idempotent, commutative, and associative. Consequently

THEOREM 1. *The linearly complete subsets of M^* can be regarded as the elements of a new system $L(M^*)$, satisfying*

(L_1) *Any a of $L(M^*)$ and b of $L(M^*)$ determine a unique "product" $a \cap b$ in $L(M^*)$.*

(L_2) *$a \cap a = a$, $a \cap b = b \cap a$, and $a \cap (b \cap c) = (a \cap b) \cap c$.*

(L_3) *There exists a "unit" i [namely, M^*] such that $a \cap i = a$ irrespective of a.*

(L_4) *In any sequence of products $a_1, a_1 \cap a_2, a_1 \cap a_2 \cap a_3, \cdots$ some two terms are equal. [Remark: This is a simple corollary of the fact that $L(M^*)$ is finite].*

Moreover distinct single elements of M^* appear as distinct single elements of $L(M^*)$.

3. *An incidental theorem.* We shall now prove a result which is obvious in the case in hand—provided we define the "join" $N \cup N'$ of N and N' directly as the product of all linearly complete subsets containing N and N'. (This definition is equivalent to defining $N \cup N'$ as the set of all elements linearly dependent on $N + N'$). The formal proof *in abstracto* has however been included, to give a neat new approach to the theory of abstract lattices.[4]

THEOREM 2. *It is an abstract consequence of Theorem 1 that any elements a and b of $L(M^*)$ determine a unique "join" $a \cup b$ with the properties*

[4] It is extremely useful in applications. For instance, it shows that if we adjoin a purely formal "unity" i, we can subsume the theory of "effective implication" as defined by E. V. Huntington in his note "Effective equality and effective implication in formal logic," *Proceedings of the USA Academy of Sciences*, May, 1935, under the theory of lattices in the finite [but not the infinite!] case. Viewed in this light, his propositions 32, 33, 34, 36, 37 appear as extensions to the infinite case of Theorems 4. 2, 6. 1, 6. 2, 6. 3 of "Subalgebras."

With regard to precedence, it was Professor Huntington's paper which suggested Theorem 2 to me originally.

(L_5) $a \cap (a \cup b) = a \cup (a \cap b) = a$.
(L_6) $a \cup a = a$, $a \cup b = b \cup a$, *and* $a \cup (b \cup c) = (a \cup b) \cup c$.

The main thing is to define $a \cup b$. To do this, note first that $i \cap a = a$ and $i \cap b = b$. Second, not that if $c \cap a = a$ and $c' \cap a = a$, *then*

$$(c \cap c') \cap a = c \cap (c' \cap a) = c \cap a = a.$$

Therefore if we set $c_0 = i$, and try to find successive elements $c_1, c_2, c_3, \cdots$ such that $c_k \cap a = a$, $c_k \cap b = b$, and yet $c_0 \cap \cdots \cap c_k$ differs from every $c_0 \cap \cdots \cap c_j$ $[j < k]$, then by (L_4) we obtain after a finite number of attempts, an element $d = c_0 \cap \cdots \cap c_n$ such that (1) $d \cap a = a$ and $d \cap b = b$ (2) if $c^* \cap a = a$ and $c^* \cap b = b$, then $c^* \cap d = c_1 \cap \cdots \cap c_k$ for some $k < n$—which, since

$$c^* \cap d = c^* \cap (d \cap d) = (c^* \cap d) \cap d = c_1 \cap \cdots \cap c_k \cap d = d$$

[by iterated use of (L_2)], means $c^* \cap d = d$.

Conditions (1)-(2) and (L_2) show that d is determined uniquely by a and b; we shall *define* $a \cup b$ as d. To prove that $a \cup b$ satisfies (L_5)-(L_6) is now merely a question of substituting from (1)-(2), and using condition (L_2).

This completes the proof, the details of which the reader should be able to fill in without difficulty.

4. *Principal results.* But by definition, an "abstract lattice" is any system which satisfies (L_1)-(L_2) and (L_5)-(L_6). Consequently, we have

THEOREM 3. *$L(M^*)$ is an abstract lattice.*

Not all lattices correspond to matroids; in particular, every lattice corresponding to a matroid contains only a finite number of elements. The facts as to which lattices correspond to matroids can however be summarized as follows.

A finite lattice corresponds to a matroid if and only if (1) if b and c are smallest distinct elements larger than a, then $b \cup c$ is a smallest element larger than b. (This is the "dual" [under the inversion of inclusion [5]] of the abstract property of composition subgroups used in the classic proof of the Theorem of Jordan-Hölder) (2) every element can be expressed as the "join" of elements of rank one.

[5] It being observed that the definition of a lattice is invariant under inversion of meet and join. Property (1) amounts in the case in hand to the property that if two linearly complete subsets of M^* of rank $(n+1)$ intersect in a subset of rank n, then the rank of their join is $(n+2)$.

Moreover the "rank" $r(N)$ of any subset N of elements $e_1, \cdots, e_k$ of M^* is equal to the length of any chain [in the sense of the Theorem of Jordan-Hölder] in $L(M^*)$, between the lattice-element corresponding to the null subset, and the "join" of the lattice-elements corresponding (under Theorem 1) to the e_i. That is,

THEOREM 4. *The structure of $L(M^*)$ determines that of M^*—and hence that of M.*

5. *Theoretical consequences.* The results of the last sections suggest trying to interpret theorems on matroids as theorems on lattices, and conversely. This can be done in at least two cases.

The less interesting of these is Whitney's Theorem 3, which may be regarded as a reappearance[6] of the first formula of Theorem 9.2 of "Subalgebras."

A more interesting correspondence concerns Whitney's Theorem 15 on separability, which is a generalization of a well-known theorem on graphs. This theorem may, in the light of the above results, be regarded as a specialization of a theorem[7] on the "strong" uniqueness of the representation of *any* lattice as the direct product of irreducible (= non-separable) components.

6. *Geometrical correspondence.* In another paper,[8] S. MacLane has shown that M^* also corresponds to a schematic configuration of dimensions $r(M^*) - 1$, and has pointed out a relationship with projective geometries. This defines an obvious correspondence *via* matroids between "matroid lattices" (or lattices satisfying conditions (1)-(2) of § 4) and schematic configurations.

In terms of this correspondence, results proved elsewhere[9] by the author directly imply

THEOREM 5. *A schematic configuration is the direct product of a finite number of projective geometries and single points, if and only if the dual [under the inversion of join and meet] of the corresponding matroid lattice, is again a matroid lattice.*

---

[6] This formula very likely has a long previous history.

[7] Proved in the author's "On the lattice theory of ideals," *Bulletin of the American Mathematical Society*, vol. 40 (1934), p. 616, Theorem 2.

[8] Cf. footnote 3, above. A much weaker result of the same sort was obtained by the author, in Theorem 5 of "On the structure of abstract algebras," to appear shortly in the *Proceedings of the Cambridge Philological Society*.

[9] "Subalgebras," Theorem 10.2, and "Combinatorial relations in projective geometries," *Annals of Mathematics*, vol. 36 (1935), pp. 743-748.

(By the " direct product " of two projective geometries P and P^* is meant the system whose elements are the couples (A, A^*) [A any m-plane of P and A^* any m^*-plane of P^*], and in which (A, A^*) lies in (B, B^*) if and only if A lies in B and A^* in B^*. This is the inverse of the notion of separation into non-separable components mentioned above).

Theorem 5 and well-known results [10] show that such schematic configurations can in general be realized by vectors in spaces whose coördinates lie in suitable finite fields.

SOCIETY OF FELLOWS,
HARVARD UNIVERSITY.

[10] O. Veblen and W. H. Bussey, "Finite projective geometries," *American Transactions*, vol. 7 (1906), pp. 241-259.

NON-SEPARABLE AND PLANAR GRAPHS*

BY
HASSLER WHITNEY

Introduction. In this paper the structure of graphs is studied by purely combinatorial methods. The concepts of rank and nullity are fundamental. The first part is devoted to a general study of non-separable graphs. Conditions that a graph be non-separable are given; the decomposition of a separable graph into its non-separable parts is studied; by means of theorems on circuits of graphs, a method for the construction of non-separable graphs is found, which is useful in proving theorems on such graphs by mathematical induction. In the second part, a dual of a graph is defined by combinatorial means, and the paper ends with the theorem that a necessary and sufficient condition that a graph be planar is that it have a dual.

The results of this paper are fundamental in papers by the author on *Congruent graphs and the connectivity of graphs*† and on *The coloring of graphs*.‡

I. Non-separable graphs

1. **Definitions.**§ A *graph G* consists of two sets of symbols, finite in number: *vertices*, $a, b, c, \cdots, f$, and *arcs*, $\alpha(ab), \beta(ac), \cdots, \delta(cf)$. If an arc $\alpha(ab)$ is present in a graph, its *end vertices* a, b are also present. We may write an arc $\alpha(ab)$ or $\alpha(ba)$ at will; we may write it also ab or ba if no confusion arises,— if there is but a single arc joining a and b in G. We say the vertices a and b are *on* the arc $\alpha(ab)$, and the arc $\alpha(ab)$ is *on* the vertices a and b. The *null graph* is the graph containing no arcs or vertices.

The obvious geometrical interpretation of such a graph, or *abstract graph*, is a *topological graph*, let us say. Corresponding to each vertex of the abstract graph, we select a point in 3-space, a vertex of the topological graph. Corresponding to each arc $\alpha(ab)$ of the abstract graph, we select an arc joining the corresponding vertices of the topological graph. An arc is here a set of points in (1, 1) correspondence with the unit interval, its end vertices corresponding

* Presented to the Society, October 25, 1930; received by the editors February 2, 1931. An outline of this paper will be found in the Proceedings of the National Academy of Sciences, vol. 17 (1931), pp. 125–127.

† American Journal of Mathematics, vol. 54 (1932), pp. 150–168.

‡ An outline will be found in the Proceedings of the National Academy of Sciences, vol. 17 (1931), pp. 122–125.

§ Compare Ste. Laguë, *Les Réseaux*, Mémorial des Sciences Mathématiques, fascicule 18, Paris, 1926.

with the ends of the interval. Moreover, we let no arc pass through other vertices or intersect other arcs. We shall consider topological graphs no further till we come to the section on planar graphs.

An *isolated vertex* is a vertex which is not on any arc. A *chain* is a set of one or more distinct arcs which can be ordered thus: ab, bc, cd, $\cdots$, ef, where vertices in different positions are distinct, i.e. the chain may not intersect itself. A *suspended chain* is a chain containing two or more arcs such that no vertex of the chain other than the first and last is on other arcs, and these two vertices are each on at least two other arcs. A *circuit* is a set of one or more distinct arcs which can be put in cyclic order, ab, bc, $\cdots$, ef, fa, vertices being distinct as in the case of the chain. A *k-circuit* is a circuit containing k arcs. Thus, the arc $\alpha(aa)$ is a 1-circuit; the two arcs $\alpha(ab)$, $\beta(ab)$ form a 2-circuit.

A graph is *connected* if any two of its vertices are joined by a chain. Obviously, if a and b are joined by a chain, and b and c are joined by a chain, then a and c are joined by a chain. Any graph consists of a certain number of *connected pieces* (one, if the graph is connected). In particular, an isolated vertex is one of the connected pieces of a graph. A graph is called *cyclicly connected* if any two of its vertices are contained in a circuit. If G_1, G_2, $\cdots$, G_m are a set of graphs, no two of which have a common vertex (or arc, therefore), we say the graph G, formed of the arcs and vertices of all these graphs, is the *sum* of these graphs. Thus, a graph is the sum of its connected pieces. A *forest* is a graph containing no circuit. A *tree* is a connected forest. A *subgraph* H of G is a graph containing a subset (in particular, all or none), of the arcs of G, and those vertices of G which are on these arcs.

2. **Rank and nullity.*** Given a graph G which contains V vertices, E arcs, and P connected pieces, we define its *rank* R, and its *nullity* (or *cyclomatic number* or first Betti number) N, by the equations

$$R = V - P,$$
$$N = E - R = E - V + P.$$

If G contains the single arc ab, it is of rank 1, nullity 0, while if it contains the single arc aa, it is of rank 0, nullity 1.

The first two theorems follow immediately from the definitions of rank and nullity:

THEOREM 1. *If isolated vertices be added to or subtracted from a graph, the rank and nullity remain unchanged.*

* These are just the rank and nullity of the matrix H_1 of Poincaré. See Veblen's Colloquium Lectures, *Analysis Situs*.

THEOREM 2. *Let the graph G' be formed from the graph G by adding the arc ab. Then*

(1) *if a and b are in the same connected piece in G, then*

$$R' = R, \quad N' = N + 1;$$

(2) *if a and b are in different connected pieces in G, then*

$$R' = R + 1, \quad N' = N.$$

THEOREM 3. *In any graph G,*

$$R \geqq 0, \quad N \geqq 0.$$

For let G_1 be the graph containing the vertices of G but no arcs. Then if R_1 and N_1 are its rank and nullity,

$$R_1 = N_1 = 0.$$

We build up G from G_1 by adding the arcs one at a time. The theorem now follows from Theorem 2.

THEOREM 4. *A forest G is a graph of nullity 0, and conversely.*

Suppose first G contained a circuit P. We shall show that the nullity of G is >0. We build up G arc by arc, adding first the arcs of the circuit P. In adding the last arc of the circuit, the nullity is increased by 1, as this arc joins two vertices already connected. (This argument holds even if the circuit is a 1-circuit.) But in adding the rest of the arcs, the nullity is never decreased, by Theorem 2. Thus the nullity of G is >0.

Now suppose G is a forest, and therefore contains no circuit. Build up G arc by arc. Each arc we add joins two vertices formerly not connected. For otherwise, this arc, together with the arcs of a chain connecting the two vertices, would form a circuit. Therefore, by Theorem 2, the nullity remains always the same, and is thus 0.

3. **Theorems on non-separable graphs.** We introduce the following

Definitions. Let H_1, which contains the vertex a_1, and H_2, which contains the vertex a_2, be two graphs without common vertices. Let us rename a_1 a, and rename the arcs of H_1 on a_1 accordingly; that is, if a_1b is an arc on a_1, we rename it ab. Rename also a_2 a, and rename the arcs of H_2 accordingly. H_1 and H_2 have now the vertex a in common; they form the graph G, say. We say G is formed by letting the vertex a_1 of H_1 *coalesce* with the vertex a_2 of H_2, or, by joining H_1 and H_2 at a vertex. Geometrically, we pull the vertices a_1 and a_2 together to form the single vertex a.

Let G be a connected graph such that there exist no two graphs H_1 and

H_2, each containing at least one arc, which form G if they are joined at a vertex. Then G is called *non-separable*. Geometrically, a connected graph is non-separable if we cannot break it at a single vertex into two graphs, each containing an arc. For example, the graph consisting of the two arcs ab, bc is separable, as is the graph consisting of the two arcs $\alpha(aa)$, $\beta(aa)$. A graph containing but a single arc is non-separable, as is the graph containing only the arcs $\alpha(ab)$, $\beta(ab)$.

If G is not non-separable, we say G is *separable*. Thus, a graph that is not connected is separable. Suppose some connected piece G_1 of G is separable. If H_1 and H_2 joined at the vertex a form G_1, we say a is a *cut vertex* of G. We have consequently

THEOREM 5. *A necessary and sufficient condition that a connected graph be non-separable is that it have no cut vertex.*

THEOREM 6. *Let G be a connected graph containing no* 1-*circuit. A necessary and sufficient condition that the vertex a be a cut vertex of G is that there exist two vertices b, c in G, each distinct from a, such that every chain from b to c passes through a.*

First suppose a is a cut vertex of G. Then, by definition, H_1 and H_2, each containing at least one arc which is not a 1-circuit, form G if they are joined at a. Let b be a vertex of H_1 and c a vertex of H_2, each distinct from a. As a is the only vertex in both H_1 and H_2, every chain from b to c in G passes through a.

Suppose now every chain from b to c in G passes through a. Remove the vertex a and all the arcs on a. The resulting graph G' is not connected, b and c being in different connected pieces. Let H_1' be that connected piece of G' containing b, and let H_2' be the rest of G'. Replace a by the two vertices a_1 and a_2. Now put back the arcs we removed, letting them touch a_1 if their other end vertices are in H_1', and letting them touch a_2 otherwise. Let H_1 and H_2 be the resulting graphs. Then H_1 and H_2 each contain at least one arc, and they form G if the two vertices a_1, a_2 are made to coalesce. Hence, by definition, a is a cut vertex of G.

THEOREM 7. *Let G be a graph containing no* 1-*circuit and containing at least two arcs. A necessary and sufficient condition that G be non-separable is that it be cyclicly connected.**

If G is not connected, the theorem is obvious. Assume therefore G is connected.

* A similar theorem has been proved for more general continuous curves by G. T. Whyburn, Bulletin of the American Mathematical Society, vol. 37 (1931), pp. 429–433.

Suppose first G is separable. Then, by Theorem 5, G has a cut vertex a, and by Theorem 6, there are two vertices b, c in G such that every chain from b to c passes through a. Hence there is no circuit in G containing b and c.

Suppose now there exist two vertices b, c in G which are contained in no circuit. Let $bd, de, \cdots, gc$ be some chain from b to c.

Case 1. There exists a circuit containing b and d. In this case, let a be the last vertex of the chain which is contained in a circuit passing also through b. Let f be the next vertex of the chain. Then every chain from f to b passes through a. For suppose the contrary. Let C be a chain from f to b not passing through a. Let P be a circuit containing b and a. Follow C from f till we first reach a vertex of P. Follow the circuit P now as far as b if b was not the vertex we reached, and continue along P till we reach a. Passing from a to f along the arc af completes a circuit containing both b and f, contrary to hypothesis. Hence, by Theorem 6, a is a cut vertex of G, and therefore G is separable.

Case 2. There exists no circuit containing b and d. Then there is but a single arc joining b and d, and they are joined by no other chain. As G is connected and contains at least two arcs, there is either another arc on b or another arc on d, say the first. The other case is exactly similar. If we add a vertex b' and replace the arc bd by the arc $b'd$, b and d are no longer joined by a chain, and hence the resulting graph G' is not connected. Let H_1 be that part of G' containing the arc $b'd$, and let H_2 be the rest of G'. As there is still an arc on b, H_2 contains at least one arc. Letting the vertices b and b' coalesce forms G, and hence G is separable. The proof is now complete.

THEOREM 8. *A non-separable graph G containing at least two arcs contains no 1-circuit and is of nullity >0. Each vertex is on at least two arcs.*

Suppose G contained a 1-circuit. Call it H_1. Let H_2 be the rest of the graph. Then H_1 and H_2 have but a single vertex in common, and thus G is separable.

Next, by Theorem 7, G is cyclicly connected. As G contains no 1-circuit, G contains at least two vertices. Containing these there is a circuit. Therefore, by Theorem 4, the nullity of G is >0.

Finally, if there were a vertex on no arcs, G would not be connected. If there were a vertex a on the single arc ab, b would be a cut vertex of G.

THEOREM 9. *Let G be a graph of nullity 1 containing no isolated vertices, such that the removal of any arc reduces the nullity to 0. Then G is a circuit.*

By Theorem 4, G contains a circuit. Suppose G contained other arcs besides. Removing one of these, the nullity remains 1, as the circuit is still present, contrary to hypothesis. There are no other vertices in G, as G contains no isolated vertices. Hence G is just this circuit.

THEOREM 10. *A non-separable graph G of nullity 1 is a circuit.*

If G contains but a single arc, it is a 1-circuit, being of nullity 1. Suppose G contains at least two arcs. By Theorem 8, it contains no 1-circuit. By Theorem 7, it is cyclicly connected. Remove any arc ab from G; a and b are still connected, and therefore, by Theorem 2, the nullity of G is reduced to 0. Hence, by Theorem 9, G is a circuit.

The converses of the last two theorems are obviously true.

4. Decomposition of separable graphs. If the graph G contains a connected piece which is separable, we may separate that piece into two graphs, these graphs having formerly but a single vertex in common. We may continue in this manner until every resulting piece of G is non-separable. We say G is separated into its *components*.

LEMMA. *Let the connected separable graph G be decomposed into the two pieces H_1 and H_2 which had only the vertex a in common in G. Then every non-separable subgraph of G is contained wholly in either H_1 or H_2.*

Suppose the contrary. Then some non-separable subgraph I of G is not contained wholly in either H_1 or H_2. Let I_1 be that part of I in H_1, and I_2 that part in H_2; I_1 and I_2 have at most the vertex a in common. I_1 and I_2 each contain at least one arc. For otherwise, if I_1, say, contained no arc, as it contains a vertex distinct from a, it would not be connected. Thus I is separable into the pieces I_1 and I_2, a contradiction again.

THEOREM 11. *Every non-separable subgraph of G is contained wholly in one of the components of G.*

This follows upon repeated application of the above lemma.

THEOREM 12. *A graph G may be decomposed into its components in a unique manner.*

Suppose we could decompose G into the components $H_1, H_2, \cdots, H_m$, and also into the components $H_1', H_2', \cdots, H_n'$. We shall show that these sets are identical. Take any H_i. It is a non-separable subgraph of G, and thus is contained in some component H_j', by Theorem 11. Similarly, H_j' is contained in some component H_k. Thus H_i is contained in H_k, and they are therefore identical. Hence H_i and H_j' are identical. In this manner we show that each H_k is identical with some H_l', and each H_l' is identical with some H_k, proving the theorem.

THEOREM 13. *Let $H_1, H_2, \cdots, H_m$ be the components of G. Let $R_1, R_2, \cdots, R_m$, and $N_1, N_2, \cdots, N_m$ be their ranks and nullities. Then*

$$R = R_1 + R_2 + \cdots + R_m,$$

$$N = N_1 + N_2 + \cdots + N_m.$$

Let G' be G separated into its components, and let R' be the rank of G'. G is formed from G' by letting vertices of different components coalesce. Each time we join two pieces, the number of vertices and the number of connected pieces are each reduced by 1, so that the rank remains the same. Thus

$$R = R'.$$

Now

$$V' = V_1 + V_2 + \cdots + V_m,$$

$$P' = P_1 + P_2 + \cdots + P_m$$

(where each $P_i = 1$). Subtracting,

$$R = R' = R_1 + R_2 + \cdots + R_m.$$

As also

$$E = E_1 + E_2 + \cdots + E_m,$$

it follows that

$$N = N_1 + N_2 + \cdots + N_m.$$

For a converse of this theorem, see Theorem 17.

THEOREM 14. *Divide the arcs of the non-separable graph G into two groups, each containing at least one arc, forming the subgraphs H_1 and H_2, of ranks R_1 and R_2. Then*

$$R_1 + R_2 > R.$$

Let the connected pieces of H_1 be $H_{11}, \cdots, H_{1m}$ (there may be but one piece, H_{11}), and let those of H_2 be $H_{21}, \cdots, H_{2n}$. Then obviously

$$R_1 = R_{11} + \cdots + R_{1m},$$

$$R_2 = R_{21} + \cdots + R_{2n},$$

whence

$$R_1 + R_2 = R_{11} + \cdots + R_{1m} + R_{21} + \cdots + R_{2n}.$$

Let G' be the sum of the graphs $H_{11}, \cdots, H_{2n}$. Then G' is of rank $R_{11} + \cdots + R_{2n}$. We form G from G' by letting vertices of the graphs $H_{11}, \cdots, H_{2n}$ coalesce. Each time we let vertices of different connected pieces coalesce, the rank is unaltered. Each time we let vertices in the same connected piece coalesce, the rank is reduced by 1. This latter operation happens at least once. For otherwise, let a_1 and a_2 be the last two vertices we let coalesce. Then a_1 and a_2 were formerly in two different pieces, I_1 and I_2. Thus

I_1 and I_2 joined at a vertex form G, and G is separable, contrary to hypothesis. Thus the rank of G is less than the rank of G', that is,

$$R < R_{11} + \cdots + R_{2n}.$$

Hence

$$R_1 + R_2 > R.*$$

Theorems 13 and 14 give

THEOREM 15. *A necessary and sufficient condition that a graph be non-separable is that there exist no division of its arcs into two groups H_1 and H_2, each containing at least one arc, so that*

$$R = R_1 + R_2.$$

5. **Circuits of graphs.** We shall say two non-separable graphs, each containing at least one arc, form a *circuit of graphs*, if they have at least two common vertices. (They may also have common arcs.) Thus the two graphs G_1: $\alpha(ab)$ and G_2: $\alpha(ab)$ (which are the same graph) form a circuit of graphs. However, the two graphs G_1: $\alpha(aa)$ and G_2: $\beta(aa)$, having but one common vertex, do not form a circuit of graphs. We shall say three or more non-separable graphs form a *circuit of graphs* if we can name them $G_1, G_2, \cdots, G_m$ in such a way that G_1 and G_2 have just the vertex a_1 in common, G_2 and G_3 have just the vertex a_2 in common, $\cdots$, G_m and G_1 have just the vertex a_m in common, these vertices are all distinct, and no other two of these graphs have a common vertex. Thus the three graphs G_1:ab, G_2:bc, G_3:ca form a circuit of graphs.

We note that there can be no 1-circuit in a circuit of graphs; also, no subset of the graphs in a circuit of graphs form a circuit of graphs. We may think of a circuit of graphs as forming a single graph.

THEOREM 16. *A circuit of graphs G is a non-separable graph.*

First suppose there are but two graphs, G_1 and G_2, present. Suppose G were separable. Then it is separable into at least two components $H_1, H_2, \cdots, H_k$. By Theorem 11, G_1 and G_2 are each contained wholly in one of these components. As G_1 and G_2 together form G, there are just two components, and they are G_1 and G_2. These, when joined at a vertex, form G. But this is contrary to the hypothesis that G_1 and G_2 have at least two vertices in common.

Next suppose there are more than two graphs present. Let C_1 be a chain in G_1 joining a_m and a_1, let C_2 be a chain in G_2 joining a_1 and $a_2, \cdots$, let C_m be a chain in G_m joining a_{m-1} and a_m. These chains taken together form a cir-

* This theorem may also be proved easily from Theorem 17.

cuit P passing through all the graphs. Now separate G into its components. By Theorem 11 (see the converse of Theorem 10), P is contained in one of these components. The same is true of each of the graphs $G_1, G_2, \cdots, G_m$, and hence these graphs are all contained in the same component. Thus G is itself this component, that is, G is non-separable.

THEOREM 17. *Let $G_1, \cdots, G_m$ be a set of non-separable graphs, each containing at least one arc, and let G be formed by letting vertices and arcs of different graphs coalesce. Then the following four statements are all equivalent*:

(1) *$G_1, \cdots, G_m$ are the components of G.*

(2) *No two of the graphs $G_1, \cdots, G_m$ have an arc in common, and there is no circuit in G containing arcs of more than one of these graphs.*

(3) *No subset of these graphs form a circuit of graphs.*

(4) *If $R, R_1, \cdots, R_m$ are the ranks of $G, G_1, \cdots, G_m$ respectively, then*

$$R = R_1 + \cdots + R_m.$$

We note that we cannot replace the word rank by the word nullity in (4). For let G be the graph containing the arcs $\alpha(ab), \beta(ab), \gamma(ab)$. Let G_1 contain α and β, and G_2, β and γ. Then the nullity of G is the sum of the nullities of G_1 and G_2, but G_1 and G_2 are not the components of G. We shall prove

(a) if (1) holds, (2) holds,

(b) if (2) holds, (3) holds,

(c) if (3) holds, (1) holds, establishing the equivalence of (1), (2) and (3);

(d) if (1) holds, (4) holds, and finally

(e) if (4) holds, (3) holds, establishing the equivalence of (4) and the other statements.

(a) If (1) holds, (2) holds. For first, in forming G from its components $G_1, \cdots, G_m$, we let vertices alone coalesce, and thus no two of the graphs have an arc in common. Also, there is no circuit in G containing arcs of more than one of the graphs; for each circuit, being a non-separable graph, is contained entirely in one of the components of G, by Theorem 11.

(b) If (2) holds, (3) holds. For suppose the contrary. If, first, some two graphs, say G_1 and G_2, form a circuit of graphs, they have at least two vertices in common, say a and b. Join a and b by a chain C in G_1 and by a chain D in G_2. By hypothesis, G_1 and G_2 have no arcs in common, and thus the arcs of C and D are distinct. From a follow along C till we first reach a vertex d of D. From d follow along D till we get back to a. We have formed thus a circuit containing arcs of both G_1 and G_2, contrary to hypothesis.

Now suppose the graphs $G_1, \cdots, G_k$, $k>2$, formed a circuit of graphs. In the proof of Theorem 16 we found a circuit passing through all the graphs of such a circuit of graphs, again contrary to hypothesis.

(c) If (3) holds, (1) holds. Assuming that no subset of the graphs $G_1, \cdots, G_m$ forms a circuit of graphs, we will show first that some one of these graphs has at most a single vertex in common with other of the graphs. For suppose each graph had at least two vertices in common with other graphs. Then G_1 has a vertex a_1 in common with some graph, say G_2. As G_2 has at least two vertices in common with other graphs, it has a vertex a_2, distinct from a_1, in common with another graph, say G_3. If we continue in this manner, we must at some point get back to a graph we have already considered.

Now starting with G_1, consider the graphs in order, and let G_i be the first one which has a vertex in common with one of the preceding graphs other than the vertex a_{i-1}, which we know already it has in common with G_{i-1}. Now of the graphs $G_{i-1}, G_{i-2}, \cdots, G_1$, let G_j be the first with which G_i has a common vertex, other than the vertex a_{i-1}. First suppose G_j is G_{i-1}. Then G_i and G_{i-1} have at least two vertices in common, and they form therefore a circuit of graphs, contrary to hypothesis. Next suppose G_j is not G_{i-1}. Then on account of the choice of G_i and G_j, G_j and G_{j+1} have just one common vertex a_j, G_{j+1} and G_{j+2} have just one common vertex $a_{j+1}, \cdots, G_i$ and G_j have just one common vertex a_i (for otherwise G_i and G_j would form a circuit of graphs), and no other two of these graphs have a vertex in common. These vertices $a_j, a_{j+1}, \cdots, a_i$ are all distinct. For, on account of the construction of the chain of graphs, two succeeding vertices a_k and a_{k+1} are distinct. a_i and a_j are distinct, for otherwise G_i and G_{j+1} would have a common vertex, etc. These graphs $G_j, G_{j+1}, \cdots, G_i$ form therefore a circuit of graphs, contrary to hypothesis.

Some graph therefore, say G_1, has at most a single vertex in common with the other graphs. Thus either it is separated from them, or we can separate it at a single vertex. Now among the graphs $G_2, \cdots, G_m$, there is also no circuit of graphs, so again we can separate one of them, say G_2. Continuing, we have finally separated G into its components $G_1, G_2, \cdots, G_m$.

(d) If (1) holds, (4) holds. This is just Theorem 13.

(e) If (4) holds, (3) holds. Let G' be the sum of the graphs $G_1, \cdots, G_m$. We form G from G' by letting vertices and arcs of different graphs coalesce. Each time we let two vertices coalesce, either (α) the two vertices were formerly in different connected pieces, in which case the rank is unchanged, or (β) the two vertices were in the same connected piece, in which case the rank is reduced by 1. Letting arcs alone coalesce (their end vertices having already coalesced) does not alter the rank. Thus in any case, the rank is never increased. To begin with, the rank of G' is $G_1 + \cdots + G_m$, and by hypothesis, the rank of G is $G_1 + \cdots + G_m$. Thus the rank is never altered, and (β) never occurs. Hence, obviously, no circuit of graphs is formed in forming G from G'. This completes the proof of the theorem.

6. **Construction of non-separable graphs.** We prove the following theorem:

THEOREM 18. *If G is a non-separable graph of nullity $N>1$, we can remove an arc or suspended chain from G, leaving a non-separable graph G' of nullity $N-1$.*

Assume the theorem is true for all graphs of nullity $2, 3, \cdots, N-1$. We shall prove it for any graph of nullity N (including the case where $N=2$). This will establish the theorem in general.

Take any non-separable graph G of nullity $N>1$. It contains at least two arcs, and therefore, by Theorem 8, it contains no 1-circuit. Remove from G any arc ab, forming the graph G_1. If G_1 is non-separable, we are through. Suppose therefore G_1 is separable, and let its components be $H_1, H_2, \cdots, H_{m-1}$. G_1 is connected, for between any two vertices c, d there exists a circuit in G by Theorem 7, and therefore there is a chain joining them in G_1.

Let H_m consist of the arc ab. By Theorem 17, no subset of the graphs $H_1, \cdots, H_{m-1}$ form a circuit of graphs, while some subset of the graphs $H_1, \cdots, H_m$ form a circuit of graphs. We shall show that the whole set of graphs $H_1, \cdots, H_m$ form a circuit of graphs. Otherwise, some proper subset, which includes H_m, form a circuit of graphs.

Let H be the graph formed from this circuit of graphs by dropping out H_m. By Theorem 16, the circuit of graphs is a non-separable graph; hence H is connected. All the arcs in G_1 not in the circuit of graphs, form a graph I. Let I_1 be a connected piece of I. Then I_1 has at most a single vertex in common with the rest of G. For suppose I_1 had the two vertices c and d in common with H. From c follow along some chain towards d in H till we first reach a vertex e in I_1. From e follow back along some chain in I_1 to c. We have formed thus a circuit containing arcs of both H and I_1. But as H consists of a certain subset of the components of G_1, this circuit contains arcs of at least two components of G_1, contrary to Theorem 17. Thus I_1 has at most a single vertex in common with the rest of G, and hence G is separable, contrary to hypothesis. Thus $H_1, \cdots, H_m$ form a circuit of graphs, that is, G is formed of a circuit of graphs.

As we assumed G_1 was separable, $m \geqq 3$. Therefore we can order the graphs so that H_1 and H_2 have just the vertex a_1 in common, $\cdots$, H_{m-1} and H_m have just the vertex $a_{m-1}=b$ in common, and H_m and H_1 have just the vertex $a_m=a$ in common. Moreover, these vertices are all distinct, and no other two of the graphs $H_1, \cdots, H_m$ have a common vertex.

As the nullity of G was >1, the nullity of G_1 is >0. By Theorem 13, this is the sum of the nullities of $H_1, \cdots, H_{m-1}$. Therefore the nullity of some one of these graphs, say H_i, is >0.

Suppose first the nullity of H_i is 1. Then, by Theorem 10, H_i is a circuit, consisting of two chains joining a_{i-1} and a_i. Remove one of these chains from G. This leaves a graph G', which again is a circuit of graphs. For the graph H_i we replace by an ordered set of non-separable graphs, each consisting of one of the arcs of the chain we have left in H_i.

Suppose next the nullity of H_i is >1. It is less than N, as H_i is contained in G_1, whose nullity is $N-1$. Therefore, by induction, we can remove an arc or a suspended chain, leaving a non-separable graph H_i' of nullity one less. If neither a_{i-1} nor a_i has thus been removed, we again have a circuit of graphs. Suppose a_i but not a_{i-1} was removed. Replace that part of the chain we removed joining a_i and a vertex of H_i distinct from a_{i-1}. Here again we have a circuit of graphs, H_i being replaced by H_i' and a set of arcs. The case is the same if a_{i-1} but not a_i was removed. If finally, both a_i and a_{i-1} were in the chain we removed, we put back all of the chain but that part between these two vertices. Here again, the resulting graph G' is a circuit of graphs.

Thus in all cases we can drop out from G an arc or suspended chain, leaving a circuit of graphs. By Theorem 16, the resulting graph G' is non-separable. As also the nullity of G' is one less than the nullity of G, the theorem is now proved.

As a consequence of this theorem, Theorem 8, and Theorem 10, we have

THEOREM 19. *We can build up any non-separable graph containing at at least two arcs by taking first a circuit, then adding successively arcs or suspended chains, so that at any stage of the construction we have a non-separable graph.*

It is easily seen that, conversely, any graph built up in this manner is non-separable. For each time we add an arc or suspended chain, these arcs, each considered as a graph, together with the non-separable graph already present, form a circuit of graphs.

II. DUALS, PLANAR GRAPHS

7. Congruent graphs. We introduce the following

Definitions. Given two graphs G and G', if we can rename the vertices and arcs of one, giving distinct vertices and distinct arcs different names, so that it becomes identical with the other, we say the two graphs are *congruent*.* (We used formerly the word "homeomorphic.")

* See the author's American Journal paper, cited in the introduction.

The geometrical interpretation is that we can bring the two graphs into complete coincidence by a (1, 1) continuous transformation.

Two graphs are called *equivalent* if, upon being decomposed into their components, they become congruent, except possibly for isolated vertices.

8. **Duals.** Given a graph G, if H_1 is a subgraph of G, and H_2 is that subgraph of G containing those arcs not in H_1, we say H_2 is the *complement* of H_1 in G.

Throughout this section, R, R', r, r', etc., will stand for the ranks of G, G', H, H', etc., respectively, with similar definitions for V, E, P, N.

Definition. Suppose there is a (1, 1) correspondence between the arcs of the graphs G and G', such that if H is any subgraph of G and H' is the complement of the corresponding subgraph of G', then

$$r' = R' - n.$$

We say then that G' is a *dual* of G.*

Thus, if the nullity of H is n, then H' (including all the vertices of G') is in n more connected pieces than G'.

THEOREM 20. *Let G' be a dual of G. Then*

$$R' = N,$$
$$N' = R.$$

For let H be that subgraph of G consisting of G itself. Then

$$n = N.$$

If H' is the complement of the corresponding subgraph of G', H' contains no arcs, and is the null graph. Thus

$$r' = 0.$$

But as G' is a dual of G,

$$r' = R' - n.$$

These equations give

$$R' = N.$$

The other equation follows when we note that $E' = E$.

THEOREM 21. *If G' is a dual of G, then G is a dual of G'.*

Let H' be any subgraph of G', and let H be the complement of the corresponding subgraph of G. Then, as G' is a dual of G,

* While this definition agrees with the ordinary one for graphs lying on a plane or sphere, a graph on a surface of higher connectivity, such as the torus, has in general no dual. (See Theorems 29 and 30.)

$$r' = R' - n.$$

By Theorem 20,

$$R' = N.$$

We note also,

$$e + e' = E.$$

These equations give

$$r = e - n = e - (R' - r') = e - N + (e' - n')$$
$$= E - N - n' = R - n'.$$

Thus G is a dual of G'.

Whenever we have shown that one graph is a dual of another graph, we may now call the graphs "dual graphs."

LEMMA. *If a graph G is decomposed into its components, the rank and nullity of any subgraph H is left unchanged.*

For each time we separate G at a vertex, H is either unchanged or is separated at a vertex. Hence neither its rank nor its nullity is altered. (See the proof of Theorem 13.)

THEOREM 22. *If G' and G'' are equivalent and G' is a dual of G, then G'' is a dual of G.*

Let H be any subgraph of G, and let H' be the complement of the corresponding subgraph of G'. Let G_1' and G_1'' be G' and G'' decomposed into their components. Then G_1' and G_1'' are congruent. H' turns into a subgraph H_1' of G'. Let H_1'' be the corresponding subgraph of G_1'', and H'' the same subgraph in G''. Then

$$r_1' = r_1''.$$

But by the above lemma,

$$r' = r_1', \quad r'' = r_1''.$$

Hence

$$r' = r''.$$

As a special case of this equation, letting H' be the whole of G', we have

$$R' = R''.$$

As G' is a dual of G,

$$r' = R' - n.$$

Therefore

$$r'' = R'' - n,$$

and G'' is a dual of G.

The converse of this theorem is not true. For define the three graphs

G: $\alpha(ab)$, $\beta(ab)$, $\gamma(ac)$, $\delta(cb)$, $\epsilon(ad)$, $\zeta(db)$;

$G': \alpha'(a'b'), \beta'(c'd'), \gamma'(a'd'), \delta'(a'd'), \epsilon'(b'c'), \zeta'(b'c')$;
$G'': \alpha''(a''b''), \beta''(b''c''), \gamma''(a''d''), \delta''(a''d''), \epsilon''(c''d''), \zeta''(c''d'')$.
G' and G'' are both duals of G, but they are not congruent.*

THEOREM 23. *Let $G_1, \cdots, G_m$ and $G_1', \cdots, G_m'$ be the components of G and G' respectively, and let G_i' be a dual of G_i, $i=1, \cdots, m$. Then G' is a dual of G.*

Let H be any subgraph of G, and let the parts of H in $G_1, \cdots, G_m$ be $H_1, \cdots, H_m$. Let H_i' be the complement of the subgraph corresponding to H_i in G_i', $i=1, \cdots, m$, and let H' be the union of $H_1', \cdots, H_m'$ in G'. Then H' is the complement of the subgraph in G' corresponding to H in G. Using the proof of Theorem 13, we find that

$$r' = r_1' + \cdots + r_m',$$

and

$$n = n_1 + \cdots + n_m.$$

As also

$$R' = R_1' + \cdots + R_m'$$

and

$$r_i' = R_i' - n_i \qquad (i = 1, \cdots, m),$$

adding these last equations gives

$$r' = R' - n,$$

and hence G' is a dual of G.

THEOREM 24. *Let $G_1, \cdots, G_m$ and $G_1', \cdots, G_m'$ be the components of the dual graphs G and G', and let the correspondence between these two graphs be such that arcs in G_i correspond to arcs in G_i', $i=1, \cdots, m$. Then G_i and G_i' are duals, $i=1, \cdots, m$.*

Let H_1 be any subgraph of G_1, let H' be the complement of the corresponding subgraph in G', and let H_1' be the complement in G'. Then $H_1', G_2', \cdots, G_m'$ form H'. By Theorem 13, we find

$$R' = R_1' + R_2' + \cdots + R_m'$$

and

$$r' = r_1' + R_2' + \cdots + R_m'.$$

Now

$$r' = R' - n_1,$$

hence

$$r_1' = R_1' - n_1,$$

and G_1' is a dual of G_1. Similarly for $G_2', \cdots, G_m'$.

* See the author's American Journal paper, however.

THEOREM 25. *Let G and G' be dual graphs, and let $H_1, \cdots, H_m$ be the components of G. Let $H_1', \cdots, H_m'$ be the corresponding subgraphs of G'. Then $H_1', \cdots, H_m'$ are the components of G', and H_i' is a dual of H_i, $i=1, \cdots, m$.*

H_1 is the subgraph of G corresponding to H_1' in G'. Its complement is I_1, the graph formed of the arcs of $H_2, \cdots, H_m$. Obviously $H_2, \cdots, H_m$ are the components of I_1. Hence, by Theorem 13, the nullity of I_1 is $n_2+n_3+\cdots+n_m$. Thus, as G' is a dual of G,

$$r_1' = R' - (n_2 + n_3 + \cdots + n_m).*$$

Similarly,

$$r_2' = R' - (n_1 + n_3 + \cdots + n_m),$$

$$\cdots\cdots\cdots\cdots\cdots\cdots$$

$$r_m' = R' - (n_1 + n_2 + \cdots + n_{m-1}).$$

Adding these equations gives

$$r_1' + r_2' + \cdots + r_m' = mR' - (m-1)(n_1 + n_2 + \cdots + n_m).$$

As $H_1, H_2, \cdots, H_m$ are the components of G,

$$N = n_1 + n_2 + \cdots + n_m.$$

Also, as G and G' are duals, by Theorem 20,

$$R' = N.$$

Hence

$$\begin{aligned} r_1' + r_2' + \cdots + r_m' &= mR' - (m-1)R' \\ &= R'. \end{aligned}$$

Let now $H_{11}', \cdots, H_{1k_1}'$ be the components of H_1' (there may be but one) and similarly for $H_2', \cdots, H_m'$. Then, by Theorem 13,

$$r_1' = r_{11}' + \cdots + r_{1k_1}',$$

$$\cdots\cdots\cdots\cdots\cdots$$

$$r_m' = r_{m1}' + \cdots + r_{mk_m}'.$$

Adding these equations gives

$$\sum_{i,j} r_{ij}' = r_1' + \cdots + r_m' = R'.$$

As the graphs $H_{11}', \cdots, H_{mk_m}'$ are non-separable, Theorem 17 tells us that they are the components of G'. Hence G' has at least as many components as

* Which equals n_1.

G. Similarly, G has at least as many components as G'. They have therefore the same number, m, of components.

There are therefore m graphs in the set $H_{11}', \cdots, H_{mk_m}'$. But there is at least one such graph in each graph $H_1', \cdots, H_m'$, and there is therefore exactly one in each. Hence each graph H_{i1}' fills out the graph H_i', and the two sets of graphs $H_{11}', \cdots, H_{mk_m}'$ and $H_1', \cdots, H_m'$ are identical, that is, $H_1', \cdots, H_m'$ are the components of G'.

The rest of the theorem follows from Theorem 24.

As a special case of this theorem, we have

THEOREM 26. *A dual of a non-separable graph is non-separable.*

9. **Planar graphs.** Up till now, we have been considering abstract graphs alone. However, the definition of a planar graph is topological in character. This section may be considered as an application of the theory of abstract graphs to the theory of topological graphs.

Definitions. A topological graph is called *planar* if it can be mapped in a (1, 1) continuous manner on a sphere (or a plane). For the present, we shall say that an abstract graph is *planar* if the corresponding topological graph is planar. Having proved Theorem 29, we shall be justified in using the following purely combinatorial definition: *A graph is planar if it has a dual.*

We shall henceforth talk about "graphs" simply, the terms applying equally well to either abstract or topological graphs.

LEMMA. *If a graph can be mapped on a sphere, it can be mapped on a plane, and conversely.*

Suppose we have a graph mapped on a sphere. We let the sphere lie on the plane, and rotate it so that the new north pole is not a point of the graph. By stereographic projection from this pole, the graph is mapped on the plane. The inverse of this projection maps any graph on the plane onto the sphere.

By the *regions* of a graph lying on a sphere or in a plane is meant the regions into which the sphere or plane is thereby divided. A given region of the graph is characterized by those arcs of the graph which form its boundary. If the graph is in a plane, the outside region is the unbounded region.

LEMMA. *A planar graph may be mapped on a plane so that any desired region is the outside region.*

We map the graph on a sphere, and rotate it so that the north pole lies inside the given region. By stereographic projection, the graph is mapped onto the plane so that the given region is the outside region.

We return now to the work in hand.

THEOREM 27. *If the components of a graph G are planar, G is planar.*

Suppose the graphs G_1 and G_2 are planar, and G' is formed by letting the vertices a_1 and a_2 of G_1 and G_2 coalesce. We shall show that G' is planar. Map G_1 on a sphere, and map G_2 on a plane so that one of the regions adjacent to the vertex a_2 is the outside region. Shrink the portion of the plane containing G_2 so it will fit into one of the regions of G_1 adjacent to a_1. Drawing a_1 and a_2 together, we have mapped G' on the sphere.* The theorem follows as a repeated application of this process.

THEOREM 28. *Let G and G' be dual graphs, and let $\alpha(ab)$, $\alpha'(a'b')$ be two corresponding arcs. Form G_1 from G by dropping out the arc $\alpha(ab)$, and form G_1' from G' by dropping out the arc $\alpha'(a'b')$, and letting the vertices a' and b' coalesce if they are not already the same vertex. Then G_1 and G_1' are duals, preserving the correspondence between their arcs.*

Let H_1 be any subgraph of G_1 and let H_1' be the complement of the corresponding subgraph of G_1'.

Case 1. Suppose the vertices a' and b' were distinct in G'. Let H be the subgraph of G identical with H_1. Then

$$n = n_1.$$

Let H' be the complement in G' of the subgraph corresponding to H. Then

$$r' = R' - n.$$

Now H' is the subgraph in G' corresponding to H_1' in G_1', except that H' contains the arc $\alpha'(a'b')$, which is not in H_1'. Thus if we drop out $\alpha'(a'b')$ from H' and let a' and b' coalesce, we form H_1'. In this operation, the number of connected pieces is unchanged, while the number of vertices is decreased by 1. Hence

$$r_1' = r' - 1.$$

As a special case of this equation, if H' contains all the arcs of G', we find

$$R_1' = R' - 1.$$

These equations give

$$r_1' = R_1' - n_1.$$

Thus G_1' is a dual of G_1.

Case 2. Suppose a' and b' are the same vertex in G'. In this case, defining H and H' as before, we form H_1' from H' by dropping out the arc $\alpha'(a'a')$. This leaves the number of vertices and the number of connected pieces un-

* Here and in a few other places we are using point-set theorems which, however, are geometrically evident.

changed. Thus two of the equations in Case 1 are replaced by the equations

$$r_1' = r_1, \quad R_1' = R_1.$$

The other equations are as before, so we find again that G_1' is a dual of G_1. The theorem is now proved.

THEOREM 29. *A necessary and sufficient condition that a graph be planar is that it have a dual.*

We shall prove first the necessity of the condition. Given any planar graph G, we map it onto the surface of a sphere. If the nullity of G is N, it divides the sphere into $N+1$ regions. For let us construct G arc by arc. Each time we add an arc joining two separate pieces, the nullity and the number of regions remain the same. Each time we add an arc joining two vertices in the same connected pieces, the nullity and the number of regions are each increased by 1. To begin with, the nullity was 0 and the number of regions was 1. Therefore, at the end, the number of regions is $N+1$.

We construct G' as follows: In each region of the graph G we place a point, a vertex of G'. Therefore G' contains $V'=N+1$ vertices. Crossing each arc of G we place an arc, joining the vertices of G' lying in the two regions the arc of G separates (which may in particular be the same region, in which case this arc of G' is a 1-circuit). The arcs of G and G' are now in (1, 1) correspondence.

G' is the dual of G in the ordinary sense of the word. We must show it is the dual as we have defined the term.

Let us build up G arc by arc, removing the corresponding arc of G' each time we add an arc to G. To begin with, G contains no arcs and G' contains all its arcs, and at the end of the process, G contains all its arcs and G' contains no arcs. We shall show

(1) each time the nullity of G is increased by 1 upon adding an arc, the number of connected pieces in G' is reduced by 1 in removing the corresponding arc, and

(2) each time the nullity of G remains the same, the number of connected pieces in G' remains the same.

To prove (1) we note that the nullity of G is increased by 1 only when the arc we add joins two vertices in the same connected piece. Let ab be such an arc. As a and b were already connected by a chain, this chain together with ab forms a circuit P. Let $a'b'$ be the arc of G' corresponding to ab. Before we removed it, a' and b' were connected. Removing it, however, disconnects them. For suppose there were still a chain C' joining them. As a' and b' are on opposite sides of the circuit P, C' must cross P, by the Jordan Theorem, that is, an arc of C' must cross an arc of P. But we removed this arc of C' when we put in the arc of P it crosses. (1) is now proved.

The total increase in the nullity of G during the process is of course just N. Therefore the increase in the number of connected pieces in G' must be at least N. But G' was originally in at least one connected piece, and is at the end of the process in $V=N+1$ connected pieces. Thus the increase in the number of connected pieces in G' is just N (hence, in particular, G' itself is connected) and therefore this number increases only when the nullity of G increases, which proves (2).

Let now H be any subgraph of G, let H' be the complement of the corresponding subgraph of G', and let H' include all the vertices of G'. We build up H arc by arc, at the same time removing the corresponding arcs of G'. Thus when H is formed, H' also is formed. By (1) and (2), the increase in the number of connected pieces in forming H' from G' equals the nullity of H, that is,

$$p' - P' = n.$$

But

$$r' = V' - p', \quad R' = V' - P',$$

as G' and H' contain the same vertices. Therefore

$$r' = R' - n,$$

that is, G' is a dual of G.

To prove the sufficiency of the condition, we must show that if a graph has a dual, it is planar. It is enough to show this for non-separable graphs. For if the separable graph G has a dual, its components have duals, by Theorem 25, hence its components are planar, and hence G is planar, by Theorem 27. This part of the theorem is therefore a consequence of the following theorem:

THEOREM 30. *Let the non-separable graph G have a dual G'. Then we can map G and G' together on the surface of a sphere so that*

(1) *corresponding arcs in G and G' cross each other, and no other pair of arcs cross each other, and*

(2) *inside each region of one graph there is just one vertex of the other graph.*

The theorem is obviously true if G contains a single arc. (The dual of an arc ab is an arc $a'a'$, and the dual of an arc aa is an arc $a'b'$.) We shall assume it to be true if G contains fewer than E arcs, and shall prove it for any graph G containing E arcs. By Theorem 8, each vertex of G is on at least two arcs.

Case 1. G contains a vertex b on but two arcs, ab and bc. As G is non-separable, there is a circuit containing these arcs. Thus dropping out one of them will not alter the rank, while dropping out both reduces the rank by 1. As G' is a dual of G, the arcs corresponding to these two arcs are each of nullity 0, while the two arcs taken together are of nullity 1. They are thus of the form $\alpha'(a'b')$, $\beta'(a'b')$, the first corresponding to ab, and the second, to bc.

Form G_1 from G by dropping out the arc bc and letting the vertices b and c coalesce, and form G_1' from G' by dropping out the arc $\beta'(a'b')$. By Theorem 28, G_1 and G_1' are duals, preserving the correspondence between the arcs. As these graphs contain fewer than E arcs,* we can, by hypothesis, map them together on a sphere so that (1) and (2) hold; in particular, $\alpha'(a'b')$ crosses ac. Mark a point on the arc ac of G_1 lying between the vertex c and the point where the arc $\alpha'(a'b')$ of G' crosses it. Let this be the vertex b, dividing the arc ac into the two arcs ab and bc. Draw the arc $\beta'(a'b')$ crossing the arc bc. We have now reconstructed G and G', and they are mapped on a sphere so that (1) and (2) hold.

Case 2. Each vertex of G is on at least three arcs. As then G contains no suspended chain, and G is not a circuit and therefore is of nullity $N>1$, we can, by Theorem 18, drop out an arc ab so that the resulting graph G_1 is non-separable. G' is non-separable, by Theorem 26, and hence the arc $a'b'$ corresponding to ab in G is not a 1-circuit. Drop it out and let the vertices a', b' coalesce into the vertex a_1', forming the graph G_1'. By Theorem 28, G_1 and G_1' are duals, and thus G_1' also is non-separable.

Consider the arcs of G' on a'. If we drop them out, the resulting graph G'' has a rank one less than that of G'. For if its rank were still less, G'' would be in at least three connected pieces, one of them being the vertex a'. Let c and d be vertices in two other connected pieces of G''. They are joined by no chain in G'', and hence every chain joining them in G' must pass through a', which contradicts Theorem 6. If we put back any arc, the rank is brought back to its original value, as a' is then joined to the rest of the graph. Hence, G' being a dual of G, the arcs of G corresponding to these arcs are together of nullity 1, while dropping out one of them reduces the nullity to 0. Therefore, by Theorem 9, these arcs form a circuit P. One of these arcs is the arc ab. The remaining arcs form a chain C. Similarly, the arcs of G corresponding to the arcs of G' on b' form a circuit Q, and this circuit minus the arc ab forms a chain D. C and D have the vertices a and b as end vertices. Also, the arcs of G_1 corresponding to the arcs of G_1' on a_1' form a circuit R. These arcs of G_1' are the arcs of G' on either a' or b', except for the arc $a'b'$ we dropped out. Thus the arcs of G_1 forming the circuit R are the arcs of the chains C and D.

As G_1 and G_1' contain fewer than E arcs, we can map them together on a

* Obviously G_1 is non-separable.

sphere so that properties (1) and (2) hold. a_1' lies on one side of the circuit R, which we call the inside. Each arc of R is crossed by an arc on a_1', and thus there are no other arcs of G_1' crossing R. There is no part of G_1' lying inside R other than a_1', for it could have only this vertex in common with the rest of G_1', and G_1' would be separable. Also, there is no part of G_1 lying inside R, for any arc would have to be crossed by an arc of G_1', and any vertex would have to be joined to the rest of G_1 by an arc, as G_1 is non-separable.

Let us now replace a_1' by the two vertices a' and b', and let those arcs abutting on a_1' that were formerly on a' be now on a', and those formerly on b', now on b'. As the first set of arcs all cross the chain C, and the second set all cross the chain D, we can do this in such a way that no two of the arcs cross each other. We may now join a and b by the arc ab, crossing none of these arcs. This divides the inside of R into two parts, in one of which a' lies, and in the other of which b' lies. We may therefore join a' and b' by the arc $a'b'$, crossing the arc ab. G and G' are now reconstructed, and are mapped on the sphere as required. This completes the proof of the theorem, and therefore of Theorem 29.

THEOREM 31. *A necessary and sufficient condition that a graph be planar is that it contain neither of the two following graphs as subgraphs:*

G_1. *This graph is formed by taking five vertices a, b, c, d, e, and joining each pair by an arc or suspended chain.*

G_2. *This graph is formed by taking two sets of three vertices, a, b, c, and d, e, f, and joining each vertex in one set to each vertex in the other set by an arc or suspended chain.*

This theorem has been proved by Kuratowski.* It would be of interest to show the equivalence of the conditions of the theorem and Theorem 29 directly, by combinatorial methods. We shall do part of this here, in the following theorem:†

THEOREM 32. *Neither of the graphs G_1 and G_2 has a dual.*

Suppose the graph G_1 had a dual. By Theorem 28, if G_1 contains a suspended chain, we can drop out one of its arcs and let the two end vertices coalesce, and the resulting graph will have a dual. Continuing, we see that the graph G_3, in which each pair of vertices of the set a, b, c, d, e are joined by an arc, must have a dual. Similarly, if G_2 has a dual, then the graph G_4, in which each vertex of the set a, b, c is joined to each vertex of the set d, e, f by an arc, must have a dual. Both of these are impossible.

* Fundamenta Mathematicae, vol. 15 (1930), pp. 271–283.

† The other half has recently been proved by the author. See Bulletin of the American Mathematical Society, abstract (38-1-39). (Note added in proof.)

(a) *The graph* G_3. To avoid subscripts, let us call it G. Suppose it had a dual, G'. Then

$$R = N' = 4,$$
$$N = R' = 6,$$
$$E = E' = 10.$$

If G' has isolated vertices, we drop them out, which does not alter its relation to G.

(1) There are no 1-circuits, 2-circuits or triangles in G'. For if there were, dropping out the corresponding arcs of G would have to reduce the rank of G. But we cannot reduce its rank without dropping out at least four arcs.

(2) G' contains at least five quadrilaterals. For if we drop out the four arcs on any vertex of G, the rank is reduced by 1, and if we put back any of these arcs, the rank is brought back to its original value; Theorem 9 now applies.

(3) At least two of these quadrilaterals have an arc in common, as there are but ten arcs in G'.

There are just two ways of forming two quadrilaterals out of fewer than eight arcs without forming any 2-circuits or triangles. One of these graphs, I_1', contains the arcs $a'b'$, $b'e'$, $a'c'$, $c'e'$, $a'd'$, $d'e'$. The other, I_2', contains the arcs $a'e'$, $e'f'$, $f'b'$, $b'a'$, $e'c'$, $c'd'$, $d'f'$. But there is no subgraph of the type I_1' in G', for this subgraph is of rank 4 and nullity 2, and there would have to be a subgraph of G of rank 2 and nullity 2, and such a graph contains a 1- or a 2-circuit, of which there are none in G. Hence G' contains a subgraph I_2'.

(4) Each vertex of G' is on at least three arcs, as there are no 1- or 2-circuits in G.

Each of the vertices a', b', c', d' of I_2' is on but two arcs. Hence there must be another arc on each of these vertices. As I_2' contains seven arcs, and G' contains but ten, one of the three arcs left must join two of these vertices. But if we add an arc $a'b'$ or $c'd'$, we would form a 2-circuit; if we add an arc $a'c'$ or $b'd'$, we would form a triangle; if we add an arc $a'd'$ or $b'c'$, we would form a graph of the type I_1'. As G' contains none of these graphs, we have a contradiction.

(b) *The graph* G_4. Let us call it G. If it has a dual G', then

$$R = N' = 5,$$
$$N = R' = 4,$$
$$E = E' = 9.$$

We proceed exactly as for the graph G_3. In outline:

(1) G' contains no 1- or 2-circuits.

(2) There is no subgraph of G' containing four vertices, each pair being joined by an arc. For this graph is of rank 3 and nullity 3, and G would have to contain a subgraph of rank 2 and nullity 1, that is, a 2-circuit.

(3) There are at least nine subgraphs of G' of rank 3 and nullity 2, and hence of the form $a'b', a'c', b'c', b'd', c'd'$, as there are nine quadrilaterals in G.

(4) As G' contains but nine arcs, two of these subgraphs have an arc in common. There is therefore a subgraph of one of the forms $I_1' : a'e', a'b', b'e', a'c', c'e', a'd', d'e'$, or $I_2' : a'e', a'b', b'e', b'c', c'e', c'd', d'e'$.

(5) Each vertex of G' is on at least four arcs.

Now each of the graphs I_1', I_2' contains seven arcs. We have but two arcs left which we must place so that each vertex of I_1' or I_2' is on at least four arcs. This cannot be done. The theorem is now proved.

Theorem 31 together with this theorem gives an alternative proof of the second part of Theorem 29. For suppose a graph G had a dual. Then it contains neither the graph G_1 nor G_2. For if it did, dropping out all the arcs of G but those forming one of these graphs, Theorem 28 tells us that this graph has a dual. But we have just seen that this is not so. Hence, by Theorem 31, G is planar.

Euler's formula. Map any connected planar graph G on a sphere, and construct its connected dual G' as described in the proof of Theorem 29. Then in each region of G there is a vertex of G'. Let F be the number of regions (or faces) in G. Then

$$R' = N,$$
$$R = V - 1,$$
$$R' = V' - 1,$$
$$V' = F,$$

and hence

$$V - E + F = R + 1 - E + N + 1 = 2,$$

which is Euler's formula.

HARVARD UNIVERSITY,
CAMBRIDGE, MASS.

Reprinted from *American Journal of Mathematics*,
volume 57, pages 800 through 804,
by permission of the Johns Hopkins University Press.

A LATTICE FORMULATION FOR TRANSCENDENCE DEGREES AND p-BASES

BY SAUNDERS MAC LANE

1. **Introduction.** The transcendence degree of an extension of a field is the cardinal number of a maximal set of independent transcendents in the extension (Steinitz [11]);[1] in an Abelian group without elements of finite order the *rank* is the cardinal number of a maximal set of rationally independent group elements (Baer [1]). Both these cardinal numbers are invariants; the proofs of these two facts are similar, so that there should be an underlying theorem generalizing these proofs and stated in terms of the lattice of subfields or of subgroups, as the case may be.[2] This paper constructs, in §2, a type of lattice, called an "exchange" lattice, in which such a theorem can be proved (cf. §3). This lattice theorem includes also some investigations of Teichmüller ([12] and [13]) on fields of characteristic p; in particular, we establish the invariance of the cardinal number of a "relative p-basis" of an inseparable algebraic extension of such a field.

The crucial axiom for our lattices is an "exchange" axiom, related to the Steinitz exchange theorem. This axiom is equivalent to one of the axioms recently used by Menger [7] in investigating the algebra of affine geometry, and also to a certain covering property used by Birkhoff [2] in an analysis of the Jordan Theorem (cf. §4). This axiom can be viewed as a weakened form of the Dedekind or modular axiom for a lattice (Ore [9], or Birkhoff [4]). The Dedekind axiom itself could not apply to the lattices of fields with which we are concerned (see §5 and §6).

Unfortunately the exchange axiom is stated in terms of the "points" of the lattice, or alternately in terms of the covering relation. It thus applies only trivially to continuous geometries or to other infinite lattices having no points. In §7 we succeed in constructing two exchange axioms which in the presence of the other axioms are equivalent to the original exchange axiom, but which themselves do not involve points or coverings. These new axioms yield most of the usual properties of the dimension function in a finite lattice. Their title to be considered as substitutes for the modular law rests chiefly on their versatility: each of the new exchange axioms is seen in §8 to be equivalent to a natural assertion about the possibility of specified types of transpositions in any given chain of the lattice.

Received March 1, 1938; presented to the American Mathematical Society, April 9, 1938.

[1] Numbers in brackets refer to the bibliography at the end of the paper.

[2] This remark is due to R. Baer (in conversation).

2. Exchange lattices. The transcendence degree of a field $\mathfrak{L}$ over a subfield $\mathfrak{K}$ can be defined in terms of those subfields $\mathfrak{M}, \mathfrak{K} \subset \mathfrak{M} \subset \mathfrak{L}$, which are relatively algebraically closed in $\mathfrak{L}$, in the sense that every element of $\mathfrak{L}$ algebraic over $\mathfrak{M}$ is contained in $\mathfrak{M}$. With respect to field inclusion, these subfields $\mathfrak{M}$ form a continuous lattice.

A set L is a *continuous lattice*[3] (in the terminology of O. Ore, a *complete structure*) if a transitive and irreflexive relation $a < b$ is so defined for elements a and b of L that for every subset $A \subset L$ there exist in L an element $\Sigma(A)$, the *union*, and an element $\Pi(A)$, the *cross-cut*, such that $c \geqq \Sigma(A)$ holds if and only if $c \geqq a$ for every a in A, while $d \leqq \Pi(A)$ if and only if $d \leqq a$ for every a in A. If A consists of two elements a and b, we denote the union or join by $a + b$, and the cross-cut or meet by $a \cdot b$. The cross-cut $0 = \Pi(L)$ is the *zero* element, the union $1 = \Sigma(L)$ the *unit* element of L. If $a > b$ in L, but $a > c > b$ is impossible for c in L, then a is said to *cover* b or to be prime over b (Ore [9]). An element p prime over 0 is called a *point* of L (Menger [7]).

In the field case, each indeterminate x over $\mathfrak{K}$ generates a relatively algebraically closed field $\mathfrak{K}(x)'$ which is a point in the lattice of fields $\mathfrak{M}$. The crucial property of algebraic dependence is: if x is algebraic over $\mathfrak{M}(y)$, but not over $\mathfrak{M}$, then y is algebraic over $\mathfrak{M}(x)$. For lattices, we state a corresponding *exchange axiom*:

(E_1) If a is in L while p and q are points of L, then $a < a + p \leqq a + q$ implies $q \leqq a + p$.

We need also the existence of points and the "finiteness" of dependence:

(G_1) If $b < a$ in L, then there is in L a point p with $b < b + p \leqq a$.

(F_1) If Q is a set of points and p a point of L with $p \leqq \Sigma(Q)$, then there exists a finite set of points $q_1, q_2, \cdots, q_n$ of Q with $p \leqq q_1 + q_2 + \cdots + q_n$.

DEFINITION. *An exchange lattice is a continuous lattice satisfying* (E_1), (G_1), *and* (F_1).

Note that (E_1) is equivalent to the assertion that $a < a + p < a + q$ is impossible, while (G_1) is equivalent to the following formally stronger axiom:

(G_2) If $a \nleqq b$, then there is a point p with $p \leqq a$, $p \nleqq b$.

For $a \nleqq b$ implies $ab < a$, hence by (G_1) there is a p such that $p \leqq a$, $p \nleqq ab$. Thus $p \nleqq b$. In the presence of (G_1), (F_1) can be shown equivalent to the following point-free statement:

(F_2) If $b \leqq \Sigma(A)$, $A \subset L$, then there are in A elements $a_1, \cdots, a_n$ with $b(a_1 + \cdots + a_n) > 0$.

To prove (F_2), let Q be the set of all points $q \leqq$ some a of A, and choose any $p \leqq b$. Then $p \leqq \Sigma(Q) = \Sigma(A)$ by (G_2), whence $p \leqq q_1 + \cdots + q_n$ by (F_1). This gives the conclusion.

Our postulates are related to the properties of algebraic dependence as formulated by van der Waerden [14]. He considers a relation "b depends on S" for an element b and a subset S of a given set D, with the following properties:

[3] This definition is in von Neumann [8]. For further literature on lattices, cf. Ore [10] or Köthe [5].

1. each b depends on the set $\{b\}$;
2. if b depends on S and $S \subset T$, then b depends on T;
3. if b depends on S, then b depends on some finite subset of S;
4. if b depends on $S_0 = \{c_1, c_2, \cdots, c_n\}$ but on no proper subset of S_0, then c_n depends on $\{c_1, \cdots, c_{n-1}, b\}$;
5. if b depends on S and if every element of S depends on T, then b depends on T.

A relation "a depends on S" with these five properties we call for the moment a *dependence relation.* The exact connection with our axioms we state without proof as follows:

THEOREM 1. *If D is a set with a dependence relation, and if a subset $S \subset D$ is said to be closed whenever S contains every b of D dependent on S, then the set $\mathfrak{L}(D)$ of all closed subsets of D is an exchange lattice if $S < T$ means that S is a proper subset of T. Conversely, if L is any exchange lattice, and if $D = \mathfrak{D}(L)$ is the set of all points of L, then the relation $p \leqq \Sigma(S)$ for $S \subset D$ is a dependence relation of p to S. Furthermore, $\mathfrak{L}(\mathfrak{D}(L))$ is isomorphic to the given lattice L, while, for a given D, the set $D' = \mathfrak{D}(\mathfrak{L}(D))$ with its dependence relation is isomorphic to the set D^* obtained from D by identifying all pairs of mutually dependent elements of D. Here an isomorphism of D' to D^* means a one-to-one correspondence of D' to D^* which leaves unchanged the dependence relation.*

In the finite case, there is a similar connection to the matroids of Whitney [15], as expressed in terms of lattices by Birkhoff [3].

3. **The basis theorem.** A transcendence basis for a field is a maximal set of algebraically independent elements. Generally, a set P of points in any lattice L is *independent* if $(\Sigma(P'))\cdot(\Sigma(P'')) = 0$ for any two disjoint subsets P' and P'' of P (von Neumann [8], Chapter II). An independent set P of points with union $\Sigma(P) = a$ is called a *basis* of the element a of L. In the transcendence basis case, an independent set corresponds exactly to a set of indeterminates irreducible in the sense that no one indeterminate depends algebraically on the others. The same definition of independence holds in other cases:

THEOREM 2. *A set P of points in an exchange lattice is independent if and only if $p \leqq p_1 + p_2 + \cdots + p_n$ is impossible for distinct points $p, p_1, \cdots, p_n$ of P. Hence a set P is independent if and only if every finite subset of P is independent.*

Proof. The necessity of the condition is immediate. Suppose conversely that $p \leqq \Sigma p_i$ is impossible, but that P is dependent. There then are disjoint subsets P' and $P'' \subset P$ with $(\Sigma(P'))\cdot(\Sigma(P'')) > 0$. There then exists a point q such that $0 < q \leqq (\Sigma(P'))\cdot(\Sigma(P''))$. By the finiteness axiom, points p_i' in P' and p_j'' in P'' can be chosen so that

$$q \leqq p_1' + p_2' + \cdots + p_m', \qquad q \leqq p_1'' + p_2'' + \cdots + p_n''.$$

If $m > 0$ is so small that $q \nleqq p_1' + \cdots + p_{m-1}'$, the exchange axiom implies that

$$p_m' \leqq p_1' + \cdots + p_{m-1}' + q \leqq p_1' + \cdots + p_{m-1}' + p_1'' + \cdots + p_n'',$$

contrary to the assumed property of P. Hence P is independent. A similar use of the exchange axiom gives also the result (cf. Menger [7], p. 462):

COROLLARY. *If P is an independent set in the exchange lattice L, and if q is a point with $q(\Sigma(P)) = 0$, then the set obtained by adjoining q to P is independent.*

THEOREM 3. *If L is an exchange lattice and if P is any independent set of points of L with $\Sigma(P) \leqq b$, then there is a set $Q \supset P$ which is a basis of b. In particular, every element of L has a basis.*

Proof. When $b = 1$, we can construct Q from P by well ordering the points of L and by applying repeatedly Theorem 2 and its corollary. A basis for any $b \neq 1$ can then be found by applying the previous case to the quotient lattice $b/0$ (Ore [9], p. 425). For any $b \geqq c$ such a *quotient lattice* b/c consists of all elements d with $b \geqq d \geqq c$.

THEOREM 4. *If $b \geqq c$ are elements of the exchange lattice L, then the quotient lattice b/c is also an exchange lattice. Its points are all the elements of L of the form $c + p$, where p is a point of L such that $c < c + p \leqq b$.*

Proof. Any such $c + p$ is a point of b/c, for otherwise $c + p > d > c$, hence by (G_1) $c + p > c + q > c$, and this is impossible by the exchange axiom. Conversely, the existence axiom for points of L shows that any point of b/c must have the form $c + p$. The axioms for an exchange lattice can now be directly verified for b/c.

The comparison of different bases depends upon an exchange process (Steinitz [11], Theorem 7, p. 115).

THEOREM 5. *If P is an independent set of points in an exchange lattice, q a point with $q \leqq \Sigma(P)$, then there is a finite subset $P_0 \subset P$ such that* (i) *$q \leqq \Sigma(P_0)$, but this statement is false if P_0 is replaced by any of its proper subsets*; (ii) *if $R \subset P$ and $q \leqq \Sigma(R)$, then $P_0 \subset R$*; (iii) *if in P any point p_1 of P_0 be replaced by q, the resulting set P' is independent, and $\Sigma(P') = \Sigma(P)$.*

Proof. The finiteness axiom yields at once the set P_0 as in (i). If $q \leqq \Sigma(R)$, as in (ii), then $q \leqq \Sigma(R_0)$ for a similar minimal finite set R_0. If $P_0 \subset R$ is false, then $P_0 \neq R_0$. Since P_0 is minimal, R_0 contains at least one point r not in P_0. If R_0' is the set R_0 with r deleted, then $q \leqq \Sigma(R_0') + r$, so that the exchange axiom proves $r \leqq \Sigma(R_0') + q \leqq \Sigma(R_0') + \Sigma(P_0)$, contrary to the independence of P. Finally P' in conclusion (iii) is independent because of (ii) and the corollary to Theorem 2.

Repeated applications of this exchange process with a transfinite induction yield, exactly as in Steinitz [11] (Note 126 and correction thereto), a proof of the fundamental invariance theorem:

THEOREM 6. *If P and Q are two bases for an element b in an exchange lattice L, then the sets P and Q have the same cardinal number.*[4] *This number we call the rank of b.*

THEOREM 7. *Every exchange lattice L has relative complements*; *that is, given $b \geqq d \geqq c$ in L there is an element d' in L with $d + d' = b$, $dd' = c$.*

[4] An equivalent abstract basis theorem has been developed by Reinhold Baer, who also found several alternative postulational bases for abstract independence (all unpublished). (Added May 9, 1938.)

Proof. In the quotient lattice b/c there is by Theorems 4 and 3 a basis P_d for d. By Theorem 3 we can adjoin to P_d a set of points Q' of b/c such that P_d and Q' together form a basis for b. Then $d' = \Sigma(Q')$ is the desired complement, because the definition of independence insures that $(\Sigma(P_d))\cdot(\Sigma(Q')) = c$.

4. **Alternative exchange axioms and the Dedekind axiom.** The exchange axiom is a weakened form of the usual modular or Dedekind law (Ore [9], p. 412; Birkhoff [4]).

(D_1) Dedekind Axiom. $a \geqq c$ implies $a(b + c) = ab + c$.

This axiom holds for the linear subspaces of a projective space, but not for an affine space. Menger showed that it could there be replaced in part by either of the two equivalent assertions, for all elements a, b and all points p of L, that

(E_2) $a \leqq b \leqq a + p$ implies $b = a$ or $b = a + p$; or that

(E_3) $p \nleqq a + b$ implies $ab = (a + p)b$

(see [7], Axiom 6^+). Birkhoff's investigations of the Jordan Theorem[5] involve the property that, for all a, b and d in L,

(E_4) if $a \geqq d$, $a + b > a$ and b covers d, then $a + b$ covers a.

The interrelations of these axioms are as follows.

Theorem 8. *In any lattice L satisfying the point-existence axiom* (G_1), *any two of the conditions* (E_1), (E_2), (E_3), (E_4) *are equivalent. Each of them is a consequence of the Dedekind axiom, but there exist exchange lattices which satisfy* (E_1), (F_1), *and* (G_1) *but not the Dedekind axiom.*

In an arbitrary lattice the implications (E_4) $\to$ (E_2) $\to$ (E_1), (E_2) $\leftrightarrow$ (E_3) (proved by Menger) and (D_1) $\to$ (E_4) (cf. Birkhoff [2], Theorem 9.1) will hold. To prove (E_4) $\to$ (E_2), set $d = 0$ and $b = p$ in (E_4) to obtain the assertion "$a + p > a$ implies that $a + p$ covers a", an alternative statement for (E_2). Similarly (E_2) with $b = a + q$ will yield (E_1).

To establish (E_4) as a consequence of (E_1) and (G_1), suppose that in the conclusion of (E_4) $a + b$ fails to cover a. There is then a c, $a + b > c > a$, and by (G_1) there are points p and q such that $a < a + p \leqq c$, $d < d + q \leqq b$. Since b covers d, $d + q = b$ and

$$a < a + p \leqq c < a + b = a + d + q = a + q.$$

The exchange axiom then makes $a + q \leqq a + p < a + q$, a contradiction. Given (G_1) we now have (D_1) $\to$ (E_4) $\to$ (E_3) $\to$ (E_2) $\to$ (E_1) $\to$ (E_4), as asserted in the theorem.

The Dedekind axiom does not hold in every exchange lattice. For instance, let C be a finite class with m elements and n an integer, $1 \leqq n < m$. Consider the set L_{mn} of all subclasses $a, b, \cdots$ of C which have not more than n elements or exactly m elements, and let $a < b$ mean that a is a proper subclass of b.

[5] Lattices having this property (E_4) have been studied by F. Klein, who calls the property "axiom (γ'')" and who obtains several interesting equivalent axioms. Cf. F. Klein, *Birkhoffsche und harmonische Verbände*, Mathematische Zeitschrift, vol. 42 (1937), pp. 58–81. (Added May 9, 1938.)

Then L_{mn} is a lattice, in which the points are the subclasses of C with just one element, while the sum $a + b$ is the whole class C or the ordinary point-set union $a \cup b$ according as $a \cup b$ has more than n elements or not. The axioms for an exchange lattice are readily verified for L_{mn}. Whenever $m \geqq n + 2 \geqq 4$, and in no other cases, L_{mn} fails to be a Dedekind lattice. For instance, if $m = 4$, $n = 2$, L_{42} contains four points p, q, r and s. If we set $a = p + q$, $b = r + s$, $c = p$, then $a \geqq c$, but

$$a(b + c) = (p + q)(p + r + s) = (p + q)\cdot 1 = p + q,$$

$$ab + c = (p + q)(r + s) + p = p \text{ and } p + q \neq p,$$

contrary to the Dedekind law.

We turn to the Jordan Theorem. A *chain* of length n joining a to b, $a < b$, is a sequence of $n + 1$ elements x_i, $a = x_0 < x_1 < \cdots < x_n = b$. The chain is *principal* if each x_i covers x_{i-1}, for $i = 1, \cdots, n$. A lattice L is said to be of *finite dimensions* if any two elements $a < b$ can be joined by a principal chain.

THEOREM 9. *If two elements $a < b$ of an exchange lattice L are joined by a principal chain of length n, then no chain joining a to b has length greater than n, so that any principal chain joining a to b has the same length n. If the exchange lattice L is of finite dimensions, then the rank (or dimension) $\rho(a)$ defined in Theorem 6 is equal to the length of any principal chain joining 0 to a, and $\rho(a)$ satisfies the inequality*

$$\rho(ab) + \rho(a + b) \leqq \rho(a) + \rho(b). \tag{1}$$

The first assertion, which states in part that when a and b are joined by a principal chain they can be joined by no "infinite" chain, can be established by induction, with repeated applications of axiom (E_4). The rest of the theorem restates known results, for the exchange axiom (E_4) implies the following condition:

(ξ) If a and b cover d and $a \neq b$, then $a + b$ covers a and b. From this condition the Jordan Theorem and (1) can be established (Birkhoff [2], Theorems 8.2 and 9.2; Ore [9], Chapter II).

The following converse to Theorems 7 and 9 gives in effect an alternative definition for finite exchange lattices.

THEOREM 10. *If L is a lattice which has for every $a > d > 0$ a "relative complement" d' with $d + d' = a$, $d\cdot d' = 0$, and if to every element a of L an integer $\rho(a)$ can be so assigned that*

(i) *$a < b$ implies $\rho(a) < \rho(b)$;*

(ii) *if b covers a, then $\rho(b) = \rho(a) + 1$;*

(iii) *the inequality* (1) *holds for all a and b;*

then L is an exchange lattice.

Proof. The finiteness of each $\rho(a)$ and the condition (i) enforce ascending and descending chain conditions (Ore [9], p. 410) and thence the existence of infinite unions and cross-cuts, as well as the finiteness axiom (F_1). To establish the existence of points, let $a > d \geqq 0$ and pick a relative complement d' of d in a.

The descending chain condition yields a point p, $p \leqq d'$, with the requisite properties, $p \leqq a$ and $p \nleqq d$.

If p is any point with $a < a + p$, then the inequality (1) and condition (ii) give

$$\rho(a + p) \leqq \rho(a) + \rho(p) - \rho(ap) = \rho(a) + \rho(p) - \rho(0) = \rho(a) + 1.$$

But $a + p > a$, so $\rho(a + p) > \rho(a)$ by (i), and we have $\rho(a + p) = \rho(a) + 1$. This means that $a + p$ covers a, and is in effect the exchange axiom (E_2). Therefore L is an exchange lattice.

A lattice L is *complemented* if for every a there is an a' with $a + a' = 1$, $aa' = 0$. If L is also modular, relative complements are known to exist, and in the finite case there is a dimension function. Therefore, we have the following

COROLLARY. *Any complemented Dedekind lattice of finite dimensions is an exchange lattice.*

The lattice of all linear subspaces of an affine space is also an exchange lattice, as can be established from Menger's axioms for affine geometry.

5. Transcendence degrees of fields. For fields $\mathfrak{L} \supset \mathfrak{K}$, the set of all relatively algebraically closed subfields $\mathfrak{M}$ between $\mathfrak{K}$ and $\mathfrak{L}$ forms an exchange lattice L. This is proved by noting that the relation "y depends algebraically on $\mathfrak{K}(S)$" is a dependence relation with the five properties used in Theorem 1 to construct an exchange lattice. The sets "closed" under this relation are exactly the subfields $\mathfrak{M}$, and the transcendence degree of $\mathfrak{M}$ over $\mathfrak{K}$ is the rank of $\mathfrak{M}$ in the lattice, and so is included in Theorem 6.

Such lattices of fields need not satisfy the Dedekind law. Consider over any field $\mathfrak{K}$ the field $\mathfrak{L} = \mathfrak{K}(x, y, z)$ of rational functions of three independent variables x, y and z. The subfields

$$\mathfrak{M} = \mathfrak{K}(x, y), \qquad \mathfrak{N} = \mathfrak{K}(z, x + yz), \qquad \mathfrak{R} = \mathfrak{K}(x) \tag{2}$$

are relatively algebraically closed in $\mathfrak{L}$, by Lüroth's theorem (cf. Steinitz [11], p. 126). The intersection of $\mathfrak{M}$ and $\mathfrak{N}$ is $\mathfrak{K}$. For let $\alpha \neq 0$ be an element of the intersection,

$$\alpha = f(x, y)/g(x, y) = r(t, z)/s(t, z), \qquad t = x + yz.$$

Then $g(x, y) \neq 0$, $s(t, z) \neq 0$, and we can assume that $r(t, z)$ and $s(t, z)$, as polynomials in t and z, have no factors in common except constants. Then $fs = gr$ is an identity in x, y and z; in it we set $z = 0$ to obtain

$$f(x, y)s(x, 0) = g(x, y)r(x, 0).$$

If $s(x, 0) = 0$, then, since $g(x, y) \neq 0$, $r(x, 0) = 0$. Since these are identities, $r(t, 0) = 0 = s(t, 0)$, which means that $r(t, z)$ and $s(t, z)$ have in common a factor z contrary to assumption. Hence $s(x, 0) \neq 0$ and

$$\alpha = f(x, y)/g(x, y) = r(x, 0)/s(x, 0),$$

so that α is in $\mathfrak{K}(x)$. If α were not in $\mathfrak{K}$, α would be transcendental over $\mathfrak{K}$ by Lüroth's theorem, so that x would be algebraic over $\mathfrak{K}(\alpha) \subset \mathfrak{N}$. Because $\mathfrak{N}$ is relatively algebraically closed, x is in $\mathfrak{N}$, and (2) shows that $\mathfrak{N}$ contains x, y, and z. This is a contradiction.

Because the intersection $\mathfrak{K} = \mathfrak{M} \cap \mathfrak{N}$ is the cross-cut $\mathfrak{M}\cdot\mathfrak{N}$ of $\mathfrak{M}$ and $\mathfrak{N}$ in the lattice L of relatively algebraically closed fields between $\mathfrak{K}$ and $\mathfrak{L}$, this lattice is not modular. For, by (2), $\mathfrak{M} > \mathfrak{R}$ and

$$\mathfrak{M}\cdot(\mathfrak{N} + \mathfrak{R}) = \mathfrak{M}\cdot\mathfrak{L} = \mathfrak{M}, \qquad \mathfrak{M}\cdot\mathfrak{N} + \mathfrak{R} = \mathfrak{K} + \mathfrak{R} = \mathfrak{R},$$

so that $\mathfrak{M}\cdot(\mathfrak{N} + \mathfrak{R}) \neq \mathfrak{M}\cdot\mathfrak{N} + \mathfrak{R}$, contrary to the Dedekind law.

This example can be extended in various ways. In any field $\mathfrak{K}(x_1, \cdots, x_n)$ with n independent variables one can construct two subfields $\mathfrak{M}$ and $\mathfrak{N}$, each of transcendence degree $n - 1$ over $\mathfrak{K}$, and with $\mathfrak{K}$ as intersection. In the field $\mathfrak{K}(x, y, z)$ one can also find a denumerable number of relatively algebraically closed subfields $\mathfrak{M}_i$, each of transcendence degree 2 over $\mathfrak{K}$, such that the intersection of any two of these subfields is $\mathfrak{K}$. These examples are typical of all fields, in the following sense:

THEOREM 11. *If $\mathfrak{K}$ is a relatively algebraically closed subfield of $\mathfrak{L}$, then the lattice of all relatively algebraically closed subfields $\mathfrak{M}$ with $\mathfrak{K} \subset \mathfrak{M} \subset \mathfrak{L}$ is a Dedekind lattice if and only if the transcendence degree of $\mathfrak{L}$ over $\mathfrak{K}$ is less than* 3.

Proof. When the transcendence degree is 1 or 2, the lattice has a simple form and the Dedekind law is trivially true. In the remaining cases, there are in $\mathfrak{L}$ three indeterminates x, y and z independent over $\mathfrak{K}$. Let $\mathfrak{M}$, $\mathfrak{N}$ and $\mathfrak{R}$ be the fields of (2) and $\mathfrak{M}'$, $\mathfrak{N}'$ and $\mathfrak{R}'$ be respectively their relative algebraic closures in $\mathfrak{L}$. Then $\mathfrak{M}' > \mathfrak{R}'$, and if we show that $\mathfrak{M}'$ and $\mathfrak{N}'$ intersect in $\mathfrak{K}$, the non-modularity follows as before.

Suppose then that w is an element common to $\mathfrak{M}'$ and $\mathfrak{N}'$, so that w is algebraic over both $\mathfrak{M}$ and $\mathfrak{N}$ and satisfies equations $f(u) = 0$ and $g(u) = 0$ irreducible over $\mathfrak{M}$ and $\mathfrak{N}$, respectively. Since $\mathfrak{K}(x, y, z)$ is a simple transcendental extension of $\mathfrak{M}$ and also of $\mathfrak{N}$, $f(u)$ and $g(u)$ must remain irreducible over the extension $\mathfrak{K}(x, y, z)$. If both f and g have the leading coefficient 1, then $f(u) \equiv g(u)$, so that w satisfies an equation $f(u) = 0$ whose coefficients are in $\mathfrak{M}\cdot\mathfrak{N}$; that is, in $\mathfrak{K}$. Therefore w, algebraic over $\mathfrak{K}$, is in $\mathfrak{K}$, and the intersection $\mathfrak{M}'\cdot\mathfrak{N}'$ is in fact $\mathfrak{K}$.

6. **Group ranks and p-bases.** Abelian groups furnish another example of exchange lattices. Let J be an additive Abelian group without elements of finite order (except 0). An element g of J is said to be dependent on a subset X of J if in X there are elements $x_1, \cdots, x_n$ and if there are integers $m, k_1, \cdots, k_n$, $m \neq 0$, with $mg = k_1x_1 + \cdots + k_nx_n$. This dependence relation has, as is readily verified, the five properties used in Theorem 1, so that the sets $H \subset J$ closed under this dependence form an exchange lattice. These closed subsets H are simply the subgroups $H \subset J$ for which the factor group J/H has no elements of finite order (except 0). A basis for J, in the lattice sense, is simply a

maximal independent set of elements of J, and the rank of J (Theorem 6) is a known invariant of J (Baer [1], Theorem 3.2). A simple proof shows also that the lattice of closed subgroups is a Dedekind lattice.

A field $\mathfrak{K}$ of characteristic p which is not perfect can be extended to a perfect field by the adjunction of p^n-th roots of a certain minimal set of elements of $\mathfrak{K}$. This minimal set is called a p-basis (Teichmüller [12], §3, and [13], p. 145, Hilfsatz 9). This notion can be generalized to any inseparable extension.[6] Let $\mathfrak{K}$ be a field of characteristic p, and $\mathfrak{L}$ a pure inseparable extension of exponent 1 over $\mathfrak{K}$; that is, an extension such that x^p is in $\mathfrak{K}$ for any x in $\mathfrak{L}$. An element y in $\mathfrak{L}$ will be called *p-dependent* on a subset X of $\mathfrak{L}$ if y is in the field $\mathfrak{K}(X)$. The properties 1, 2, 3 and 5 of a dependence relation are immediately verified. As for property 4, let y be in $\mathfrak{K}(x_1, \cdots, x_n)$, but not in $\mathfrak{K}(x_1, \cdots, x_{n-1})$. Since x_n^p is in $\mathfrak{K}$, $y = f(x_1, \cdots, x_n)$, where f is a polynomial with coefficients in $\mathfrak{K}$ and of degree less than p in x_n. Then x_n must actually occur in some term of f, so x_n is algebraic of degree $\leqq p - 1$ over $\mathfrak{K}_1 = \mathfrak{K}(y, x_1, \cdots, x_{n-1})$. But x_n also satisfies the inseparable equation $z^p - x_n^p = 0$ over $\mathfrak{K}_1$, so that x_n must be in $\mathfrak{K}_1$, and is p-dependent on $y, x_1, \cdots, x_{n-1}$, as asserted in property 4.

The exchange lattice corresponding to this dependence relation is simply the lattice of all subfields $\mathfrak{M}$ with $\mathfrak{K} \subset \mathfrak{M} \subset \mathfrak{L}$. The points of the lattice are the subfields of degree p over $\mathfrak{K}$, so that Theorem 6 becomes

THEOREM 12. *If $\mathfrak{L}$ is a pure inseparable extension of exponent 1 of a field $\mathfrak{K}$ of characteristic p, then there exists a set of subfields $\mathfrak{M}_\sigma$ of $\mathfrak{L}$, each of degree p over $\mathfrak{K}$, such that the adjunction of all $\mathfrak{M}_\sigma$ to $\mathfrak{K}$ gives $\mathfrak{L}$, while no $\mathfrak{M}_\sigma$ is a subfield of the field obtained by adjoining the remaining $\mathfrak{M}$'s to $\mathfrak{K}$. The cardinal number of subfields $\mathfrak{M}_\sigma$ is an invariant of $\mathfrak{L}/\mathfrak{K}$, called the relative degree of imperfection.*

Each subfield $\mathfrak{M}_\sigma$ has the form $\mathfrak{M}_\sigma = \mathfrak{K}(\sqrt[p]{x_\sigma})$ for some x_σ in $\mathfrak{K}$. The set of these x_σ's, one for each $\mathfrak{M}_\sigma$, can be called a *p-basis* of $\mathfrak{L}$ over $\mathfrak{K}$. This concept will apply to any inseparable algebraic extension, for such an extension $\mathfrak{F}$ of $\mathfrak{K}$ can be uniquely decomposed into $\mathfrak{K} \subset \mathfrak{L}_0 \subset \mathfrak{L}_1 \subset \mathfrak{L}_2 \subset \cdots \subset \mathfrak{F}$, where each $\mathfrak{L}_n$ contains all elements of $\mathfrak{F}$ of exponent not more than n over $\mathfrak{K}$, so that each $\mathfrak{L}_n$ is a pure inseparable extension of exponent 1 over $\mathfrak{L}_{n-1}$.

The lattice of subfields $\mathfrak{M}$ of $\mathfrak{L}$ is not always modular. Consider $\mathfrak{L} = \mathfrak{P}(x, y, z)$, $\mathfrak{K} = \mathfrak{P}(x^p, y^p, z^p)$, where x, y and z are independent indeterminates over the perfect field $\mathfrak{P}$ of characteristic p. $\mathfrak{L}$ over $\mathfrak{K}$ is pure inseparable of degree p^3, is obtained from $\mathfrak{K}$ by adjoining the independent p-th roots x, y and z, and so has the p-basis $\{x, y, z\}$ (cf. Teichmüller [12], Theorem 18). The subfields $\mathfrak{M} = \mathfrak{K}(x, y)$, $\mathfrak{N} = \mathfrak{K}(z, x + yz)$ are each of degree p^2 over $\mathfrak{K}$. To disprove the Dedekind law, it will suffice as in §5 to show that $\mathfrak{M}$ and $\mathfrak{N}$ intersect in $\mathfrak{K}$. Let them have in common the element

$$(3) \qquad \alpha = f(x, y) = g(z, t) \equiv \sum_{i,j=0}^{p-1} a_{ij} z^i t^j, \qquad t = x + yz,$$

[6] The possibility of such an extension is indicated by Teichmüller [12], §3.

where f and g are polynomials with coefficients in $\mathfrak{K}$ and of degree less than p in any one variable. The polynomial

$$(4) \qquad h(x, y, z) \equiv f(x, y) - g(z, x + yz)$$

must be zero in $\mathfrak{L}$ and is of degree at most $2p - 2$ in the variable z. If each term z^{p+s} be replaced by dz^s, with coefficient $d = z^p$ in $\mathfrak{K}$, then $h(x, y, z)$ is equal in $\mathfrak{L}$ to a new polynomial $h'(x, y, z)$ of degree less than p in any one variable. But the power products $x^q y^r z^s$ with exponents less than p form a basis for the algebraic extension $\mathfrak{L}/\mathfrak{K}$, so that $h'(x, y, z) = 0$ in $\mathfrak{L}$ implies that h' is identically zero. In h' terms in z^{p-1} never arise from a replacement $z^{p+s} \to dz^s$, but come only from terms $z^i t^j$ in g with $i + j \geqq p - 1$. These terms are, by expansion of (3),

$$\left(\sum_{j=0}^{p-1} \sum_{i=p-1-j}^{p-1} \binom{j}{p-1-i} a_{ij} x^{j+i-(p-1)} y^{p-1-i}\right) z^{p-1}.$$

These terms involve distinct power products $x^q y^s$ and have binomial coefficients not zero, so $h' \equiv 0$ implies $a_{ij} = 0$ for $i + j \geqq p - 1$. Thus (4) actually involves no term of degree p or more in x, in y or in z, and $h(x, y, z) \equiv 0$. But z arises only from $g(z, x + yz)$, so only the constant term of g can differ from 0, and the element $\alpha = g(0, 0)$ of $\mathfrak{M} \cdot \mathfrak{N}$ is in fact in $\mathfrak{K}$.

7. **Exchange axioms free of points.** A central feature of von Neumann's continuous geometry is the use of the modular law, which makes no reference to the points of the geometry. Similarly, Wilcox [16] has shown that affine geometry as developed algebraically by Menger can also be axiomatized without the use of points. His treatment depends on certain properties of a relation of modularity which do not hold in all exchange lattices. Nevertheless, our exchange axiom can be replaced by conditions which make no use of points or of covering relations.

To modify Menger's exchange axiom (E_3), which asserts that $p \nleqq a + c$ implies $(a + p)c = ac$, replace the point p by an arbitrary element b and the conclusion $(a + p)c = ac$ by the assertion that $(a + b_1)c = ac$ for some non-trivial part b_1 of b. No generality is lost if we require $a < c$; the hypothesis $p \nleqq a + c$ or $p(a + c) = 0$ might then become $bc < a$, and our modified statement is[7]

(E_5) $bc < a < c < b + c$ implies that there exists b_1 such that $bc < b_1 \leqq b$ and $(a + b_1)c = a$.

A similar elimination of points from the exchange axiom (E_1) leads eventually to the laws

(E_6) $bc < a < c < b + c$ implies that there exists b_1 such that $bc < b_1 \leqq b$ and $(a + b_1)c < c$;

(E_7) $bc < a < c < b + a$ implies the conclusion of (E_6).

The second assertion $(a + b_1)c < c$ of this conclusion is equivalent to $c \nleqq a + b_1$, which in turn is equivalent to $a + b_1 < c + b_1$.

[7] The statement that this law (or other similar laws) holds is to mean that it holds for all elements a, b, and c of the lattice.

These three point-free laws do not, like the exchange axiom (E_1), hold trivially in any lattice without points (as for instance in the lattice of all real numbers between 0 and 1, where $a < b$ has its usual meaning). The rôle of these laws can be stated thus:

THEOREM 13. *In any lattice,* (E_5) *implies* (E_6) *and* (E_6) *is equivalent to* (E_7).

THEOREM 14. *In a lattice satisfying the point-existence axiom* (G_1), *the exchange axiom* (E_1) *is equivalent to* (E_5) *and also to* (E_6). *Hence* (E_5) *(or* (E_6)) *can replace* (E_1) *in the definition of an exchange lattice.*

THEOREM 15. *In any lattice,* (E_7) *(and hence* (E_6) *or* (E_5)) *implies the covering law* (E_4). *Therefore a dimension function* $\rho(a)$ *can be defined in any complete lattice of finite dimensions satisfying* (E_7); *in other words, the conclusions of Theorem 9 hold for such a lattice.*

Proof. (E_5) $\to$ (E_6) $\to$ (E_7) is immediate. Conversely, to prove (E_7) $\to$ (E_6), let $bc < a < c < b + c$ as in (E_6). If $a + b > c$, then the hypothesis of (E_7) holds and yields the desired conclusion. If $a + b = c$, then $b \leqq c$, $b + c = c < b + c$, a contradiction. In the remaining case, $(a + b)c < c$, which states that the conclusion of (E_6) holds with $b_1 = b$. This gives Theorem 13.

For Theorem 15 we need only prove (E_7) $\to$ (E_4). Given the hypothesis of (E_4), the conclusion could be false only if $a + b > c > a$ for some c. Then $d \leqq bc \leqq b$, whence $bc = b$ or $bc = d$. In the former case, $b \leqq c$, $a + b \leqq a + c = c < a + b$, a contradiction. Therefore $bc = d$. Omitting the trivial case $d = a$, we have $bc < a < c < a + b$, as in (E_7), so that $(a + b_1)c < c$ for $bc < b_1 \leqq b$. Because b covers bc, $b_1 = b$ and $c = (a + b)c = (a + b_1)c < c$, a contradiction.

Since (E_7) $\to$ (E_4), Theorem 14 now needs only a proof that (E_5) holds whenever (E_1) and (G_1) do. Let $bc < a < c < b + c$. Since $b \nleqq c$, the property (G_2) of §2 furnishes a point p with $p \leqq b$, $p \nleqq c$. Hence $p \nleqq a + c = c$, so that $(a + p)c = ac$ by the exchange axiom (E_3), which is known to hold (Theorem 8). If we set $b_1 = bc + p$, then $bc < b_1 \leqq b$, while $(a + b_1)c = (a + bc + p)c = (a + p)c = ac = a$, as in the conclusion of (E_5).

Any one of these three point-free axioms can be viewed as a weaker form of the modular law (D_1), for in any lattice this modular law is equivalent to the following assertion, of a form similar to (E_5),

(D_2) $bc < a < c < b + c$ implies $(a + b)c = a$.

This is a direct consequence of the modular law. Conversely, the modular law requires that $c' = c(b + a)$ be equal to $a' = cb + a$ for any $c \geqq a$. In any event, $c' \geqq a'$ and $bc' \leqq a'$. If either $bc' = a'$ or $b + c' = c'$, the conclusion $c' = a'$ will follow readily. Hence the conclusion can be false only with $bc' < a' < c' < b + c'$, exactly as in the hypothesis of (D_2), whence we obtain the conclusion

$$cb + a = a' = (a' + b)c' = (cb + a + b)c(b + a)$$
$$= (a + b)c(b + a) = c(b + a),$$

which is the modular law.

The complexity of the axioms (E_5) to (E_7), attendant on the existence assertion for b_1, is unavoidable. Specifically, suppose that an axiom (E_M) could be found which, like (E_5), is equivalent to the exchange axiom (E_1) in the presence of the axioms (F_1) and (G_1), but which, unlike (E_5), contains no existence assertions. In other words, (E_M) is built up from certain elements $a_1, \cdots, a_n$ of the lattice L by any combinations using $<$, $=$, $+$, $\cdot$, "and", "implies" and "not", and is to be asserted for all choices of the elements a_i. Any such axiom (E_M) true in a lattice L is automatically true in any sublattice L' of L, for $<$, $+$, and $\cdot$ have the same meaning in L' as in L. But then (E_M) cannot be equivalent to (E_1). For the non-modular lattice L_{42} of §4 contains four points p, q, r and s and has a sublattice L' composed of 0, p, r, $p + r$, $q + s$ and 1. In L' axioms (F_1) and (G_1) hold, but axiom (E_1) fails, because in L' p and $q' = q + s$ are points, $p \leqq r + q'$, $p \nleqq r$, so that, according to (E_1), $q' = q + s \leqq r + p$ should hold, contrary to the construction of L_{mn}. But (E_M), true in L, is also true in L', and so is not equivalent to (E_1) in the presence of (F_1) and (G_1).

8. **Transposition axioms and modularity**. The point-free exchange axiom (E_5) states in part that the chain in its hypothesis can be subjected to the following successive transpositions,

$$bc < a < c < b + c, \tag{5}$$

$$bc < a < a + b_1 < b + c, \tag{6}$$

$$bc < b_1 < a + b_1 < b + c. \tag{7}$$

Note that the inclusions in (6) are all proper inclusions, for $a + b_1 = a$ would give $b_1 \leqq a$, $bc < b_1 \leqq ab \leqq cb$, a contradiction to ($E_5$), while $a + b_1 = b + c$ implies $(a + b_1)c = (b + c)c = c$, although $(a + b_1)c = a$ in (E_5). Similarly $b_1 < a + b_1$ in (7).

To investigate the relations of the exchange axiom to the possibility of such transpositions, consider any chain of length n,

$$C : a_0 < a_1 < a_2 < \cdots < a_n . \tag{8}$$

A *transposition* of C is the operation of replacing an element a_k by a new element a_k' between a_{k-1} and a_{k+1} ($a_{k-1} < a_k' < a_{k+1}$). We call the transposition *primary* if $a_k \cdot a_k' = a_{k-1}$, *proper* if $a_k \cdot a_k' < a_k$ and $a_k \cdot a_k' < a_k'$. A primary transposition is always proper, but not conversely. In these terms can be stated the following alternative modular and exchange axioms:

(T_5) If C is a chain of length $n \geqq 2$ and if $a_0 < b < a_n$ and $a_1 b = a_0$, then[8] one can find $n - 1$ or fewer successive primary transpositions of C which yield a chain $a_0 < b_1 < \cdots < a_n$ with second term $b_1 \leqq b$.

[8] The hypothesis $a_1 b = a_0$ is not highly restrictive, since in the contrary case $a_1 \geqq a_1 b > a_0$, so a simple insertion of $b_1 = a_1 b$ in C yields a new chain of the desired form. If in (T_5) the hypothesis $a_1 b = a_0$ were replaced by $a_{n-1} b = a_0$, and the conclusion "$n - 1$ or fewer" by "exactly $n - 1$", the resulting statement is equivalent to the original (T_5), by the arguments used for Theorem 16.

(T_6) Statement as in (T_5), with "primary" replaced by "proper".

(T_D) If C is a chain of length $n \geqq 2$ and if $a_0 < b < a_n$ and $a_{n-1}b = a_0$, then one can find $n - 1$ successive primary transpositions of C which yield a chain $a_0 < b < \cdots < a_n$ with second term b.

THEOREM 16. *In any lattice L, the Dedekind law is equivalent to* (T_D), *while the exchange axioms* (E_5) *and* (E_6) *are equivalent respectively to* (T_5) *and* (T_6).

We prove (E_5) $\rightarrow$ (T_5) by induction. If $n = 2$, C is $a_0 < a_1 < a_2$, and the transposition to $a_0 < b < a_2$ is primary by the hypothesis $a_1 b = a_0$. Let (T_5) be true for all chains in L of length $k < n$. If $a_{n-1}b > a_0$ in the given chain C, then $a_{n-1} > a_{n-1}b > a_0$, so we apply the induction assumption to $b' = a_{n-1}b$ and the chain $a_0 < \cdots < a_{n-1}$ of length $n - 1$. In the remaining case, $a_{n-1}b = a_0$, $b \nleqq a_{n-1}$, so the four-element chain $a_0 < a_{n-2} < a_{n-1} < a_{n-1} + b$ has the form of the hypothesis of (E_5). There exists a b_1, $a_0 < b_1 \leqq b$, with $(a_{n-2} + b_1) a_{n-1} = a_{n-2}$, which is to say that the transposition $a_{n-1} \rightarrow (a_{n-2} + b_1)$ in C is primary. The induction assumption applied to the chain $a_0 < a_1 < \cdots < a_{n-2} < a_{n-2} + b_1$ and to the element b_1 therefore gives $n - 2$ more transpositions leading to the desired type of chain.

As to the converse, (T_5) $\rightarrow$ (E_5), the chain (5) in the hypothesis of (E_5) is by (T_5) reducible by one or two primary transpositions to the form $bc < b_1 < c_1 < b + c$, with $b_1 \leqq b$. The first of these transpositions could not have introduced b_1, for then $b_1 < c$, which entails the contradiction $b_1 = bb_1 \leqq bc < b_1$. The first transposition is therefore $c \rightarrow c_1$; it is primary, so $cc_1 = a$. But $a < c_1$, $b_1 < c_1$, $a + b_1 \leqq c_1$ so $c(a + b_1) \leqq a$, and therefore $c(a + b_1) = a$, which is the conclusion of (E_5). This argument depends essentially on the circumstance that (T_5) allows only *two* transpositions on the chain (5).

An analogous proof that (T_6) $\leftrightarrow$ (E_6) is possible because (E_6) again gives two transpositions (5) $\rightarrow$ (6) and (6) $\rightarrow$ (7) in the chain (5) of length 3. The second transposition is primary, hence proper. The first transposition is proper because (E_6) asserts that $(a + b_1)c < c$ and because $(a + b_1)c < a + b_1$, for otherwise $a + b_1 \leqq c$ would give the paradox $bc \geqq b_1 > bc$.

To show that the Dedekind law implies (T_D), observe that the chain (8) can be subjected to the successive transpositions $a_k \rightarrow (a_{k-1} + b)$ $(k = n - 1, \cdots, 1)$. Each one is primary, for by the Dedekind law, $(a_{k-1} + b)a_k = a_{k-1} + ba_k \leqq a_{k-1} + ba_{n-1} = a_{k-1} + a_0 = a_{k-1}$. The final chain $a_0 < b < a_1 + b < \cdots < a_n$ has the specified form. The converse assertion that (T_D) $\rightarrow$ (D_1) can be readily checked as above by using the Dedekind law in the form (D_2) of §7.

BIBLIOGRAPHY

1. R. BAER, *The subgroup of elements of finite order of an Abelian group*, Annals of Mathematics, vol. 37(1936), p. 768.
2. G. BIRKHOFF, *On the combination of subalgebras*, Proceedings of the Cambridge Philosophical Society, vol. 29(1933), pp. 441–464.
3. G. BIRKHOFF, *Abstract linear dependence and lattices*, American Journal of Mathematics, vol. 57(1935), pp. 800–804.
4. G. BIRKHOFF, *Combinatorial relations in projective geometries*, Annals of Mathematics, (2), vol. 36(1935), pp. 743–748.

5. G. Köthe, *Die Theorie der Verbände*, Jahresbericht der deutschen Mathematiker-Vereinigung, vol. 47(1937), pp. 125–144.
6. S. Mac Lane, *Some interpretations of abstract linear dependence in terms of projective geometry*, American Journal of Mathematics, vol. 58(1936), pp. 236–240.
7. K. Menger, *New foundations of projective and affine geometry; algebra of geometry*, Annals of Mathematics, (2), vol. 37(1936), pp. 456–482.
8. J. von Neumann, *Continuous Geometry*, Notes of lectures at the Institute for Advanced Study, 1936.
9. O. Ore, *On the foundation of abstract algebra*. I, Annals of Mathematics, (2), vol. 36 (1935), pp. 406–437.
10. O. Ore, *On the decomposition theorems of algebra*, Comptes Rendus du Congrès International des Mathématiciens, 1936.
11. E. Steinitz, *Algebraische Theorie der Körper*, edited by R. Baer and H. Hasse, 1930.
12. O. Teichmüller, *p-Algebren*, Deutsche Mathematik, vol. 1(1936), pp. 362–388.
13. O. Teichmüller, *Diskret bewertete perfekte Körper mit unvollkommenem Restklassenkörper*, Journal für die Mathematik, vol. 176(1936), pp. 141–152.
14. B. L. van der Waerden, *Moderne Algebra*, vol. I, p. 204.
15. H. Whitney, *On the abstract properties of linear dependence*, American Journal of Mathematics, vol. 57(1935), pp. 509–533.
16. L. R. Wilcox, *Modularity in the theory of lattices*, Bulletin of the American Mathematical Society, vol. 44(1938), p. 50 (abstract 44-1-79).

University of Chicago.

Reprinted from *American Journal of Mathematics*,
volume 57, pages 800 through 804,
by permission of the Johns Hopkins University Press.

Strong Maps of Geometries

D. A. HIGGS

Faculty of Mathematics, University of Waterloo, Waterloo, Ontario, Canada

Communicated by the Editor-in-Chief

ABSTRACT

Strong maps, the analogs for geometry of continuous functions in topology, are studied using a certain "lift" construction. It is shown that, in the category of geometries and strong maps, the ordered set of quotient objects of a given object satisfies the Jordan-Dedekind chain condition and, as a further application, a proof of a result due to Edmonds (4) is given (Theorem III below).

INTRODUCTION

In this paper, a *geometry* is a geometric lattice of finite length, that is, a (non-empty) lattice G which contains a finite maximal chain and which satisfies the condition : $x \prec y$ if and only if there exists a point p of G such that $x \wedge p = 0$ and $x \vee p = y$.

Let G and H be geometries. A function σ from G into H is a *strong map* if and only if σ is $\bigvee$-preserving and for each point p of G either $p\sigma = 0_H$ or $p\sigma$ is a point of H. As a simple consequence we have:

(1) *If σ is a strong map then $x \leqslant y$ imples $x\sigma \leqslant y\sigma$.*

Geometries and strong maps form a category $\mathscr{G}_s$. As a preliminary to the study of $\mathscr{G}_s$, the following lattice-theoretic remarks are helpful. If $\sigma : G \to H$ is a $\bigvee$-preserving function from a geometry G into a geometry H define the $\bigwedge$-preserving function $\sigma^\Delta : H \to G$ by $y\sigma^\Delta = \bigvee \{x \mid x\sigma \leqslant y\}$; and if $\tau : H \to G$ is a $\bigwedge$-preserving function define the $\bigvee$-preserving function $\tau^\Delta : G \to H$ dually. In particular, if Q is a $\bigwedge$-closed subset of G, define $\xi : G \to Q$ to be $^\Delta$ of the inclusion function $i : Q \to G$. Then any $\bigvee$-preserving function $\sigma : G \to H$ may be factorized as

$$G \xrightarrow{\xi} H\sigma^\Delta \cong G\sigma \xrightarrow{i} H,$$

the isomorphism $H\sigma^\Delta \cong G\sigma$ being the appropriate restriction of σ.

This suggests the following definitions: call a $\vee$-closed subset S of a geometry G a *subgeometry* of G if S is a geometry under the restriction of the order in G to S and each point of S is a point of G (so that $i : S \to G$ is a strong map); and call a $\wedge$-closed subset Q of G a *quotient* of G if Q is a geometry under the restriction of the order in G to Q and for each point p of G either $p \leqslant 0_Q$ or $p \leqslant$ some point of Q (so that $\xi : G \to Q$ is a strong map). In particular, if a is any element of G then $G \cdot a = \{s \in G \mid s \leqslant a\}$ is a subgeometry of G and $G/a = \{t \in G \mid a \leqslant t\}$ is a quotient of G.

(2) *If* $\sigma : G \to H$ *is a strong map then* $S = G\sigma$ *is a subgeometry of* H *and* $Q = H\sigma^{\Delta}$ *is a quotient of* G.

PROOF: Since S is $\vee$-closed in H, S is a non-empty lattice with $\vee_S = \vee_H$. Also S contains a finite maximal chain and the points of S are the points of H which are in S. Let $x \lessdot_S y$. Then there is a point p of G such that $p\sigma \nleqslant x$ and $p\sigma \leqslant y$. Hence $q = p\sigma$ is a point of S for which $x \wedge_S q = 0_S$ and $x \vee_S q = y$. Conversely let x be in S and let q be a point of S such that $q \nleqslant x$. Then $x \lessdot_H x \vee_H q$ and hence $x \lessdot_S x \vee_S q$. It follows that S is a subgeometry of H. Now Q is a $\wedge$-closed subset of G and, since $Q \cong S$, Q is a geometry. Also for each point p of G, $p\xi = p\sigma\zeta^{-1}$, where ζ is the isomorphism $\zeta : Q \cong S$, so that either $p\xi = 0_Q$ or $p\xi$ is a point of Q. It follows that Q is a quotient of G. Q.E.D.

Call a strong map an *injection* if it is 1-1 and a *surjection* if it is onto.

COROLLARY. *Every strong map be factorized as a surjection followed by an injection.*

This factorization is unique to within isomorphisms. A less obvious factorization of strong maps is given in Theorem III below.

For any geometry G, it is easily seen that the ordered set of subobjects in $\mathcal{G}_s$ of G and the ordered set of quotient objects in $\mathcal{G}_s$ of G may be identified with, respectively, the set $\mathcal{S}_s(G)$ of subgeometries of G and the set $\mathcal{Q}_s(G)$ of quotients of G, each ordered by inclusion. The structure of $\mathcal{S}_s(G)$ is very simple—it is just that of the set of all subsets of the set of points of G. $\mathcal{Q}_s(G)$ is not so simple—for instance it is not even a lattice when $\lambda(G)$ exceeds 2. (Here $\lambda(G)$ denotes the rank of G; if x is an element of G we will write $\lambda_G(x)$ for $\lambda(G \cdot x)$.)

THE LIFT CONSTRUCTION

Let Q be a quotient of a geometry G and write $I = I(Q, G)$ for the set of elements x of G for which $\xi : G \to Q$ is 1-1 on $G \cdot x$. It is clear that 0_G is in I and that if x is I and $y < x$ then y is in I.

(3) *For all x in G the following conditions are equivalent:*

(i) *x is in I;*

(ii) $\lambda_Q(x\xi) = \lambda_G(x)$;

(iii) *$y \lessdot_G x$ implies $y\xi \lessdot_Q x\xi$ for all y in G.*

PROOF: By virtue of (1), (i) implies (ii) and (ii) implies (iii). Suppose that (i) fails and that y and z are distinct members of $G \cdot x$ such that $y\xi = z\xi$. Since ξ is $\vee$-preserving, we may assume that $y < z$ and hence that $y \lessdot_G z$. Let w be a minimal complement of z in $G \cdot x$. Then $y \vee_G w \lessdot_G z \vee_G w = x$ and $(y \vee_G w)\xi = y\xi \vee_Q w\xi = z\xi \vee_Q w\xi = (z \vee_G w)\xi = x\xi$, so that (iii) fails. Thus (iii) implies (i). Q.E.D.

COROLLARY. $I\xi = Q$.

PROOF: Let t be in Q and let x be a minimal element of G such that $x\xi = t$. Then x satisfies (3) (iii) and hence x is in I. Q.E.D.

Now write $L = L(Q, G) = Q \cup I(Q, G)$.

THEOREM I. *L is a quotient of G; if $Q \neq G$ then $\lambda(L) = \lambda(Q) + 1$.*

PROOF: It is clear that L is $\wedge$-closed in G, so that L is a non-empty lattice with $\wedge_L = \wedge_G$. Also L contains a finite maximal chain and the points of L are the points of G which are not $\leqslant 0_Q$, together with 0_Q itself if $0_Q \neq 0_G$. Note further that $0_L = 0_G$ and that, for all x, y in L, $x \vee_L y = x \vee_G y$ if $x \vee_G y$ is in I and $x \vee_L y = (x \vee_G y)\xi$ if $x \vee_G y$ is not in I.

Let $x \lessdot_L y$. We wish to find a point q of L such that $q \nleqslant x$ and $q \leqslant y$. If $0_Q \nleqslant x$ and $0_Q \leqslant y$ then we may take $q = 0_Q$. If not, then there is a point p of G such that $p \nleqslant 0_Q$, $p \nleqslant x$, and $p \leqslant y$ — and we may then take $q = p$.

Now let x be in L and let q be a point of L such that $q \nleqslant x$. We wish to show that $x \lessdot_L x \vee_L q$. If $x \vee_G q$ is in I then q is a point of G, so that $x \lessdot_G x \vee_G q$ and hence $x \lessdot_L x \vee_L q$. If $x \vee_G q$ is not in I and x is in I then $x \vee_L q = (x \vee_G q)\xi = x\xi$ (for q a point of G, the second equality follows from the equivalence of (i) and (ii) in (3); for $q = 0_Q$ it is obvious). It follows that $x \lessdot_L x \vee_L q$ in this case also (for if $x < y < x\xi$ then $y\xi = x\xi$ so that y is neither in Q nor in I). Suppose that x is not in I. Then x is in Q and q must be a point of G. By (1) we have $x\xi \lessdot_Q (x \vee_G q)\xi$ and this is the same as $x \lessdot_L x \vee_L q$, the only members of L greater than x being those in Q.

We have now shown that L is a geometry and since each point of G is $\leqslant$ some point of L, L is a quotient of G.

To find $\lambda(L)$, let x be an element of I for which $x\xi = 1$ (such elements exist by (3), Corollary). Then by the last parenthetical remark we have $x \prec_L 1$. It follows that $\lambda(L) = \lambda_L(x) + 1 = \lambda_G(x) + 1 = \lambda(Q) + 1$.

Q.E.D.

COROLLARY. *Let Q_1 and Q_2 be quotients of G. Then $Q_1 \prec Q_2$ if and only if $Q_1 \leqslant Q_2$ and $\lambda(Q_2) = \lambda(Q_1) + 1$.*

(The relations here between Q_1 and Q_2 refer to the order in $\mathcal{Q}_s(G)$.)

PROOF: If $Q_1 \prec Q_2$ then by Theorem I, $Q_1 \leqslant L(Q_1, Q_2) \leqslant Q_2$ and $\lambda(L(Q_1, Q_2)) = \lambda(Q_1) + 1$. Hence $L(Q_1, Q_2) = Q_2$ so that $\lambda(Q_2) = \lambda(Q_1) + 1$. The converse follows from the fact that if $Q_1 < Q_2$ then $\lambda(Q_1) < \lambda(Q_2)$.

Q.E.D.

As an immediate consequence of this corollary we have:

THEOREM II. *For each geometry G, $\mathcal{Q}_s(G)$ satisfies the Jordan-Dedekind chain condition and the height of a quotient Q of G in $\mathcal{Q}_s(G)$ is $\lambda(Q)$.*

A FACTORIZATION THEOREM FOR STRONG MAPS

Define the *kernel*, ker σ, of a strong map $\sigma : G \to H$ to be the element $0_H\sigma^{\Delta}$ of G and call σ a *contraction* if it is a surjection, 1-1 on $G/\ker \sigma$. In particular, if a is any element of G, then $\xi : G \to G/a$ is a contraction.

THEOREM III. *Every strong map be factorized as an injection followed by a contraction.*

In proving this it is convenient, given a surjection $G \to H$, to define the quotients $L^i(G \to H)$, $i = 0, 1, 2, ——$, of G by: $L^0(G \to H) = H\sigma^{\Delta}$, $L^{i+1}(G \to H) = L(L^i(G, H), G)$. Then $L^n(G \to H) = G$ for $n = \lambda(G) - \lambda(H)$ and $G \to H$ may be factorized as:

$$G = L^n(G \to H) \to L^{n-1}(G \to H) \to —— \to L^1(G \to H) \to L^0(G \to H) \cong H.$$

(4) *Suppose that in the commutative diagram* (α):

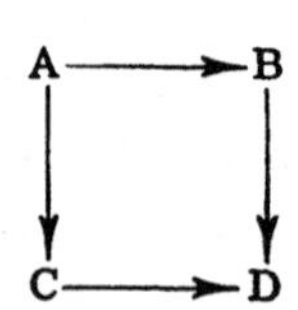

of $\mathcal{G}_s$, the horizontal maps are surjections and the vertical maps are injections.

Then there is an injection from $L = L^1(A \to B)$ *into* $M = L^1(C \to D)$ *which may be inserted into* (α) *so that the resulting diagram:*

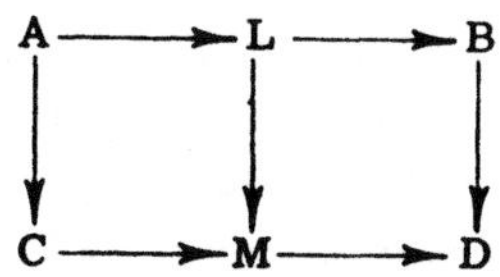

commutes.

PROOF: We may without loss of generality regard A as a subgeometry of C, D as a quotient of C, and B both as a subgeometry of D and as a quotient of A. Conversely, any such situation gives rise to a commutative diagram (α)—to see this it is sufficient, on account of the $\vee$-preserving property of strong maps, to verify only that the diagram commutes on points, and this is straightforward. It follows that (4) will be obtained if we show that L is a subgeometry of M. Write $I = I(B, A)$ and $J = I(D, C)$. Then clearly $I \subseteq J$ and hence $L \subseteq M$. Also $0_L = 0_A = 0_B = 0_M$ and $0_C = 0_D$. Let x and y be in L. If $x \vee_A y$ is in I then $x \vee_L y = x \vee_A y$ is in I and hence $x \vee_L y$ is in M. If $x \vee_A y$ is not in I then $x \vee_L y$ is the image of $x \vee_A y$ under the map $A \to B$, and this is the same as the image of $x \vee_B y$ under the map $C \to D$. Hence $x \vee_L y$ is in M in this case also. It follows that L is a $\vee$-closed subset of M. Since clearly every point of L is a point of M, L is a subgeometry of M. Q.E.D.

PROOF OF THEOREM III: Consider first a surjection $G \to H$. Let $\lambda(G) - \lambda(H) = n$ and let E be any geometry of rank n. Form (in $\mathcal{G}_s$) the sum $G + E \to H$ of $G \to H$ and $E \to O \to H$ (sums in $\mathcal{G}_s$ are obtained by taking lattice products and the obvious maps; O is the zero object of $\mathcal{G}_s$). Write $L^n(G + E \to H) = L$ and note that $L^n(G \to H) = G$. Applying (4) n times to the diagram:

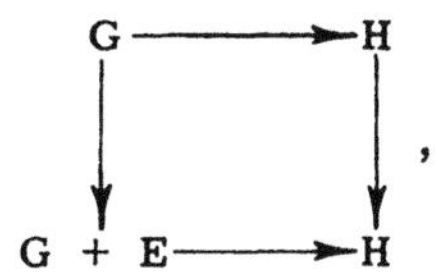

where $G \to G + E$ is the natural injection of G into the sum $G + E$ and $H \to H$ is the identity on H, we deduce that $G \to L = G \to G + E \to L$ is an injection; likewise for $E \to L = E \to G + E \to L$. This latter fact shows that if a is the image of 1_E under $E \to L$ then $\lambda_L(a) = n$. Now the image of a under $L \to H$ is 0_H and we can therefore factorize $L \to H$ as $L \xrightarrow{\varepsilon} L/a \to H$, $L/a \to H$ being the restriction of $L \to H$ to L/a. Since

$\lambda(L/a) = \lambda(L) - n = \lambda(H)$ by Theorem I, we must have $L/a \cong H$, and $L \to H$ is a contraction. Thus $G \to H = G \to L \to H$ is an injection followed by a contraction.

To deal with the case of a strong map $G \to H$ which is not a surjection, consider the sum $G + H \to H$ of $G \to H$ and the identity $H \to H$. This is a surjection and by what has already been proved we can express it as an injection followed by a contraction, say $G + H \to H = G + H \to L \to H$. It follows that $G \to H = G \to G + H \to L \to H$, where $G \to G + H$ is the natural injection, is also an injection followed by a contraction. Q.E.D.

Remarks

1. To reconcile the present notions with those of [5], we shall call a function σ from a geometry G into a geometry H a *weak map* if, for all elements x of G,

(i) $x\sigma = \bigvee \{ p\sigma \mid p \text{ a point of } G \cdot x\}$,

(ii) $\lambda(x\sigma) \leqslant \lambda(x)$.

If $x\sigma = 0_H$ for $x = 0_G$ only, let us say that σ is *non-singular*. Then the maps of [5], so far as finite rank geometries are concerned, are the same as the non-singular weak maps. Geometries and weak maps form a category $\mathscr{G}_w$, provided that we define the product of two weak maps $\sigma : G \to H$ and $\tau : H \to K$, not by ordinary composition as we may (and do) in the case of strong maps, but rather by the equation $x(\sigma\tau) = \bigvee \{(p\sigma)\tau \mid p \text{ a point of } G \cdot x\}$.

Every strong map is evidently also a weak map. If Q is a quotient of a geometry G then $I(Q, G)$ coincides with the set $\mathscr{K}_1(\xi) = \mathscr{K}_2(\xi)$ of [5], where $\xi = \xi : G \to Q$.

2. Crapo [3] has independently set up a theory of strong maps, equivalent to that given here. I am grateful to him for many helpful discussions which have considerably influenced the present paper.

3. Quotients and thus strong maps are closely related to the linear representations of Bruck and Bose [1]. Indeed, if Σ is a Desarguesian projective geometry, then the presentations

$$\pi = \pi(\Sigma, \mathscr{C}_p, \mathscr{C}_l)$$

which satisfy the lattice axiom (4) of [1] and the condition: each point of Σ is contained in some member of $\mathscr{C}_p$, are, with the appropriate 0_π and 1_π added, precisely the same as the rank 3 projective quotients of Σ. For such presentations, the axioms ($N1$), ($N1'$), and ($N2'$) of [1] necessarily hold; ($N2$) and the completion axioms (5) and (6) are each equivalent to the condition that the strong map $\xi : \Sigma \to \pi$ be non-singular (see 1 above); and the condition that each member of $\mathscr{C}_p$ be a point of Σ is equivalent to the condition that $\xi : \Sigma \to \pi$ be monic in $\mathscr{G}_s$ (a strong map $\sigma : G \to H$ is monic in $\mathscr{G}_s$ if and only if it is 1-1 on the points-and-0 of G.)

4. The injections and contractions defined here correspond respectively to the reductions and contractions of Tutte [6].

5. I would like to express my thanks to Jack Edmonds for his generosity in communicating Theorem III to me and for several valuable remarks concerning an earlier version of this paper.

6. The particular case of Theorem III in which $G \to H$ is a surjection with $\lambda(G) = \lambda(H) + 1$ may be regarded as a portion ([2], p. 59) of Crapo's theorem on single-element extensions of matroids, the modular cuts $C \neq \phi, G$ of G corresponding to the quotients Q of G with $\lambda(Q) = \lambda(G) - 1$ via the relation $G - C = I(Q, G)$.

References

1. R. H. Bruck and R. C. Bose, Linear Representations of Projective Planes in Projective Spaces, *J. Algebra* **4** (1966), 117–172.
2. H. H. Crapo, Single-Element Extensions of Matroids, *J. Res. Nat. Bur. Standards Sect. B* **69** (1965), 55–65.
3. H. H. Crapo, Structure Theory for Geometric Lattices, *Rend. Sem. Mat. Univ. Padova* **38** (1967), 14–22.
4. J. Edmonds, letter dated May 5, 1966.
5. D. A. Higgs, Maps of Geometries, *J. Lond. Math. Soc.* **41** (1966), 612–618.
6. W. T. Tutte, Lectures on Matroids, *J. Res. Nat. Bur. Standards Sect. B* **69** (1965), 1–47.

Chapter II
Linear representation of matroids

1. Non-representable matroids

The axioms of matroid theory are abstracted from the properties of linear dependence. Thus, a natural question is whether every matroid can be represented by a set of vectors with coordinates over some field $\boldsymbol{k}$ so that abstract dependence in the matroid coincides with linear dependence of the vectors. If such a set of vectors exists, the matroid is said to be *representable* over the field $\boldsymbol{k}$. This question was first considered by Whitney in [I.1], §16. There he gave the Fano plane as an example of a matroid representable only over a field of characteristic 2, and hence, not representable over the real or complex numbers. The question whether there exists a matroid not representable over any field is answered in Mac Lane's paper, "Some interpretation of abstract linear dependence in terms of projective geometry," the first paper reprinted in this chapter.

The basic idea in this paper is to construct non-representable matroids by taking a "suitable" theorem of projective geometry, drawing the geometric configuration that it implies, and modifying the configuration so that it remains a matroid but no longer satisfies the theorem. (The natural problem of characterizing the theorems of projective geometry for which this construction works seems to be a difficult one. Indeed, one does not even know how to formulate the problem precisely.) An important example is obtained by taking the bundle theorem. This theorem states

> *Let $\ell_1, \ell_2, \ell_3, \ell_4$ be lines, no three on a plane. If five of the six pairs $\{\ell_i, \ell_j\}$, $i \neq j$, are coplanar, then so is the sixth.*

Letting $\ell_i = a_i \vee b_i$, we obtain the three-dimensional configuration with eight points $a_1, b_1, a_2, b_2, a_3, b_3, a_4, b_4$ with no three points collinear and the following six four-element subsets of points (and no other) coplanar:

$$a_1, b_1, a_2, b_2; \quad a_1, b_1, a_3, b_3; \quad a_1, b_1, a_4, b_4;$$
$$a_2, b_2, a_3, b_3; \quad a_2, b_2, a_4, b_4; \quad a_3, b_3, a_4, b_4.$$

By decreeing one of the above four-element subsets non-coplanar, we obtain a matroid, the *failed bundle configuration,* which is not representable over any field. The failed bundle configuration is one of two non-representable matroids with eight points. They are the non-representable matroids with the smallest number of points. (The failed bundle configuration first appeared, it seems, in Higgs [66] and Vámos [71]; it is sometimes called the *twisted cube* or the *Vámos matroid.*) In §3 of Mac Lane's paper, Pascal's theorem for a hexagon inscribed in a degenerate conic consisting of two lines and Desargues' theorem are used in this way to obtain non-representable matroids. In addition, the failed Pascal configuration is generalized to yield an infinite family of rank 3 non-representable matroids.

The last section of Mac Lane's paper is concerned with representing matroids over algebraic extensions of the rationals. There are two results. The second result (Theorem 4) is that if a (finite) matroid is representable over the complex numbers, then it is also representable over a finite algebraic extension of the rationals. The first (Theorem 3) is that given a finite algebraic extension K of the rationals R, one can construct a matroid M which is representable over an extension K' of R only if K' contains K. The proof is "by example" and a matroid associated with the extension $R(2^{1/2})$ is constructed. To aid the reader, we shall perform this construction in more detail.

The construction is based on von Staudt's algebra of throws. This technique was used by von Staudt to prove that Desarguesian projective geometries are representable over some (skew) field and is a distant descendant of the "geometric algebra" of the Greeks. What we wish to construct is a matroid reflecting the fact that

$$(2^{1/2})(2^{1/2}) = 1 + 1.$$

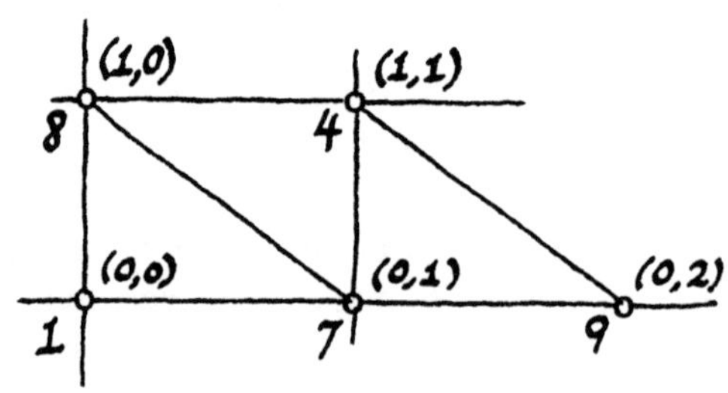

Figure 1. $1 + 1 = 2$

We first take care of $1+1$. This is done using the configuration in Figure 1. (The points in Figures 1 and 2 are labelled according to Mac Lane's description of the matroid; in Figure 1, the following pairs of lines are parallel: 18, 47; 78, 49; 48, 179.)

Next, we add points $\alpha=(0, 2^{1/2})$ and $0=(2^{1/2}, 0)$, thus obtaining the configuration in Figure 2. (In Figure 2, the lines 78, 49, 0α, are parallel; 8α, 90 are also parallel.)

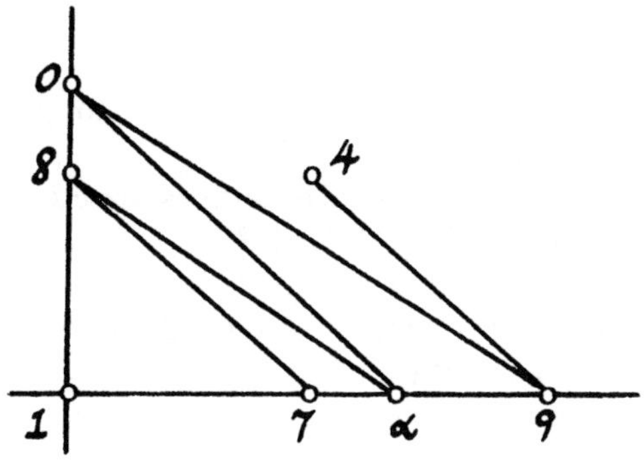

Figure 2. $(2^{1/2})(2^{1/2})=2$

Finally, we add points of intersection on the line at infinity for every pair or triple of parallel lines in the following way: 2 is the point of intersection of 48, 179; 3 is the point of intersection of 180, 47; 5 is the point of intersection of 78, 49 and 0α; 6 is the point of intersection of 90 and 8α. Note that by construction, the four points 2356 are collinear. The points 1, 2, ..., 0, α with the incidence relations given above form precisely the matroid given in §4.

We end with a list of related papers. Of particular importance are two papers. The first is Ingleton [71] which sets out in detail many useful examples of geometric configurations. The second is Kahn [82] which shows that for every finite set of primes $\mathcal{P}$, there exists a matroid M such that $\mathcal{P}$ is the set of characteristics of the fields over which M is representable.

General: Ingleton [71].
Configurations: Mason [71, 77].
Non-representable matroids: Dowling [73], §5, Higgs [67], Ingleton [59], Lazarson [58], Vámos [71]. See also Fournier [71].
Characteristic sets: Brylawski [82], Gordon [84], Ingleton [71], Kahn [82], Reid [80].

Computational Complexity of deciding representability: Jensen and Korte [82], Seymour and Walton [81], Truemper [82].

Necessary condtions for representability: Ingleton [71, 71′], Rado [57], Vámos [78].

Ring theoretic approaches to representation: Fenton [84], Vámos [71], White [75, 80].

References

BRYLAWSKI, T.: Finite prime-field characteristic sets for planar configurations, *Linear Algebra Appl.* *46*(1982), 155-176.

DOWLING, T.A.: A class of geometric lattices based on finite groups, *J. Combin. Theory Ser. B 14*(1973), 61-86.

FOURNIER, J.-C.: Représentation sur un corps des matroïds d'ordre $\leqslant 8$, *Théorie des Matroïdes* (*Rencontre Franco-Britannique, Brest,* 1970), pp. 50-61, Lecture Notes in Math., Vol. 211, Springer-Verlag, Berlin, 1971.

FENTON, N.E.: Matroid representations – an algebraic treatment, *Quart. J. Math. Oxford Ser. (2) 35*(1984), 263-280.

GORDON, G.: Matroids over F_p which are rational excluded minors, *Discrete Math. 52*(1984), 51-65.

GORDON, G.: Constructing prime-field planar configurations, *Proc. Amer. Math. Soc. 91*(1984), 492-502.

HIGGS, D.A.: *Geometry,* Lecture notes, Univ. Waterloo, Waterloo, Ont., 1967.

INGLETON, A.W.: Note on independence functions and rank, *J. London Math. Soc. 34*(1959), 49-56.

INGLETON, A.W.: Representations of matroids, *Combinatorial Mathematics and its Applications (Proc. Conf., Oxford, 1969),* pp. 149-167, Academic Press, London, 1971.

INGLETON, A.W.: Conditions for representability and transversality of matroids, *Théorie des matroïdes (Rencontre Franco-Britannique, Brest, 1970),* pp. 62-66, Lecture Notes in Math., Vol. 211, Springer-Verlag, Berlin, 1971.

JENSEN, P.M. and KORTE, B.: Complexity of matroid property algorithms, *SIAM J. Comput. 11*(1982), 184-190.

KAHN, J.: Characteristic sets of matroids, *J. London Math. Soc. (2) 26*(1982), 207-217.

LAZARSON, T.: The representation problem for independence functions, *J. London Math. Soc. 33*(1958), 21-25.

MASON, J.H.: Geometric realization of combinatorial geometries, *Proc. Amer. Math. Soc. 30*(1971), 15-21.
MASON, J.H.: Matroids as the study of geometric configurations, *Higher Combinatorics* (M . Aigner, ed.), Reidel, Dordrecht, 1977.
RADO, R.: Note on independence functions, *Proc. London Math. Soc. (3) 7*(1957), 300-320.
REID, R.: Obstructions to representations of combinatorial geometries, unpublished; appeared as an appendix to T. Brylawski and D.G. Kelly, *Matroids and Combinatorial Geometries,* Univ. North Carolina Press, Chapel Hill, N.C., 1980.
SEYMOUR, P.D. and WALTON, P.N.: Detecting matroid minors, *J. London Math. Soc. (2) 23*(1981), 193-203.
TRUEMPER, K.: On the efficiency of representability tests for matroids, *Europ. J. Combin. 3*(1982), 275-291.
VÁMOS, P.: Linearity of matroids over division rings (Notes by G. Roulet), *Möbius Algebras (Proc. Conf., Univ. Waterloo, Waterloo, Ont., 1971),* pp. 170-174, Univ. Waterloo, Waterloo, Ont., 1971.
VÁMOS, P.: A necessary and sufficient condition for a matroid to be linear, *Möbius Algebras (Proc. Conf., Univ. Waterloo, Waterloo, Ont., 1971),* pp. 162-169, Univ. Waterloo, Waterloo, Ont., 1971.
VÁMOS, P.: The missing axiom of matroid theory is lost forever, *J. London Math. Soc. (2) 18*(1978), 403-408.
WHITE, N.L.: The bracket ring of a combinatorial geometry. I and II. *Trans. Amer. Math. Soc. 202*(1975), 79-95 and *214*(1975), 233-248.
WHITE, N.L.: The transcendence degree of a coordinatization of a combinatorial geometry, *J. Combin. Theory Ser. B 29*(1980), 168-175.

2. Homotopy

A major problem in the theory of linear representations of matroids is to find the forbidden minors for representability over a given field (or a given set of fields). The first forbidden minor theorem is Kuratowski's theorem characterizing planar graphs, here stated in a weaker form.

Theorem 1 (Kuratowski [30]): A graphic matroid is the cycle matroid of a planar graph if and only if it does not contain the cycle matroids of K_5 and $K_{3,3}$ as minors.

This theorem was greatly extended by Tutte.

Theorem 2 (Tutte [58]): A matroid is the cycle matroid of a graph if and only if it does not contain the four-point line or the dual of the cycle matroid of K_5 or $K_{3,3}$ as minors. Moreover, a matroid is the cycle matroid of a planar graph if and only if it does not contain the four-point line, the cycle matroid of K_5 or $K_{3,3}$, or their duals as minors.

In addition, Tutte also discovered forbidden minor theorems for two classes of representable matroids.

Theorem 3 (Tutte): A matroid is binary (i.e. representable over GF(2)) if and only if it does not contain the four-point line as a minor.

Theorem 4 (Tutte): A matroid is regular (i.e. representable over every field) if and only if it does not contain the four-point line, the Fano plane or its dual as minors.

Theorems 3 and 4 appeared in Part II of the paper "A homotopy theorem for matroids." Both parts of this paper are reprinted in this chapter. (It may be worth remarking that there does not exist a finite set of forbidden minors for representability over some field. This follows, for example, from the existence of the infinite family of generalized failed Pascal configurations given in Mac Lane [II. 1].)

In order to prove these theorems, it is necessary to study "paths" of copoints in geometric lattices. In Part I of Tutte's paper, a notion of homotopy for such paths is introduced and it is proved that all re-entrant paths are homotopic to paths made up of elementary paths of four specific kinds. The homotopy theorem is then used in Part II to prove the forbidden minor theorems.

To read Tutte's paper, it is crucial to understand his terminology. Since most readers are more familiar with the terminology of geometric lattices and closed sets, we shall explain the relation between this terminology and Tutte's terminology. In §1 of Part I, a matroid is defined in terms of its collection ***M*** of circuits. A lattice is then defined whose elements are unions of circuits. This lattice is essentially the lattice of closed sets of the dual matroid, except that rather than considering closed sets, we consider their complements, and the lattice is inverted. (Here, we use the fact that the cocircuits, or complements of copoints, are the circuits of the dual matroid. See, for example, Theorem 2 of [II. 1].) Thus, if the matroid is of nullity n, a $(k-1)$-flat (in Tutte's terminology) is the complement of a closed set in the dual matroid whose rank (in the dual matroid) is $n-k$. In particular, the points corresponds to the copoints, the lines to the colines, and so on. Elements of the matroid are called cells.

Once Tutte's terminology is understood, the paper becomes quite accessible. §2 is concerned with the (inverted) lattice of flats and several basic properties (such as the existence of a relative modular complement (2.4) and the (lower) semimodularity of the dimension function (2.5)) are proved. §3 deals with connectivity and §4 the disconnected line. It should be noted that a flat U is connected if and only if the upper interval $[U^C, 1]$ is connected in the lattice of closed sets of the dual. (Here, U^C is the complement of U.)

§5 is concerned with convex subclasses (also called linear subclasses) and paths. A convex subclass of circuits (defined in §1) can be thought of as the set of all circuits not containing a given cell, although that cell may not be in the matroid (see Crapo [65]). A path (also defined in §1) is a sequence of points or 0-flats or circuits $(X_1, \ldots, X_n)$ such that $X_i \cup X_{i+1}$ is a line containing at least three points. (The reason for considering such paths will be made clear in Part II.) If $X_1 = X_n$, the path is said to be re-entrant. The path theorem (5.1) which states that in a connected matroid, there exist paths avoiding a given convex subclass, is proved. Next, four kinds of elementary re-entrant paths are described and using them, the notion of a path being null-homotopic is defined.

The homotopy theorem (6.1) is stated and proved in §6. The proof is in two parts. First, it is proved that certain paths of the form $Q = (W, X, Y, Z, W)$ are null-homotopic (the lemma in §6). The proof proceeds as follows. It is first observed that any two paths of the given form are homotopic. Using the notion of a transversal and the path theorem, the possible cases for Q are reduced progressively to two. One of these cases is that Q is an elementary reentrant path of the fourth kind. In both cases, Q is found to be null-homotopic. Returning to the proof of the homotopy theorem, suppose that there is a re-entrant path P (off a convex subclass) which is not null-homotopic. Choose one for which the union $F(P)$ of all the points (or circuits) in P has minimum dimension n. As $F(P)$ is connected, there exists a connected flat E of dimension $n-1$ lying between $F(P)$ and X_0, the origin of the re-entrant path P. Let R be a re-entrant path $(X_0, X_1, \ldots, X_m, X_0)$ on the flat $F(P)$. Let $u(R)$ be the number of points X_i in R not on E and $v(R)$ be the dimension of the flat $X_{i-1} \cup X_i \cup X_{i+1}$ where X_i is the first point in R not on E, if such a point exists, and let $v(R)$ be 0 otherwise. Now choose R so that it is homotopic to P, $u(R)$ is a minimum, and $v(R)$ is a minimum. Suppose $u(R) > 0$. There are three cases to consider, $v(R) = 1$, $v(R) = 2$, and $v(R) > 2$. The first case is relatively simple; the second and third use the lemma proved in the first part. All three cases lead to a contradiction and so $u(R) = v(R) = 0$. But this means the path R is on E, a flat of smaller dimension than $F(P)$. By the choice of P, R is null-homotopic and hence, P is also null-homotopic. This completes the proof of the homotopy theorem. Part I ends with an interpretation of the homotopy theorem for graphs.

Two other notions of homotopy can be found in Lovász [77] and Maurer [73].

References

Crapo, H.H.: Single-element extensions of matroids, *J. Res. Nat. Bur. Standards Sect. B 69B*(1965), 55-65.

Kuratowski, K.: Sur les problèmes des courbes gauches en Topologie, *Fundamenta Math. 15*(1930), 271-283.

Lovász, L.: A homology theory for spanning trees of a graph, *Acta Math. Acad. Sci. Hungar. 30*(1977), 241-251.

Maurer, S.B.: Matroid basis graphs.I, *J. Combin. Theory Ser.B 14*(1973), 216-240.

Tutte, W.T.: Matroids and graphs, *Trans. Amer. Math. Soc. 90*(1959), 527-552.

3. Binary and regular matroids

The forbidden minor theorems for binary and regular matroids are proved in Part II of Tutte's paper.

The first two sections are concerned with chain groups. The relation between the matroid of a chain group and the chain group is essentially the relation between a matrix and its circuit matrix described in Part III of Whitney [I. 1]. Let M be a set of column vectors over a field or an integral domain R, thought of as the columns of a matrix. The collection $\boldsymbol{M}$ of minimal linearly dependent sets in M is the collection of circuits of a matroid on M. If X is a circuit, then there exists a function $f(X)$ or f_X (both notation are used) defined from M to R such that

(1) For $a \in M$, $f_X(a) \neq 0$ if and only if $a \in X$ and

(2) $\sum_{a \in M} f_X(a)\, a = 0$.

The function f_X is unique up to a non-zero multiple and gives the coefficients of the minimal linear relation between the vectors in X. The function f_X is called an *elementary chain* and X is called its *domain*. The *circuit matrix* of M

is the matrix with rows indexed by $\boldsymbol{M}$ and columns indexed by M whose Xa-entry is $f_X(a)$. The *chain group* associated with $\boldsymbol{M}$ is the abelian group (under addition of functions) consisting of sums and differences of elementary chains. Thus, if R is a finite field or the integers, the chain group is the row space of the circuit matrix. From the chain group, the circuits $\boldsymbol{M}$ of the matroid can be reconstructed: they are the domains of elementary chains (defined intrinsically) in the chain group (see (2.2) and (2.3) for details). From the circuit matrix, a representation of the matroid $\boldsymbol{M}$ as a set of vectors can be constructed by taking the subspace orthogonal to the row space (see [I. 1], Part III). Thus, to represent a matroid, it suffices to construct the circuit matrix, or, equivalently, the elementary chains of a chain group. (The representation can also be constructed directly by taking the cocircuit matrix, i.e. the circuit matrix of the dual.)

A very effective way to construct the elementary chains is given by the Tutte coordinatization theorem (2.5). This theorem says that to check that a set of chains f_X indexed by the circuits of a matroid $\boldsymbol{M}$ is the set of elementary chains of a chain group representing $\boldsymbol{M}$, it suffices to check two conditions:

(1) $f_X(a) \neq 0$ if and only if $a \in X$;

(2) If X, Y and Z are three circuits on a connected line, then the chains f_X, f_Y and f_Z are linearly dependent.

Thus, it suffices only to check the conditions for elementary chains at the bottom of the lattice of flats (or the top of the lattice of closed sets of the dual). This theorem is proved using the path theorem. From the coordinatization theorem, the forbidden minor theorem for binary matroids follows almost immediately (see (2.6) and (2.7)).

The rest of Part II is devoted to proving the forbidden minor theorem for regular matroids. In §3, some preliminary results about restrictions and contractions are assembled. Once again, it is important to understand the terminology used. Let $\boldsymbol{M}$ be the collection of circuits of a matroid on the set M and let $S \subseteq M$. There are two matroids defined on S.

(a) $\boldsymbol{M} \times S$. This is the collection of all circuits in $\boldsymbol{M}$ contained in S. (Thus, this is usually called the"restriction" of $\boldsymbol{M}$ to S.)

(b) $\boldsymbol{M} \cdot S$. This is the collection of all minimal non-empty members of $\boldsymbol{M}_S = \{X \cap S\colon\ X \in \boldsymbol{M}\}$. (Thus, this is usually called the "contraction" of $\boldsymbol{M}$ by $M - S$.)

The *carrier* of $\boldsymbol{M}\cdot S$ in $\boldsymbol{M}$ is defined to be a subset Z, $S \subseteq Z \subseteq M$, such that $(\boldsymbol{M}\times Z)\cdot S = \boldsymbol{M}\cdot S$. In particular, if $Z - S$ is an independent spanning set of $M - S$, then Z is a carrier of $\boldsymbol{M}\cdot S$. Carriers are used to define an injection θ of flats of $\boldsymbol{M}\cdot S$ into the lattice of flats of $\boldsymbol{M}$. (The injection θ is perhaps easier to define if one observe that the lattice of flats is the inverted lattices of closed sets of the dual and that the dual of $\boldsymbol{M}\cdot S$ is a submatroid of the dual of $\boldsymbol{M}$.)

§4 begins with more preliminary results. First, it is shown that a regular matroid is binary (4.4). (Note that a regular matroid is defined in §1 to be the matroid of a chain group over the integers whose elementary chains take only the values -1, 0 or 1. This is equivalent to being representable over every field.) Next, the figures of Type BI (the Fano plane) and Type BII (the dual of the Fano plane) are described. A figure is an "upper minor" in the lattice of closed sets of the dual.

Finally, the main theorem (4.5) is stated: A binary matroid is regular if and only if it contains no figures of Type BI or BII. The implication is easy, as the Fano plane and its dual are not regular. The converse is proved by induction. Let $\boldsymbol{M}$ be a binary matroid not containing figures of Type BI or BII and let a be an element of $\boldsymbol{M}$. The smaller matroid $\boldsymbol{M}' = \boldsymbol{M}\cdot(\boldsymbol{M} - \{a\})$ is regular; by induction, we can construct a chain group whose elementary chains g_U, where U is a circuit of $\boldsymbol{M}'$, take only the values -1, 0 or 1. (In the paper, $g_U(a)$ is sometimes written as $s(U, a)$.) This chain group will be extended to a chain group for $\boldsymbol{M}$.

To do so, three preliminary definitions are needed. We may as well assume that $\{a\}$ is independent in $\boldsymbol{M}$ and that a carrier for $\boldsymbol{M}'$ is the entire set M. Let θ be the injection of the flats of $\boldsymbol{M}'$ into the flats of $\boldsymbol{M}$ defined in §3. The injection θ acts on a circuit T of $\boldsymbol{M}'$ in the following way: if T is also a circuit of $\boldsymbol{M}$, then $\theta T = T$; if T is not a circuit of $\boldsymbol{M}$, then $T \cup \{a\}$ is a circuit of $\boldsymbol{M}$ and $\theta T = T \cup \{a\}$. The convex subclass $\boldsymbol{C}$ is the collection of all the circuits in $\boldsymbol{M}'$ such that $a \notin \theta T$, or, equivalently, $\theta T = T$.

Now let X and Y be distinct circuits of $\boldsymbol{M}'$ not in $\boldsymbol{C}$ on the same connected line. The function $t(X, Y)$ is defined by

$$t(X, Y) = g_X(d)\, g_Y(d),$$

where g_X and g_Y are the elementary chains with domain X and Y, and d is any element in $X \cap Y$. This definition is independent of the choice of d. If $R = (X_1, \ldots, X_k)$ is a path off the convex subclass $\boldsymbol{C}$, then set

$$u(R) = \prod_{i=1}^{k-1} t(X_i, X_{i+1}).$$

Using the homotopy theorem and the fact that $\boldsymbol{M'}$ contains no figures of Type BI and BII, it is shown that $u(R) = 1$ for any re-entrant path R. This fact is use in the next definition.

The circuits in $\boldsymbol{M'} - \boldsymbol{C}$ are partitioned into equivalence classes $K_1, \ldots, K_q$ by the relation:

X and Y are in the same class K_i if there is a path off $\boldsymbol{C}$ from X to Y.

For each equivalence class K_i, choose a representative T_i and set $\nu(T_i) = 1$. For any other circuit X in the same class K_i, set

$$\nu(X) = u(Q)\,\nu(T_i) = u(Q),$$

where Q is a path off $\boldsymbol{C}$ from T_i to X. As $u(R) = 1$ for any re-entrant path, this definition is independent of the choice of the path Q.

All is now ready for the final step: to extend the chain group for $\boldsymbol{M'}$ to a chain group for $\boldsymbol{M}$. Let U be a circuit in $\boldsymbol{M}$. The elementary chain f_U is defined in the following way:

- If $a \in U$, then $X = U - \{a\}$ is a circuit in $\boldsymbol{M'}$. Set $f_U = g_X$ on $\boldsymbol{M} - \{a\}$ and $f_U(a) = \nu(X)$.
- $a \notin U$, then there are two possibilities:
  1. U is a circuit in $\boldsymbol{M'}$. Set $f_U = g_U$ on $M - \{a\}$ and $f_U(a) = 0$.
  2. U is the disjoint union of two circuits in $\boldsymbol{M'}$. In this case, U is the symmetric difference of two circuits V and W in $\boldsymbol{M}$, with $V \cap W = \{a\}$. As a is contained in V and W, f_V and f_W are already defined. Set

     $$f_U = f_V(a) f_W - f_W(a) f_V.$$

The proof of (4.5) is now completed by checking that the elementary chains so defined satisfy the conditions of the Tutte coordinatization theorem (2.5).

The forbidden minor theorem for regular matroids follows from the forbidden minor theorem for ternary matroids (see §3 of [II. 3]). Several extensions of the forbidden minor theorems for binary and regular matroids can be found in Bixby [74, 76] and Seymour [77, 81]. A more leisurely exposition of the homotopy and forbidden minor theorems can be found in Tutte [65] (see also [71]). Other related papers are Brylawski [75] and White [70].

References

BIXBY, R.E.: ℓ-matrices and a characterization of binary matroids, *Discrete Math.* *8*(1974), 139-145.

BIXBY, R.E.: A strengthened form of Tutte's characterization of regular matroids, *J. Combin. Theory Ser. B* *20*(1976), 216-221.

BRYLAWSKI, T.: A note on Tutte's unimodular representation theorem, *Proc. Amer. Math. Soc.* *52*(1975), 499-502.

SEYMOUR, P.D.: A note on the production of matroid minors, *J. Combin. Theory Ser. B* *22*(1977), 289-295.

SEYMOUR, P.D.: On minors of non-binary matroids, *Combinatorica* *1*(1981), 387-394.

TUTTE, W.T.: Lectures on matroids, *J. Res. Nat. Bur. Standards Sect. B* *69B*(1965), 1-47.

TUTTE, W.T.: *Introduction to the theory of matroids,* American Elsevier, New York, 1971.

WHITE, N.L.: Coordinatization of combinatorial geometries, *Combinatorial Mathematics and its Applications (Proc. Conf., Univ. North Carolina, Chapel Hill, N.C., 1970),* pp. 484-486, Univ. North Carolina, Chapel Hill, N.C., 1970.

4. Ternary matroids

After binary and regular matroids have been characterized by forbidden minors, the next step is to do the same for ternary matroids (i.e. matroids representable over $GF(3)$). Here, an immediate conjecture is that the forbidden minors are the five-point line, the Fano plane, and their duals. A proof of this conjecture was found by Reid around 1971 but he never published his proof. Two proofs were independently discovered later by Bixby and Seymour. Bixby's proof (see [79]) used the homotopy theorem and was very similar to Tutte's proof of the forbidden minor theorem for regular matroids. Seymour's proof used entirely different techniques and appeared in his paper "Matroid representation over $GF(3)$." This paper is the last paper reprinted in this chapter.

Seymour's proof rests on two results proved in §§2 and 3. The first (Lemma 2.7) states that if two matroids M_1 and M_2 on the same set are "almost the same" and satisfy certain connectivity conditions, then one of them has a five-point line or the Fano plane as a minor. This is proved using inductive arguments. The second result (Lemma 3.2) is that a ternary matroid is uniquely representable, i.e. any two representations of a ternary matroid are projectively equivalent.

The proof of the forbidden minor theorem is contained in §4. The plan of the proof is as follows. Let M be a matroid not containing any of the four forbidden minors. The first step is to dispose of the relatively easy cases when M or $M \setminus \{z\}$ (for some element z) is disconnected. Next, a connectivity lemma (Lemma 2.3) is used to show that there exist elements a and b such that $M \setminus \{a, b\}$ is connected. By induction, the matroids $M \setminus \{a\}$ and $M \setminus \{b\}$ can be represented over $GF(3)$. By Lemma 3.2, these representations can be chosen so that they coincide on $M \setminus \{a, b\}$. These two representations are now combined (by taking the chain group generated by the two chain groups associated with the representations) to yield a ternary matroid. The proof is completed by observing that, by Lemma 2.7, the ternary matroid must be the same as the matroid M.

In addition, it is also proved that a matroid is regular if and only if it is representable over $GF(2)$ and $GF(3)$ (Lemma 3.4). Using this, Tutte's theorem for regular matroids is derived as a corollary.

Since Seymour's paper, two other proofs of the forbidden minor theorem for ternary matroids have appeared. These can be found in Kahn [84] and Truemper[82]. See also Truemper [84].

References

Bixby, R.E.: On Reid's characterization of the ternary matroids, *J. Combin. Theory Ser. B 26*(1979), 174-204.

Kahn, J.: A geometric approach to forbidden minors for $GF(3)$, *J. Combin. Theory Ser. A 37*(1984), 1-12.

Truemper, K.: Alpha-balanced graphs and matrices and $GF(3)$-representability of matroids, *J. Combin. Theory Ser. B 32*(1982), 112-139.

Truemper, K.: Partial matroid representations, *Europ. J. Combin.*, *5*(1984), 377-394.

Reprinted from *American Journal of Mathematics*,
volume 57, pages 800 through 804,
by permission of the Johns Hopkins University Press.

SOME INTERPRETATIONS OF ABSTRACT LINEAR DEPENDENCE IN TERMS OF PROJECTIVE GEOMETRY.[1]

By Saunders MacLane.

1. *Introduction.* The abstract theory of linear dependence, in the form recently developed by Whitney,[2] is closely related to the study of projective configurations. For any matroid (that is, any finite system of elements for which a suitably restricted notion of "linear dependence" is given) can be interpreted as a schematic geometric figure. Such a schematic figure, like a schematic configuration, is composed of a number of points, lines, planes, etc., with certain combinatorially defined incidences. The problem of representing a matroid by a matrix then becomes simply the problem of realizing a schematic figure by some geometric figure—and the impossibility of always finding such a representation turns out to be a simple consequence of Pascal's theorem! Even when such representation is possible, it depends essentially upon the field from which the elements of the representing matrix are taken. However, only algebraic fields need be used, and hence arises a connection between certain matroids and the algebraic fields in which they can be best represented.

Matroids will be defined by axioms on "rank," as in Whitney's paper. Without loss of generality we can also assume the following two axioms:

R_4: The rank of a single element is always 1.

R_5: The rank of a pair of elements is always 2.

For example, an element e which does not satisfy R_4 may be dropped from or added to a matroid M without otherwise altering the structure of M. These two axioms are equivalent to the following conditions on "bases":

B_3: Every element belongs to at least one base.

B_4: There is no pair of elements (e_1, e_2) such that every base containing e_1 remains a base when e_1 is replaced by e_2.

These conditions on M are in turn equivalent to the following conditions on the dual matroid M^*:

C^*_3: Every element is omitted from at least one circuit complement.

[1] Presented to the American Mathematical Society, December 28, 1934.

[2] H. Whitney, "On the abstract properties of linear dependence," *American Journal of Mathematics*, vol. 57 (1935), pp. 509-533.

C^*_4: For every pair of elements (e_1, e_2) there is a circuit complement containing e_1 but not e_2.

2. *Schematic geometric figures.* A rectilinear plane figure consists of a number of points and of all the lines joining these points in pairs. The combinatorial structure of such a figure can be specified by giving for each line L the set of all those points of the figure which lie on L. These sets satisfy the following axioms:

F_1: Any pair of points belongs to one and only one line.
F_2: Every line contains at least two points.
F_3: No line contains all the points.
F_4: There are at least two points.

A system consisting of a finite number of "points" and certain sets of these points, called "lines," and satisfying these axioms will be called a *schematic plane figure*; it may or may not correspond to some actual figure. In the same way a *schematic space figure* would involve "points," "lines," and "planes," satisfying F_1 to F_4 and the following axioms:

S_1: Every triple of points belonging to no line belongs to one and only one plane.
S_2: Every plane contains three points not on a line.
S_3: No plane contains all the points.
S_4: If a plane contains two points of a line, it contains all the points of that line.

The definition of a schematic n-dimensional figure is similar; it involves "points" and "k-dimensional planes" for $k = 1, \cdots, n-1$.

The equivalence of schematic figures and matroids may be formulated as follows:

THEOREM 1. *Every schematic n-dimensional figure is a matroid of rank $n+1$ if the rank of a set of points A is defined as the smallest r such that all the points of A are contained in some $(r-1)$-plane. Conversely, every matroid of rank $n+1$ becomes a schematic n-dimensional figure if the k-planes are taken as maximal sets of elements of rank $k+1$. This translation sets up a one-one correspondence between matroids and schematic figures.*[3]

From this theorem it follows that a schematic n-dimensional figure is

[3] The conditions R_4 and R_5 of § 1 are necessary here to exclude the geometrically meaningless cases of a point of dimension -1 or of two coincident points.

determined once the corresponding matroid M or its dual M^* is given. By Whitney's results, this dual M^* is completely determined by its circuit complements. The circuit complements in M^* correspond to maximal sets of rank $r(M)-1$ in M. Hence

THEOREM 2. *A schematic n-dimensional figure is completely determined if its* $(n-1)$*-planes are given. If a set of "points" and certain subsets of this set are given, these subsets will be the* $(n-1)$*-planes of some figure if and only if they are the circuit complements of a matroid* M^**; that is, if and only if these subsets satisfy the above axioms* C^*_3 *and* C^*_4*, while their complements satisfy Whitney's axioms* C_1 *and* C_2 *for circuits.*

These results also show that matroids form a direct generalization of schematic configurations. A *schematic plane configuration*[4] $p_\gamma g_\pi$ consists of p "points" and g "lines," with each point on γ lines and each line on π points. Such a configuration becomes a schematic figure in the above sense if those pairs of points not already joined by lines are joined by new "diagonal" lines. Similar transformations are possible for space configurations.

3. *Matrix representations of matroids.* The columns of a matrix stand in relations of rank and thus form a matroid. The question whether every matroid can be represented in this way by a matrix is clarified by the equivalence of matroids and schematic figures. Thus Whitney has constructed a matroid of rank 3 which cannot be represented as a matrix. This matroid has 9 elements $1, 2, \cdots, 9$ and the following 20 circuit complements:

712, 814, 923, 734, 836, 945, 756, 825;
16, 19, 69, 13, 15, 24, 26, 35, 46, 78, 79, 89.

Any attempt to represent this matroid yields a figure in which the lines 16, 19, and 69 coalesce into one line 169. A geometric representation reveals at once that this is simply Pascal's theorem for the hexagon 723845 inscribed in the degenerate conic composed of the two lines 743 and 825. The points 1, 6, and 9 are the intersections of opposite sides of the hexagon. In exactly the same way the theorem of Desargues may be used to construct a matroid with ten elements which has no matrix representation. Furthermore, the matroid arising from Pascal's theorem can be generalized to the case of $2m+3$ elements, which we denote by $1, 2, \cdots, 2m, \alpha, \beta, \gamma$. The circuit complements are:

[4] F. Levi, *Geometrische Konfigurationen*, p. 4.

$$12\alpha,\ 34\alpha,\cdots,(2m-3,\ 2m-2,\ \alpha),\ (2m-1,\ 2m,\ \alpha),$$
$$14\beta,\ 36\beta,\cdots,(2m-3, 2m, \beta),\ (2m-1, 2, \beta),$$
$$23\gamma,\ 45\gamma,\cdots,(2m-2,\ 2m-1,\ \gamma).$$

together with all the pairs of elements not included in one of these triples. No matrix representation is possible, for any attempt to construct one yields a matroid with the additional circuit complement $(2m, 1, \gamma)$.

These matroids fail to be matrices because of the presence of too few circuit complements. Failure is also possible for the opposite reason. Thus the plane figures (matroids) formed by finite projective geometries [5] can be represented only by matrices of elements from a finite field. Another important special case of the matrix representation of matroids is the problem of constructing a geometric realization for schematic plane configurations. Here it is well-known that a configuration (p_γ, g_π) cannot in general be realized if [6]

$$2(p+g)-p\gamma-8<0.$$

The use of geometric figures also simplifies the investigation of the conditions for the representability of individual matroids. Thus for a matroid M of rank 3 we need only find three homogeneous coördinates for each element (point) of the matroid, such that when three points lie on a line (i. e., are contained in a circuit complement of the dual matroid), then the determinant of the corresponding coördinates is zero, and conversely. This application of the usual theorems of analytic geometry can replace Whitney's Theorem 32.

4. *Representation in finite algebraic fields.* The configuration of eight elements which can be represented in the complex but not in the real plane [7] suggests that the representability of a matroid depends essentially on the field from which the elements of the representing matrix are taken. Another similar example can be constructed for the field $R(2^{1/2})$, where R is the field of rational numbers. We need only take a point with coördinates $(1, 2^{1/2}, 0)$ and carry out the constructions in the von Staudt algebra of throws corresponding to

$$(2^{1/2})(2^{1/2})=1+1.$$

The resulting figure (matroid) consists of 11 points, $1, 2, \cdots, 9, 0, \alpha$, the following sets of points being collinear:

$$1279\alpha,\ 2356,\ 1380,\ 248,\ 347,\ 578,\ 549,\ 690,\ 50\alpha,\ 68\alpha.$$

[5] Veblen and Young, *Projective Geometry*, vol. I, p. 3 and p. 201.

[6] E. Steinitz, *Encyklopädie der mathematischen Wiss.*, III A B 5a, p. 485.

[7] F. Levi, *loc. cit.*, pp. 98-102.

Any attempt to represent this matroid by a matrix leads to a matrix whose elements generate a field containing $R(2^{1/2})$. This matroid is thus a sort of geometric analog of the irreducible equation for $2^{1/2}$. Generalization yields:

THEOREM 3. *Let K be a finite algebraic field over the field of rational numbers. Then there exists a matroid M of rank 3 which can be represented by a matrix with elements in K, while any other representation of M by a matrix with elements in a number-field K_1 requires $K_1 \supset K$.*

Such finite fields are sufficient for the representation of all representable matroids, in the following sense.

THEOREM 4. *Let the matroid M be representable by a matrix of complex numbers. Then M can also be represented by a matrix with elements from an algebraic field of finite degree.*

For let the matroid M have rank n and consist of p points, and let these points be assigned the indeterminate homogeneous coördinates a_{ij}, for $i = 1, \cdots, n$; $j = 1, \cdots, p$. Each circuit complement of the dual of M requires the vanishing of a number of determinants of the a_{ij}, and thus corresponds to a number of algebraic equations for these quantities. The set of all values of the a_{ij} giving at least the required circuit complements thus constitutes an algebraic manifold N_1 in the np-dimensional space of all coördinates a_{ij}. The set of those coördinates yielding additional undesired circuit complements forms another manifold N_2. Since the original matrix was representable, there exists a point of $N_1 - N_2$. The parametric representation of irreducible algebraic manifolds[8] thus makes possible the construction of a point with algebraic coördinates in $N_1 - N_2$.

HARVARD UNIVERSITY.

[8] B. L. van der Waerden, *Moderne Algebra*, vol. 2, p. 51 ff.

Reprinted from *American Journal of Mathematics*,
volume 57, pages 800 through 804,
by permission of the Johns Hopkins University Press.

A HOMOTOPY THEOREM FOR MATROIDS, I

BY
W. T. TUTTE

1. **Introduction.** By a *matroid* on a finite set M we understand a class $\boldsymbol{M}$ of non-null subsets of M which satisfies the following axioms.

AXIOM I. *No member of* $\boldsymbol{M}$ *contains another as a proper subset.*

AXIOM II. *If* $(X, Y) \in \boldsymbol{M}$, $a \in X \cap Y$ *and* $b \in X - (X \cap Y)$, *then there exists* $Z \in \boldsymbol{M}$ *such that* $b \in Z \subseteq (X \cup Y) - \{a\}$.

Such systems were introduced by Hassler Whitney [1].

As an example let $\boldsymbol{L}$ be any class of subsets of M forming a group under mod 2 addition, and let $\boldsymbol{M}$ be the class of all minimal non-null members of $\boldsymbol{L}$. Then it is easily verified that $\boldsymbol{L}$ satisfies Axiom II and that each non-null member of $\boldsymbol{L}$ is a sum of non-null members of $\boldsymbol{M}$. It follows that $\boldsymbol{M}$ satisfies both axioms and is thus a matroid. Such a matroid we call *binary*.

In particular $\boldsymbol{M}$ may be the set of edges of a finite graph G and $\boldsymbol{L}$ may be the class of 1-cycles mod 2 of G. Then it is found that the members of $\boldsymbol{M}$ are those sets of edges of G which define circuits. In this case we call $\boldsymbol{M}$ the *circuit-matroid* of G.

Given a matroid $\boldsymbol{M}$ let $\boldsymbol{Q}$ be the class of all unions of members of $\boldsymbol{M}$. Then each element of $\boldsymbol{Q}$ is a subset of M. We partition $\boldsymbol{Q}$ into disjoint classes $\boldsymbol{Q}_{-1}, \boldsymbol{Q}_0, \boldsymbol{Q}_1, \boldsymbol{Q}_2, \cdots$ according to the following rules.

(i) *The null subset* $\emptyset$ *of* M, *considered as an empty union, is the only member of* $\boldsymbol{Q}_{-1}$.

(ii) *When* $\boldsymbol{Q}_r$ *has been determined for* $-1 \leqq r \leqq k$ *we define* $\boldsymbol{Q}_{k+1}$ *as the class of all minimal members of*

$$\boldsymbol{P}_k = \boldsymbol{Q} - \bigcup_{r=-1}^{k} \boldsymbol{Q}_r.$$

That is $\boldsymbol{Q}_{k+1}$ consists of all members of $\boldsymbol{P}_k$ which have no other members of $\boldsymbol{P}_k$ as a subset. The members of $\boldsymbol{Q}$ are the *flats* of $\boldsymbol{M}$. Those belonging to $\boldsymbol{Q}_d$ are the flats of *dimension* d, or *d-flats*.

At the end of §2 of this paper we interpret the dimensions of the flats of a circuit-matroid in terms of graph theory.

We shall see that the flats of a matroid $\boldsymbol{M}$ on a set M have some properties resembling those of the elements of a projective geometry. Because of this analogy we refer to the 0-flats, 1-flats and 2-flats of $\boldsymbol{M}$ as its *points*, *lines* and *planes* respectively. The points are simply the members of the class $\boldsymbol{M}$.

We have to recognize one distinction which has no analogue in projective

Received by the editors October 26, 1956.

geometry. A flat F is *disconnected* if it can be represented as the union of two disjoint non-null subsets F' and F'' of M such that each point X of $\boldsymbol{M}$ satisfying $X \subseteq F$ satisfies also either $X \subseteq F'$ or $X \subseteq F''$. If no such representation is possible then F is *connected*. Thus the points of $\boldsymbol{M}$ and its (-1)-flat are connected.

A *path* in $\boldsymbol{M}$ is a finite sequence $P=(X_1, \cdots, X_k)$ of one or more points of $\boldsymbol{M}$, not necessarily all distinct, such that any two consecutive terms are distinct points of $\boldsymbol{M}$ which are subsets of the same connected line. The first and last terms of P are its *origin* and *terminus* respectively. If they are the same point we call P *re-entrant*. If P has only one term we call it *degenerate*.

If $P=(X_1, \cdots, X_k)$ and $P'=(X_k, \cdots, X_m)$ are paths of $\boldsymbol{M}$ such that the origin of P' is the terminus of P then we define their *product* PP' as the path $(X_1, \cdots, X_k, \cdots, X_m)$. Multiplication of paths is clearly associative. It is therefore permissible to write a path $(PQ)R$ or $P(QR)$ simply as PQR.

Suppose we have two paths PR and PQR where Q is either (i) of the form (X, Y, X) or (ii) of the form (X, Y, Z, X) with X, Y and Z subsets of the same plane. Then we say that each of PR and PQR can be derived from the other by an *elementary deformation*. Two paths P_1 and P_2 are *homotopic* if they are identical or if one can be derived from the other by a finite sequence of elementary deformations. Homotopy is clearly an equivalence relation. A path homotopic to a degenerate path is said to be *null-homotopic*.

In this paper we show that every re-entrant path in a matroid is null-homotopic. Actually we prove a more general theorem, as the result just stated is not sufficient for the purposes of Paper II. We first agree to call a subclass $\boldsymbol{C}$ of $\boldsymbol{M}$ *convex* if it has the following property: if two distinct members X and Y of $\boldsymbol{C}$ are subsets of the same line L then every point of $\boldsymbol{M}$ which is a subset of L is a member of $\boldsymbol{C}$. Given a convex subclass $\boldsymbol{C}$ of $\boldsymbol{M}$ we say that a path P is *off* $\boldsymbol{C}$ if no term of P is a point of $\boldsymbol{C}$. We then enquire into the condition that a path P off $\boldsymbol{C}$ can be transformed into a degenerate path by a finite sequence of elementary deformations so that all the intermediate paths are off $\boldsymbol{C}$. In this paper we show how the idea of an elementary deformation must be generalized so as to make this transformation possible for every re-entrant path P off $\boldsymbol{C}$.

It is hoped that the technique here developed for the study of matroids will be found useful in graph theory when applied to the circuit-matroids of graphs.

2. **Flats.** Let $\boldsymbol{M}$ be a matroid on a set M. We refer to the elements of M as the *cells* of the matroid. If S and T are subsets of M we use the symbol $S \subset T$ to denote that S is a proper subset of T. We write $\langle S \rangle$ for the union of all the points of $\boldsymbol{M}$ which are subsets of S. If S is a flat of $\boldsymbol{M}$ we denote its dimension by dS.

(2.1) *If S is a flat of $\boldsymbol{M}$ and k is an integer satisfying $-1 \leqq k < dS$ then there exists a flat T of $\boldsymbol{M}$ such that $dT=k$ and $T \subset S$.*

Proof. If the theorem fails let k be the greatest integer satisfying -1

$\leqq k < dS$ such that no k-flat T of $\boldsymbol{M}$ satisfies $T \subset S$. Clearly $k > -1$ and therefore $dS \geqq 1$.

By the definition of k there exists a $(k+1)$-flat T' of $\boldsymbol{M}$ such that $T' \subseteq S$. By the definition of the classes $\boldsymbol{Q}_r$ we have

$$T' \in \boldsymbol{Q} - \bigcup_{r=-1}^{k} \boldsymbol{Q}_r \subset \boldsymbol{Q} - \bigcup_{r=-1}^{k-1} \boldsymbol{Q}_r = \boldsymbol{P}_{k-1}.$$

But T' is not a minimal member of $\boldsymbol{P}_{k-1}$, since dT' is not k. Hence there exists a minimal member T of $\boldsymbol{P}_{k-1}$ such that $T \subset T'$. But then $dT = k$. Since $T' \subseteq S$ this contradicts the definition of k. The theorem follows.

(2.2) *If S and T are flats of $\boldsymbol{M}$ such that $S \subset T$, then $dS < dT$.*

Proof. Since T is non-null we have $dT > -1$. If $dS \geqq dT$ there is a flat U of $\boldsymbol{M}$ such that $U \subseteq S$ and $dU = dT$, by (2.1). But then T is not a minimal member of $\boldsymbol{Q}_{dT}$, contrary to the definition of this class.

It follows from (2.2) that $\langle M \rangle$ has a greater dimension than any other flat of $\boldsymbol{M}$.

It is convenient to say that a flat S is *on* a flat T if either $S \subseteq T$ or $T \subseteq S$. If S and T are distinct we can distinguish between the two cases by comparing dimensions.

(2.3) *If S is a flat of $\boldsymbol{M}$ and $a \in S$, then $d\langle S - \{a\}\rangle = dS - 1$.*

Proof. If possible choose S and a so that $d\langle S - \{a\}\rangle \neq dS - 1$ and so that dS has the least value consistent with this.

By (2.1) there is a flat T of $\boldsymbol{M}$ such that $T \subseteq S$ and $dT = dS - 1$. Choose $b \in S - T$. Then $T \subseteq \langle S - \{b\}\rangle \subset S$. Hence $d\langle S - \{b\}\rangle = dS - 1$, by (2.2).

Suppose $a \notin \langle S - \{b\}\rangle$. Then $\langle S - \{b\}\rangle \subseteq \langle S - \{a\}\rangle \subset S$. Hence $d\langle S - \{a\}\rangle = dS - 1$, by (2.2). But this is contrary to the choice of S and a.

We deduce that $a \in \langle S - \{b\}\rangle$. Hence there exists $X \in \boldsymbol{M}$ such that $X \subseteq S$, $a \in X$ and $b \notin X$. Since $b \in S$ there exists $Y \in \boldsymbol{M}$ such that $Y \subseteq S$ and $b \in Y$. It follows by Axiom II that there exists $Z \in \boldsymbol{M}$ such that $Z \subseteq S$, $a \notin Z$ and $b \in Z$. ($Z = Y$ if $a \notin Y$). These results imply

$$\langle\langle S - \{a\}\rangle - \{b\}\rangle \subset \langle S - \{a\}\rangle \subset S, \tag{2.3a}$$

$$\langle\langle S - \{b\}\rangle - \{a\}\rangle \subset \langle S - \{b\}\rangle \subset S. \tag{2.3b}$$

We also have

$$\langle\langle S - \{a\}\rangle - \{b\}\rangle = \langle\langle S - \{b\}\rangle - \{a\}\rangle, \tag{2.3c}$$

since each side of this equation represents the union of those points of $\boldsymbol{M}$ which include neither a nor b.

Since $d\langle S - \{b\}\rangle = dS - 1$ it follows from (2.3b) and the choice of S and a that $d\langle\langle S - \{b\}\rangle - \{a\}\rangle = dS - 2$. Hence $d\langle S - \{a\}\rangle = dS - 1$, by (2.2), (2.3a) and (2.3c). This contradiction establishes the theorem.

(2.4) *Let S and T be flats of $\boldsymbol{M}$ such that $S \subseteq T$. Then there exists a flat U of $\boldsymbol{M}$ such that $U \subseteq T$, $\langle U \cap S\rangle = \varnothing$ and $dU = dT - dS - 1$.*

Proof. Write $S_0=S$, $T_0=T$. If possible choose $a_0\in S_0$ and write $S_1=\langle S_0-\{a_0\}\rangle$, $T_1=\langle T_0-\{a_0\}\rangle$. Observe that $S_1\subseteq T_1$. If possible choose $a_1\in S_1$ and write $S_2=\langle S_1-\{a_1\}\rangle$, $T_2=\langle T_1-\{a_1\}\rangle$. Then $S_2\subseteq T_2$. Continue this process until it terminates. By (2.3) this will be with S_k and T_k, where $k=dS+1$ and $S_k=\varnothing$. Applying (2.3) to the sequence of the T_i we find that $dT_k=dT-k=dT-dS-1$. We note that $\langle T_k\cap S\rangle\subseteq\langle S-\{a_0,\cdots,a_{k-1}\}\rangle=S_k=\varnothing$. Hence the theorem is satisfied with $U=T_k$.

(2.5) *If S and T are any flats of* $\boldsymbol{M}$ *then* $d(S\cup T)+d\langle S\cap T\rangle\geqq dS+dT$.

Proof. Write $S_0=S$. If possible choose $a_0\in S_0-(S_0\cap T)$ and write $S_1=\langle S_0-\{a_0\}\rangle$. If possible choose $a_1\in S_1-(S_1\cap T)$ and write $S_2=\langle S_1-\{a_1\}\rangle$, and so on. By (2.3) the process terminates with $S_k=\langle S\cap T\rangle$, where $k=dS-d\langle S\cap T\rangle$. We now have $T=S_k\cup T\subset S_{k-1}\cup T\subset\cdots\subset S_0\cup T=S\cup T$. Hence $d(S\cup T)-dT\geqq k=dS-d\langle S\cap T\rangle$, by (2.2).

Many "geometrical" results can be deduced from (2.2) and (2.5). For example any two distinct lines L_1 and L_2 on a plane P have a unique common point. To prove this we first use (2.2) to show that $d\langle L_1\cap L_2\rangle<dL_1=1$ and $L_1\cup L_2=P$. Then $d\langle L_1\cap L_2\rangle\geqq 0$, by (2.5). Hence $d\langle L_1\cap L_2\rangle=0$ and $\langle L_1\cap L_2\rangle$ is a single point of $\boldsymbol{M}$. We can prove in the same way that if P_1 and P_2 are distinct planes on the same 3-flat E of $\boldsymbol{M}$, then $\langle P_1\cap P_2\rangle$ is a line on E. Similarly if P is a plane and L a line on the same 3-flat E, and L is not on P, then $\langle P\cap L\rangle$ is a point on E.

Not all the axioms of projective geometry are valid for matroids. For example two points are not necessarily on a common line. In general matroids are like geometrical figures but not like complete geometries.

Suppose $\boldsymbol{M}$ is the circuit-matroid of a graph G. If $S\in\boldsymbol{Q}$ we write $G\cdot S$ for the subgraph of G made up of the edges of S and their incident vertices. We see that the flats S of $\boldsymbol{M}$ correspond to those subgraphs $G\cdot S$ in which each edge belongs to some circuit of the subgraph. In virtue of (2.3) $dS+1$ is the least number of edges which must be removed from $G\cdot S$ in order to destroy all its circuits, that is $dS+1$ is the rank or first Betti number of $G\cdot S$. The subgraph $G\cdot S$ is nonseparable if and only if the flat S is connected.

3. Connected flats. We begin this section with a study of the line.

(3.1) *Any line L of* $\boldsymbol{M}$ *is on at least two points. If X and Y are distinct points on L then* $L=X\cup Y$. *Moreover* $X\cap Y$ *is non-null if and only if L is connected.*

Proof. Choose $a\in L$. Then $\langle L-\{a\}\rangle$ is a point on L, by (2.3). Choose $b\in\langle L-\{a\}\rangle$. Then $\langle L-\{b\}\rangle$ is a point on L, by (2.3), which is distinct from $\langle L-\{a\}\rangle$.

Let X and Y be distinct points on L. Then $X\subset X\cup Y\subseteq L$. Hence $X\cup Y=L$, by (2.2). If $X\cap Y$ is non-null then L is clearly connected. If $X\cap Y$ is null then either L is disconnected or there exists $Z\in\boldsymbol{M}$ such that $Z\subseteq X\cup Y$ and Z meets both X and Y. In the latter case $X\subset X\cup Z\subset X\cup Y=L$, by Axiom I. This is impossible, by (2.2).

(3.2) *A disconnected line is on just two points, and a connected line is on at least three points.*

Proof. By (3.1) any two distinct points on a disconnected line L are disjoint and have L as their union. Hence L has at most two points, and therefore just two by (3.1).

By (3.1) any connected line L has two distinct points X and Y, and we can find $a \in X \cap Y$. By (2.3) $\langle L - \{a\} \rangle$ is a point on L distinct from X and Y.

We shall need the following general theorems on connected flats.

(3.3) *Let S and T be connected flats of $\boldsymbol{M}$ such that $S \subset T$. Then there exists a connected $(dS+1)$-flat U of $\boldsymbol{M}$ which is on both S and T.*

Proof. Since T is connected we can find $X \in \boldsymbol{M}$ such that $X \subseteq T$ and X meets both S and $T-S$. Choose such an X so that $S \cup X$ has the least possible number of cells. Clearly $S \cup X$ is a connected flat of $\boldsymbol{M}$. Its dimension exceeds dS, by (2.2).

Suppose $d(S \cup X) > dS+1$. Choose $a \in (S \cup X) - S$. Then $d\langle (S \cup X) - \{a\} \rangle \geq dS+1$, by (2.3). Hence there exists $Y \in \boldsymbol{M}$ such that $Y \subseteq (S \cup X) - \{a\}$ and Y meets $(S \cup X) - S$. But $Y \cap S$ is null, by the choice of X. Hence $Y \subset X$, which is impossible by Axiom I. We deduce that $d(S \cup X) = dS+1$. Hence the theorem is true with $U = S \cup X$.

(3.4) *Let S be a connected d-flat on a connected $(d+2)$-flat T of $\boldsymbol{M}$. Then there exist distinct connected $(d+1)$-flats U and V of $\boldsymbol{M}$ such that $S = \langle U \cap V \rangle$ and $T = U \cup V$.*

Proof. By (3.3) there is a connected $(d+1)$-flat U which is on both S and T. Choose $a \in U - S$ and write $W = \langle T - \{a\} \rangle$. By (2.3) W is another $(d+1)$-flat on S and T. By (2.4) there is a line L on T having no point in common with S. It meets U and W in points X and Z respectively, by (2.5). (See Figure I.) By (2.2) we have $S \cup X = U$ and $S \cup Z = W$. Hence Z is not on U and therefore $U \cup Z = T$, by (2.2).

Assume W is not connected. Then $S \cap Z = \emptyset$.

Suppose $U \cap Z = \emptyset$. By the connection of T there exists $Z' \in \boldsymbol{M}$ such that $Z' \subseteq T$ and Z' meets both U and Z. Then $U \subset U \cup Z' \subset U \cup Z = T$, by Axiom I. This is impossible by (2.2). We deduce that $U \cap Z \neq \emptyset$. A similar argument in which X, S and U replace Z, U and T respectively shows that $X \cap S \neq \emptyset$. Choose $b \in Z \cap U$ and $c \in X \cap S$.

Write $V = \langle T - \{b\} \rangle$. By (2.3) V is a $(d+1)$-flat. It is on S since $b \in Z$ and $S \cap Z = \emptyset$. By (2.5) it has a common point Y with L. Clearly V is distinct from U and W and therefore Y is distinct from X and Z, since $V = S \cup Y$ by (2.2).

Now $c \in S \cap X \subseteq S \cap L = S \cap (Y \cup Z) = S \cap Y$, by (3.1). Hence V is connected.

If instead W is connected we write $V = W$.

We now have two connected $(d+1)$-flats U and V of $\boldsymbol{M}$ each of which is on both S and T. Hence $S \subseteq \langle U \cap V \rangle \subset U \subset U \cup V \subseteq T$, since U and V are

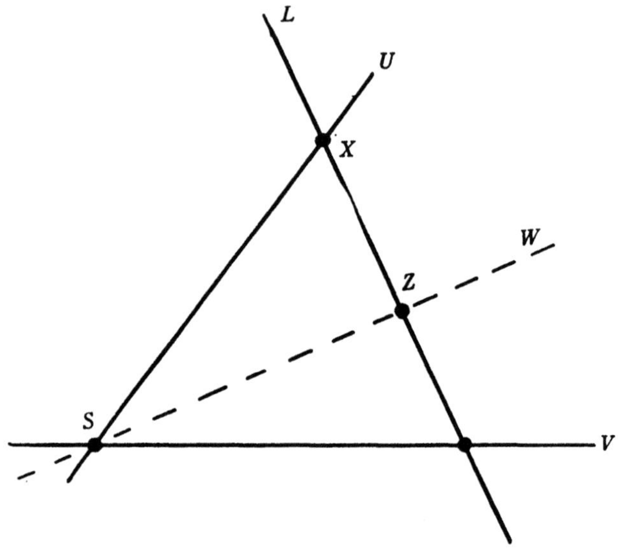

FIG. I

distinct and, by (2.2), neither is a subset of the other. In view of (2.2) this is possible only if $S=\langle U\cap V\rangle$ and $T=U\cup V$.

(3.5) *Let S, T and U be flats of $\boldsymbol{M}$ such that S and T are connected, $S\cup U\subseteq T$ and $\langle S\cap U\rangle=\emptyset$. Then there exists a connected flat R of $\boldsymbol{M}$ such that $S\subseteq R\subseteq T$, $\langle R\cap U\rangle=\emptyset$ and $dR=dT-dU-1$.*

Proof. If possible choose S, T and U so that the theorem fails and dU has the least value consistent with this. Then $dU>-1$ since otherwise the theorem holds with $T=R$. Let W be a connected flat of $\boldsymbol{M}$ of greatest possible dimension such that $S\subseteq W\subseteq T$ and W does not contain U. Then $dW=dT-1$ since otherwise, by (3.3) and (3.4), there exist distinct connected $(dW+1)$-flats K and L of $\boldsymbol{M}$ on T such that $\langle K\cap L\rangle=W$, and these cannot both contain U. By the choice of S, T and U there is a connected flat R of $\boldsymbol{M}$ such that $S\subseteq R\subseteq W\subset T$, $\langle R\cap U\rangle=\emptyset$ and $dR=dW-d\langle U\cap W\rangle-1$. But then $dR\geqq dT-dU-1$, by (2.2), and therefore $dR=dT-dU-1$, by (2.2) and (2.5). This contradiction establishes the theorem.

The foregoing results can be applied to circuit-matroids to obtain rather simple theorems about graphs. Thus from (3.5) with $S=\emptyset$ we find that if a nonseparable graph G has rank r and a subgraph $G\cdot U$ has a rank s then there is a nonseparable subgraph $G\cdot R$ of G of rank $r-s$ having no circuit in common with $G\cdot U$.

4. The disconnected line. By a *separation* $\{S_1, S_2\}$ of a disconnected flat S of $\boldsymbol{M}$ we mean a pair of complementary non-null subsets of S such that each $X\in\boldsymbol{M}$ satisfying $X\subseteq S$ satisfies either $X\subseteq S_1$ or $X\subseteq S_2$.

(4.1) *If $\{S_1, S_2\}$ is a separation of a flat S of $\boldsymbol{M}$ and X_1 and X_2 are* *points of **M** such that $X_1 \subseteq S_1$ and $X_2 \subseteq S_2$, then $X_1 \cup X_2$ is a disconnected line of **M**.*

Proof. Suppose Y is a point on $X_1 \cup X_2$ distinct from X_1 and X_2. Since $Y \subseteq S$ we have $Y \subseteq S_1$ or $Y \subseteq S_2$. Hence $Y \subset X_1$ or $Y \subset X_2$, contrary to Axiom I. Thus the only subsets of $X_1 \cup X_2$ which are flats of **M** are $\emptyset$, X_1, X_2 and $X_1 \cup X_2$. Hence, by the definition of dimension, $X_1 \cup X_2$ is a line of M having $\{X_1, X_2\}$ as a separation.

(4.2) *Let L be a disconnected line on a connected d-flat S of **M**, where $dS > 1$. Then there exists a connected plane P of M such that $L \subset P \subseteq S$.*

Proof. Let the two points on L be X and Y. Let P be a connected flat of **M** of least possible dimension such that $L \subset P \subseteq S$. Assume $dP > 2$.

Suppose first that there is a disconnected line L', distinct from L, on X and P. Let its point other than X be Z. By (3.5) there is a connected $(dP-2)$-flat U on Y and P having no point in common with L'. By (3.4) there are distinct connected $(dP-1)$-flats V and W of **M** on P such that $\langle V \cap W \rangle = U$. By (2.2) and (2.5) V and W meet L' in distinct points. Since there are only two points on L' we may suppose X is on V. But then L is on V and the definition of P is contradicted. A similar argument applies if there is a disconnected line distinct from L on Y and P.

In the remaining case we choose $a \in P - L$ and write $R = \langle P - \{a\} \rangle$. Then $L \subseteq R$. Moreover $dR = dP - 1$, by (2.3). By the definition of P the flat R is disconnected. But there is no disconnected line on R, other than L, which is on either X or Y. Hence, by (4.1), the only possible separation of R is $\{X, Y\}$. Accordingly $R = L$ and $dP = 2$, contrary to assumption. From this contradiction we deduce that P is a plane.

(4.3) *Let L be a disconnected line on a connected plane P of **M**. Let X and Y be the two points of L and let Z be any other point on P. Then $X \cup Z$ and $Y \cup Z$ are connected lines. Moreover they are the only lines of **M** which are on both Z and P.*

Proof. Any line on Z and P has a common point with the line $X \cup Y$. Hence, by (3.1), the only flats on Z and P which can be lines are $X \cup Z$ and $Y \cup Z$. By (3.4) both these flats must be connected lines.

(4.4) *Let L be a disconnected line on a connected plane P of **M**. Then every line on P other than L is connected.*

Proof. Let L' be any such line. By (3.1) it is on a point Z distinct from X and Y. Hence, by (4.3) it is one of the connected lines $X \cup Z$ and $Y \cup Z$.

5. **Convex subclasses.** Convex subclasses of a matroid **M** were defined in the Introduction. As an example we may take the class of all points of **M** on a given flat. The convexity of this class follows from (3.1).

Consider any path $P = (X_1, X_2, \cdots, X_k)$ of **M**. We say P is a path *from X_1 to X_k*. Any two consecutive terms of P have a non-null intersection, by (3.1). Hence the flat $X_1 \cup X_2 \cup \cdots \cup X_k$ is connected. We denote this flat by $F(P)$. If S is any flat of **M** such that $F(P) \subseteq S$ we say that P is a path *on S*.

(5.1) *Let **C** be any convex subclass of **M**. Let S be a non-null connected flat* *of **M** and let X and Y be points on S such that $Y \notin \mathbf{C}$. Then there exists a path P from X to Y on S such that no term of P other than the first is a point of* **C**.

Proof. If possible choose S, X and Y so that the theorem fails and dS has the least value consistent with this. Clearly $dS>1$. By (3.3) and (3.4) there is a connected $(dS-2)$-flat U and two distinct connected $(dS-1)$-flats V and W on S such that $X \subseteq U = \langle V \cap W \rangle$. Now Y is not on V or W, for otherwise there would be a path from X to Y on V or W of the kind required. By (3.5) there is a connected line L on S and Y such that $\langle L \cap U \rangle = \emptyset$. This meets V and W in distinct points $Z(V)$ and $Z(W)$ respectively, by (2.2) and (2.5). At least one of these, say $Z(V)$, belongs to $\mathbf{M}-\mathbf{C}$ since **C** is convex. By the choice of S, X and Y there is a path Q from X to $Z(V)$ on V such that no term of Q other than the first is a point of **C**. Adjoining Y to Q we obtain a path P from X to Y on S of the kind required. This contradiction establishes the theorem.

We now distinguish four kinds of re-entrant paths of **M** as *elementary* with respect to a given convex subclass **C** of **M**. The first kind consists of all paths off **C** of the form (X, Y, X). The second consists of all paths off **C** of the form (X, Y, Z, X) such that $d(X \cup Y \cup Z) \leqq 2$.

Suppose P is a plane of **M** on which there are two distinct points A and B of **C** such that each connected line on P is on either A or B. Then any path off **C** on P of the form (X, Y, Z, T, X) such that X, Y, Z and T are distinct, the lines $X \cup Y$ and $Z \cup T$ are on A, and the lines $Y \cup Z$ and $T \cup X$ are on B is an elementary re-entrant path of the third kind with respect to **C**.

Suppose E is a 3-flat of **M** on which there are three points A, B and C such that $A \cup B$, $B \cup C$ and $C \cup A$ are disconnected lines. Let there be just six connected planes on E, two on each of these disconnected lines. Suppose A, B and C are all in $\mathbf{M}-\mathbf{C}$ but there are two distinct members of **C** on each of the six connected planes. Then any path off **C** of the form (A, X, B, Y, A), where X and Y are on distinct connected planes on $A \cup B$ and E, is an elementary re-entrant path of the fourth kind with respect to **C**.

In studying the preceding case it is convenient to use the following notation. We write Z_1, Z_2 and Z_3 for A, B and C. We enumerate the six connected planes as $P_1, \cdots, P_6$ in such a way that $\langle P_i \cap P_{i+3} \rangle = Z_j \cup Z_k$, where $1 \leqq i \leqq 3$ and (i, j, k) is a permutation of $(1, 2, 3)$. In general we write $\langle P_i \cap P_j \rangle = L_{ij}$ for $1 \leqq i < j \leqq 6$. If $j = i+3$ then L_{ij} is the disconnected line $(Z_1 \cup Z_2 \cup Z_3) - Z_i$. If $j \neq i+3$ let k be that integer 1, 2 or 3 which is not congruent to i or j mod 3. Then L_{ij} is on Z_k and it meets P_k and P_{k+3} in two distinct points. It is therefore connected, by (3.2). Clearly it is on no connected plane on E other than P_i and P_j. The 12 lines L_{ij}, $j \neq i+3$, are the only connected lines on E, for by (3.4) any connected line on E is on two distinct connected planes on E.

We write $\langle P_i \cup P_j \cup P_k \rangle = X_{ijk}$ for $1 \leqq i < j < k \leqq 6$. Then X_{ijk} is a point of **M**, being identical with $\langle L_{ij} \cap P_k \rangle$. If two of the suffices i, j and k are congruent mod 3 then X_{ijk} is one of the points Z_1, Z_2 and Z_3. The remaining eight points X_{ijk} are all distinct, for on any one of them there can be only three

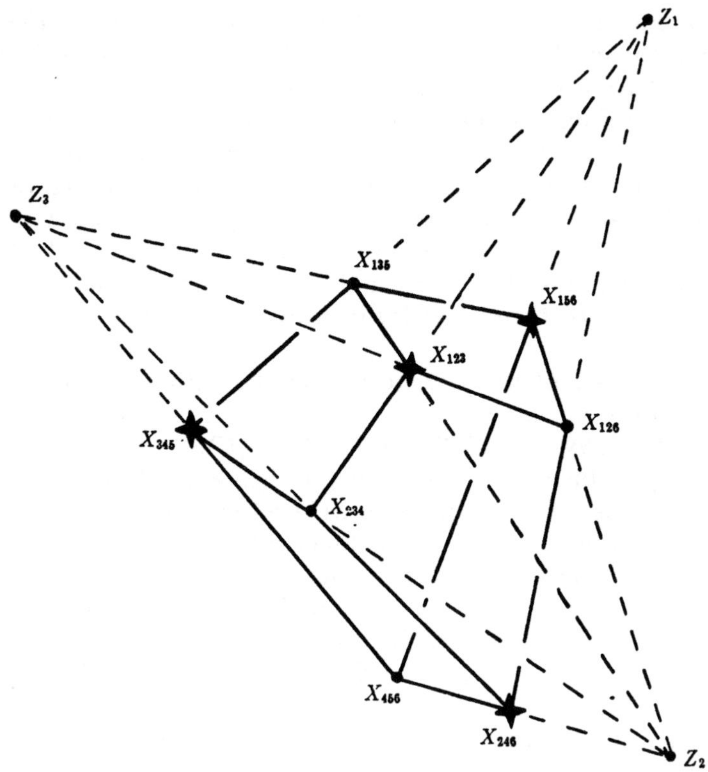

FIG. II

distinct planes such that each is on one of the lines $Z_1 \cup Z_2$, $Z_2 \cup Z_3$ and $Z_3 \cup Z_1$. These eight points, together with Z_1, Z_2 and Z_3, are the only points on E. For any point on E is on three distinct connected planes on E, by two applications of (3.4). (See Figure II.)

Consider the plane P_1. The only points on it are Z_2, Z_3, X_{123}, X_{126}, X_{135} and X_{156}. We may adjust the notation so that $X_{123} \in \mathbf{C}$. The other point of $\mathbf{C}$ on P_1 can have no common connected line with X_{123} and must therefore be X_{156}. We now find that $X_{246} \in \mathbf{C}$ since this is the only point on P_2 having no common connected line with X_{123}. Proceeding in this way we find that $X_{ijk} \in \mathbf{C}$ if and only if no two of the suffices are congruent mod 3 and the number of suffices less than 4 is odd. In Figures II, III and IV we represent points of $\mathbf{C}$ by four-pointed stars.

To construct a matroid having the structure just described we may use a method based on (2.3). We take M to be a set of six cells in 1-1 correspondence with the planes P_i. Any point X_{ijk} is represented by the set of those cells not corresponding to planes on X_{ijk}.

Suppose we have two paths PQR and PR off $\mathbf{C}$, where Q is an elementary re-entrant path of the kth kind with respect to $\mathbf{C}$. Then we call the process of deriving one of the paths PQR and PR from the other an *elementary deformation* of the kth kind with respect to **C**. We say that two given paths P' and P'' off **C** are *homotopic* with respect to **C** (written $P' \sim P''$ (**C**)) if they are identical or if one can be derived from the other by a finite sequence of elementary deformations with respect to **C**. Homotopy with respect to **C** is an equivalence relation. A path P homotopic to a degenerate path with respect to **C** is said to be *null-homotopic* with respect to **C** (written $P \sim 0$ (**C**)).

The null subset of $\boldsymbol{M}$ is clearly convex. If **C** is null we have only elementary deformations of the first and second kind to consider, and homotopy with respect to **C** becomes identical with the homotopy defined in the Introduction.

If P is any path of $\boldsymbol{M}$ we write P^{-1} for the path obtained by taking the terms of P in reverse order.

(5.2) *If P is any path off* **C** *then* $PP^{-1} \sim 0$ (**C**).

Proof. If possible choose P so that the theorem fails and P has the least number s of terms consistent with this. If $s>1$ we can write $P=QR$, where Q and R have each fewer than s terms. Since RR^{-1} and QQ^{-1} can be converted into degenerate paths by elementary deformations we have $PP^{-1}=QRR^{-1}Q^{-1} \sim QQ^{-1} \sim 0$ (**C**). If $s=1$ then PP^{-1} is an elementary re-entrant path of the first kind, and so $PP^{-1} \sim 0$ (**C**). The theorem follows.

(5.3) *If PUR and PVR are paths off* **C** *such that* $UV^{-1} \sim 0$ (**C**), *then* $PUR \sim PVR$ (**C**).

Proof. By (5.2) we have $V^{-1}V = V^{-1}(V^{-1})^{-1} \sim 0$ (**C**). Hence $PUR \sim PUV^{-1}VR \sim PVR$ (**C**).

(5.4) *Let* **C** *be any convex subclass of a matroid* $\boldsymbol{M}$. *Let S be a d-flat of* $\boldsymbol{M}$ *on a $(d+1)$-flat T of* $\boldsymbol{M}$. *Suppose all the points on S and at least one other point on T are members of* **C**. *Then all the points on T are members of* **C**.

Proof. Suppose the theorem false. Then we can find points $X \in \mathbf{C}$ and $Y \notin \mathbf{C}$, both on T but not on S. The flat $X \cup Y$ is connected since otherwise $S \subset S \cup X \subset T$, contrary to (2.2). By (5.1) there is a path from X to Y on $X \cup Y$ whose second term, X' say, is not a member of **C**. By (2.5) the line $X \cup X'$ has a point X'' in common with S. But $X'' \notin \mathbf{C}$, by the definition of a convex subclass. This is contrary to hypothesis.

6. **Proof of the main theorem.**

(6.1) *Let* **C** *be any convex subclass of a matroid* $\boldsymbol{M}$ *and let P be any re-entrant path of* $\boldsymbol{M}$ *off* **C**. *Then* $P \sim 0$ (**C**).

Proof. Assume the theorem false. Let P be any re-entrant path off **C** which is not null-homotopic with respect to **C**, and for which $dF(P)$ has the least value, n say, consistent with this condition. For an arbitrary path Q of $\boldsymbol{M}$ we call $dF(Q)$ the *dimension* of Q.

By far the most difficult part of the proof is that covered by the following lemma.

LEMMA. *Suppose $n \geqq 3$. Let $Q=(W, X, Y, Z, W)$ be a path off* **C** *of dimension*

n such that $W \cup X \cup Y$ and $Y \cup Z \cup W$ are connected planes and $W \cup Y$ is a disconnected line. Then $Q \sim 0$ (**C**).

Proof. Write $F_1 = W \cup X \cup Y$ and $F_2 = Y \cup Z \cup W$.

We note that if $Q' = (W, X', Y, Z', W)$ is a path off **C** such that X' is on F_1 and Y' on F_2, then

(6.1a) $$Q' \sim Q \text{ } (\mathbf{C}).$$

For

$$Q' \sim (W, X', Y)(Y, X, W)(W, X, Y)(Y, Z, W)(W, Z, Y)(Y, Z', W) \text{ } (\mathbf{C})$$

by (5.2). But $(W, X', Y)(Y, X, W)$ and $(W, Z, Y)(Y, Z', W)$ are re-entrant paths off **C** of dimension $<n$ and are therefore null-homotopic with respect to **C**. Hence $Q' \sim (W, X, Y)(Y, Z, W) = Q(\mathbf{C})$.

A *transversal of dimension* $n-1$ is a connected $(n-1)$-flat of M which is on $F(Q)$ but not on both W and Y. By (2.2) and (2.5) such a transversal meets each of F_1 and F_2 in a line. These two lines are connected, by (4.4).

A *transversal of dimension* $n-2$ is a connected $(n-2)$-flat of M which is on $F(Q)$ but not on W or Y. By (2.2) and (2.5) the transversal has just one point in common with each of F_1 and F_2. We call these two points the *poles* of the transversal.

Let B be any transversal of dimension $n-2$, with poles X' on F_1 and Z' on F_2. Then B is on two distinct connected $(n-1)$-flats of M on $F(Q)$, by (3.4). Using (2.5) we find that each of these is on one, but not both, of W and Y. Hence, by (2.2) they are $B \cup W$ and $B \cup Y$. They are transversals of dimension $n-1$. The flats $X' \cup W$, $X' \cup Y$, $Z' \cup W$ and $Z' \cup Y$ are their connected lines of intersection with F_1 and F_2. We note that a path (W, X', Y, Z', W) exists.

Assume that Q is not null-homotopic with respect to **C**.

Suppose B is a transversal of dimension $n-2$ with poles X' on F_1 and Z' on F_2. Suppose further that neither X' nor Z' belongs to **C**. Then, by (5.1) there is a path R off **C** from X' to Z' on B. Now $(W, X')R(Z', W)$ and $(X', Y, Z')R^{-1}$ are paths on the $(n-1)$-flats $B \cup W$ and $B \cup Y$ respectively. Hence their dimensions are less than n and so they are null-homotopic with respect to **C**. Using (6.1a), (5.3) and (5.2) we find $Q \sim (W, X', Y, Z', W) = (W, X')(X', Y, Z')(Z', W) \sim (W, Z')R^{-1}R(Z', W) \sim 0$ (**C**). This is contrary to assumption. We deduce that each transversal of dimension $n-2$ has at least one pole in **C**.

By (3.5) there is a transversal A of dimension $n-1$ which is not on Y. Let its lines of intersection with F_1 and F_2 be L_1 and L_2 respectively. They are connected lines on W. By (3.2) there is a point X' of $M - \mathbf{C}$ other than W on L_1. By (3.5) there is a connected $(n-2)$-flat B of M which is on A and X' but not on W. Now B is a transversal of dimension $n-2$. Let its pole on L_2 be U_2. Then $U_2 \in \mathbf{C}$. Similarly there is a transversal B' of dimension $n-2$ on A having a point Z' of $M - \mathbf{C}$ as its pole on L_2 and a point U_1 of **C**

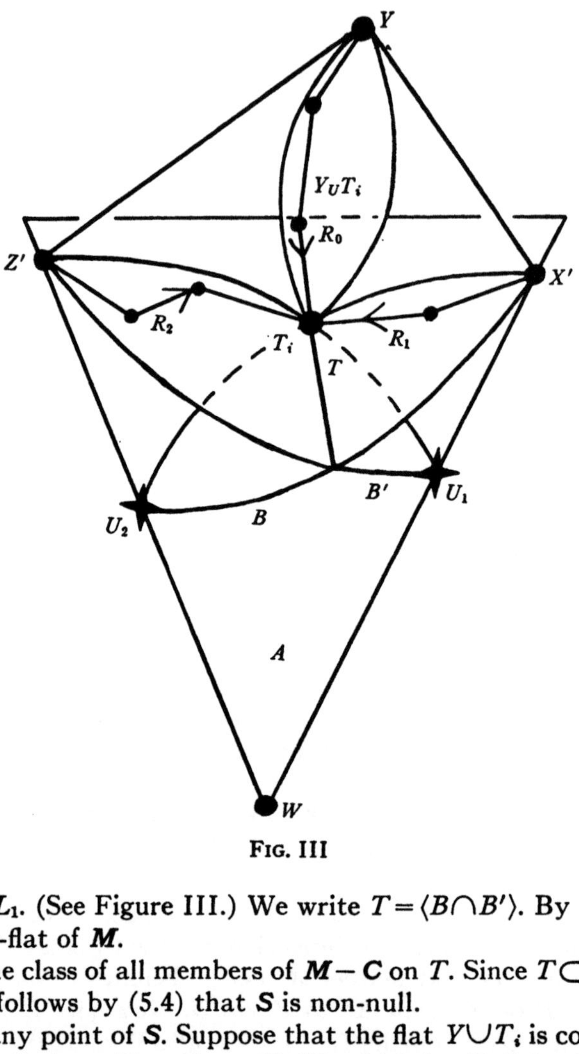

FIG. III

as its pole on L_1. (See Figure III.) We write $T = \langle B \cap B' \rangle$. By (2.2) and (2.5) T is an $(n-3)$-flat of $\boldsymbol{M}$.

Let $\boldsymbol{S}$ be the class of all members of $\boldsymbol{M} - \boldsymbol{C}$ on T. Since $T \subset B$, $X' \in \boldsymbol{M} - \boldsymbol{C}$ and $U_2 \in \boldsymbol{C}$ it follows by (5.4) that $\boldsymbol{S}$ is non-null.

Let T_i be any point of $\boldsymbol{S}$. Suppose that the flat $Y \cup T_i$ is connected. Then there is a path R_0 from Y to T_i on $Y \cup T_i$ which is off $\boldsymbol{C}$, by (5.1). Similarly there is a path R_1 from X' to T_i on B and a path R_2 from Z' to T_i on B', both R_1 and R_2 being off $\boldsymbol{C}$. Now $(X', Y)R_0R_1^{-1}$ is a re-entrant path on the transversal $B \cup Y$ of dimension $n-1$, and $(Y, Z')R_2R_0^{-1}$ is a re-entrant path on the transversal $B' \cup Y$ of dimension $n-1$. Hence both these paths are null-homotopic with respect to $\boldsymbol{C}$. Applying (6.1a), (5.3) and (5.2) we find $Q \sim (W, X', Y, Z', W) = (W, X')(X', Y)(Y, Z')(Z', W) \sim (W, X')R_1R_0^{-1}R_0R_2^{-1} \cdot (Z', W) \sim (W, X')R_1R_2^{-1}(Z', W)$ $(\boldsymbol{C})$. But the last path is on the $(n-1)$-flat A and is therefore null-homotopic with respect to $\boldsymbol{C}$. Hence $Q \sim 0$ $(\boldsymbol{C})$, contrary to assumption. We deduce that $Y \cup T_i$ has a separation $\{Y, T_i\}$. That is $Y \cup T_i$ is a disconnected line, and Y and T_i are the two points on it.

We can repeat the above argument with $B \cup Y$ replacing A. Instead of

B' we then obtain a transversal B'' of dimension $n-2$ on $B\cup Y$ with its pole on $U_2\cup Y$ a member of $\boldsymbol{M}-\boldsymbol{C}$ and its pole on $X'\cup Y$ a member of $\boldsymbol{C}$. We denote the $(n-3)$-flat $\langle B\cap B''\rangle$ by T'. We find that any point T_j' of $\boldsymbol{M}-\boldsymbol{C}$ on T' is such that $W\cup T_j'$ is a disconnected line. But, by (2.5), B'' has a point in common with the disconnected line $Y\cup T_i$, and this point can only be T_i. Hence T_i is one of the points of $\boldsymbol{M}-\boldsymbol{C}$ on T'.

We conclude that any point T_i of $\boldsymbol{S}$ is such that $W\cup T_i$ and $Y\cup T_i$ are disconnected lines.

Let E be a connected flat of $\boldsymbol{M}$ on F_1 and $F(Q)$ which is on some point of $\boldsymbol{S}$ and has the least dimension consistent with this property. It is clear that either $F(Q)$ or one of its subsets satisfies these conditions. We have

$$n = dF(Q) \geqq dE \geqq 3, \tag{6.1b}$$

$$d\langle E\cap T\rangle \geqq dE - 3, \tag{6.1c}$$

by (2.2) and (2.5). Choose a point N on $\langle E\cap T\rangle$, taking $N\in\boldsymbol{C}$ if this is possible. By (3.5) there is a connected $(dE-1)$-flat E' of $\boldsymbol{M}$ on F_1 and E but not on N. By (2.2) and (2.5) $\langle E'\cap T\rangle$ is a $(d\langle E\cap T\rangle-1)$-flat on $\langle E\cap T\rangle$. All the points of $\boldsymbol{M}$ which are subsets of $\langle E'\cap T\rangle$ belong to $\boldsymbol{C}$, by the definition of E. By the choice of N this implies that either $d\langle E'\cap T\rangle=-1$ or $N\in\boldsymbol{C}$. But in the latter case all the points on $\langle E\cap T\rangle$ belong to $\boldsymbol{C}$, by (5.4), contrary to the definition of E. Hence $d\langle E'\cap T\rangle=-1$ and therefore $d\langle E\cap T\rangle=0$. Hence $dE=3$, by (6.1b) and (6.1c). Henceforth we use the symbol T_i to denote the single point $\langle E\cap T\rangle$, which must be in $\boldsymbol{S}$.

Suppose $n\geqq 4$. Then F_2 is not on E. By (3.5) there is a connected $(n-1)$-flat E'' of $\boldsymbol{M}$ on F_2 and $F(Q)$ but not on T_i. Write $F_3=\langle E''\cap E\rangle$. Then F_3 is a plane on E and $W\cup Y$, by (2.2) and (2.5). By (3.5) there is a connected line L on E and T_i having no common point with $W\cup Y$. Let its common points with F_1 and F_3 be X_1 and X_3 respectively. Neither of these is T_i. We note that $F_3=W\cup Y\cup X_3$, by (2.2). Now $X_1\cup W$ and $X_1\cup Y$ are connected lines by (4.3). Hence $X_1\cap W$ and $X_1\cap Y$ are both non-null, by (3.1). But we have shown that $W\cup T_i$ and $Y\cup T_i$ are disconnected lines. Hence $T_i\cap W$ and $T_i\cap Y$ are both null. But $L=X_1\cup T_i=X_3\cup T_i$, by (3.1). Hence $X_3\cap W$ and $X_3\cap Y$ are both non-null and therefore F_3 is connected.

By (5.1) there is a path R from Y to W on F_3 which is off $\boldsymbol{C}$. The re-entrant paths $(W, X, Y)R$ and $(Y, Z, W)R^{-1}$ are on E and E'' respectively and so have dimensions $<n$. Hence they are null-homotopic with respect to $\boldsymbol{C}$. Using (5.3) and (5.2) we find $Q=(W, X, Y)(Y, Z, W)\sim R^{-1}R\sim 0$ $(\boldsymbol{C})$, contrary to assumption.

We deduce, using (6.1b), that $n=3$. This implies $dT=0$. Hence T is a point of $\boldsymbol{M}$, identical with T_i and therefore a member of $\boldsymbol{M}-\boldsymbol{C}$. The three flats $W\cup Y$, $Y\cup T$ and $T\cup W$ are disconnected lines. The flat $W\cup Y\cup T$ is not a connected plane of $\boldsymbol{M}$ by (4.3).

Any plane on $F(Q)$ has a point in common with each of the disconnected lines $W \cup Y$, $Y \cup T$ and $T \cup W$. It is therefore on one of these lines. Each line of $F(Q)$ is on a plane of $F(Q)$, by (2.2) and (2.3). It follows that each line on $F(Q)$ is on one of the points W, Y and T.

Let F be any transversal of dimension 2. It meets F_1 and F_2 in connected lines L_1 and L_2 respectively. Let the points on L_1 other than W or Y be $X_1, \cdots, X_k$. By (3.4) there is a transversal B_i of dimension 1 on F and X_i for each i. The line B_i must be on T. Hence $B_i = X_i \cup T$, and B_i is uniquely determined for each i. Let X_i' denote the point of intersection of B_i and L_2. Since $B_i = T \cup X_i'$ for each i the k points $X_1', \cdots, X_k'$ are all distinct. But at most one point on each of the lines L_1 and L_2 belongs to $\mathbf{C}$, and no transversal of dimension 1 has both its poles in $\mathbf{M} - \mathbf{C}$. Applying (3.2) we deduce that $k=2$. Moreover we can adjust the notation so that $(X_1, X_2') \in \mathbf{C}$ and $(X_2, X_1') \in \mathbf{M} - \mathbf{C}$.

Distinct lines on T and F meet L_1 in distinct points, by (3.1). Hence the only connected lines on T and F are B_1 and B_2. But each point of F is on two connected lines on F, and one of these is on T. Hence X_1 and X_2' are the only points of $\mathbf{C}$ on F. We thus prove that each connected plane on $F(Q)$ not on $W \cup Y$ is on just two points of $\mathbf{C}$.

Any connected line on F_1 is on a transversal of dimension 2, by (3.4). Hence it is on just three points, one of which is in $\mathbf{C}$. Distinct lines on F_1 and W (or Y) meet a given connected line on F_1 and Y (or W) in distinct points, by (3.1). It follows that on F_1 there are just two connected lines on each of the points W and Y. As each point on F_1 is on two connected lines on F_1 we deduce that F_1 is on just two points of $\mathbf{C}$. Analogous results hold for F_2. Two distinct transversals of dimension 2 are both on T and therefore meet F_1 in distinct lines, by (2.2). Accordingly there are just two connected planes on $F(Q)$ and $T \cup W$, and just two on $F(Q)$ and $Y \cup T$ (since $W \cup Y \cup T$ is not a connected plane).

It follows from these results that either Q is an elementary re-entrant path of the fourth kind with respect to $\mathbf{C}$ or there is a third connected plane F_3 on $F(Q)$ and $W \cup Y$. The first alternative must be rejected since it implies $Q \sim 0$ ($\mathbf{C}$).

Let the points of intersection with F_3 of B_1 and B_2 be X_1'' and X_2'' respectively. These are both in $\mathbf{M} - \mathbf{C}$. Each is on two connected lines on F_3, one on W and the other on Y, by (4.3). We have $(W, X, Y, X_1'', W) = (W, X, Y) \cdot (Y, X_1'', W) \sim 0$ ($\mathbf{C}$). For otherwise we can repeat the first parts of the preceding proof with (W, X, Y, X_1'', W) replacing Q and obtain a contradiction, for the transversal B_2 then has both poles in $\mathbf{M} - \mathbf{C}$. Similarly $(Y, Z, W, X_1'', Y) = (Y, Z, W)(W, X_1'', Y) \sim 0$ ($\mathbf{C}$). Applying (5.3) and (5.2) we find $Q = (W, X, Y)(Y, Z, W) \sim (W, X_1'', Y)(Y, X_1'', W) \sim 0$ ($\mathbf{C}$), contrary to assumption. The lemma follows.

We return to the path P defined at the beginning of this proof. We note that $n = dF(P) \geqq 1$ since otherwise P would be trivially null-homotopic with respect to $\mathbf{C}$. We choose a connected $(n-1)$-flat E of $\mathbf{M}$ which is on $F(P)$

and the origin X_0 of P. This choice is possible, by (3.3).

Let $R=(X_0, \cdots, X_m, X_0)$ be any re-entrant path with the same origin as P on $F(P)$. We write $u(R)$ for the number of terms of R, counting repetitions, which are not on E. If $u(R)>0$ we write X_i for the first term of R which is not on E. We then write $v(R)=d(X_{i-1}\cup X_i\cup X_{i+1})$, taking $X_{m+1}=X_0$ if $i=m$. If $u(R)=0$ we write $v(R)=0$.

Henceforth we suppose R chosen so as to satisfy the following conditions:

(i) $R\sim P$ ($\mathbf{C}$),

(ii) $u(R)$ has the least value consistent with (i),

(iii) $v(R)$ has the least value consistent with (i) and (ii).

We consider first the case $u(R)>0$. Then $v(R)>0$. We may conveniently write R in the form $R_1(X_{i-1}, X_i, X_{i+1})R_2$, noting that R_1 is a path on . We write also $F=X_{i-1}\cup X_i\cup X_{i+1}$.

Suppose $v(R)=1$. Then F is a connected line. If $X_{i+1}=X_{i-1}$ we have $R\sim R_1R_2$ ($\mathbf{C}$), by an elementary deformation of the first kind. This is impossible since $u(R_1R_2)<u(R)$. If $X_{i+1}\neq X_{i-1}$ then $(X_{i-1}, X_i, X_{i+1}, X_{i-1})$ is an elementary re-entrant path of the second kind with respect to $\mathbf{C}$. Applying (5.3) we find $R\sim R_1(X_{i-1}, X_{i+1})R_2$ ($\mathbf{C}$). This is impossible since

$$u(R_1(X_{i-1}, X_{i+1})R_2) < u(R).$$

Suppose $v(R)=2$. Then F is a connected plane on $F(Q)$. It meets E in a line L, by (2.2) and (2.5). Let Z be the point of intersection of the lines L and $X_i\cup X_{i+1}$ on F. We discuss first the case $Z\in \mathbf{M}-\mathbf{C}$. In this case we define Q as the degenerate path (Z) if $Z=X_{i+1}$ and as the path (Z, X_{i+1}) otherwise. Then $(X_i, X_{i+1})Q^{-1}(Z, X_i)$ is an elementary re-entrant path of the first or second kind. Hence $R\sim R_1(X_{i-1}, X_i, Z)QR_2$ ($\mathbf{C}$), by (5.3). If L is connected we have $(X_{i-1}, X_i, Z, X_{i-1})\sim 0$ ($\mathbf{C}$) and therefore $R\sim R_1(X_{i-1}, Z)QR_2$ ($\mathbf{C}$), by (5.3). If L is not connected it is on a connected plane F' of $\mathbf{M}$ on E, by (4.2). We can find a connected line L' on X_{i-1} and F', and a point T of $\mathbf{M}-\mathbf{C}$ distinct from X_{i-1} on L'. Then $T\cup X_{i-1}$ and $T\cup Z$ are connected lines, by (4.3). Using the lemma and the definition of n we find $(X_{i-1}, X_i, Z, T, X_{i-1})\sim 0$ ($\mathbf{C}$). Hence $R\sim R_1(X_{i-1}, T, Z)QR_2$ ($\mathbf{C}$), by (5.3). So whether L is connected or not we have $R\sim R_3QR_2$ ($\mathbf{C}$), where R_3 is on E. This is impossible since $u(R_3QR_2)<u(R)$.

We go on to the case $Z\in\mathbf{C}$, illustrated in Figure IV. By (3.4) there is a connected line L' other than $X_i\cup X_{i+1}$ on X_{i+1} and F. If L' is on X_{i-1} we have $R\sim R_1(X_{i-1}, X_{i+1})R_2$ ($\mathbf{C}$), using (5.3) with the elementary re-entrant path $(X_{i-1}, X_i, X_{i+1}, X_{i-1})$ of the second kind. This is impossible since $u(R_1(X_{i-1}, X_{i+1})R_2)<u(R)$. Hence L' must meet the lines $X_{i-1}\cup X_i$ and L in distinct points U and V respectively. Since $Z\in\mathbf{C}$ and $X_{i-1}\in\mathbf{M}-\mathbf{C}$ we have $V\in\mathbf{M}-\mathbf{C}$.

Suppose $U\in\mathbf{M}-\mathbf{C}$. Using (5.3) with elementary re-entrant paths of the first and second kinds we find

$R \sim R_1(X_{i-1}, U, X_i, X_{i+1})R_2 \sim R_1(X_{i-1}, V, U, X_{i+1})R_2 \sim R_1(X_{i-1}, V, X_{i+1})R_2 (\mathbf{C})$.
This is impossible since $u(R_1(X_{i-1}, V, X_{i+1})R) < u(R)$.

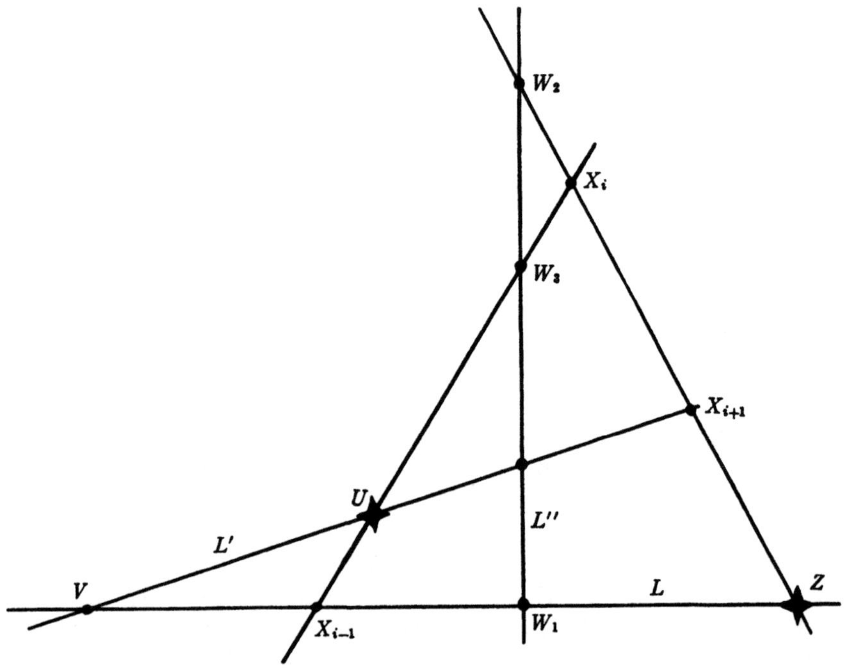

FIG. IV

Suppose $U \in \mathbf{C}$. It may happen that each connected line on F is on either U or Z. Then $(X_{i-1}, X_i, X_{i+1}, V, X_{i-1})$ is an elementary re-entrant path of the third kind with respect to $\mathbf{C}$. Using (5.3) we have $R \sim R_1(X_{i-1}, V, X_{i+1})R_2$ $(\mathbf{C})$, which is impossible, as before. Hence there is a connected line L'' on F which is not on U or Z. If L'' is on X_{i+1} we can substitute it for L' in the preceding argument and so reduce to the case $U \in \boldsymbol{M} - \mathbf{C}$. We may therefore suppose L'' is not on X_{i+1}.

If L'' is on X_i it meets L in a point W_1 distinct from X_{i-1} and Z. Writing $R' = R_1(X_{i-1}, W_1, X_i, X_{i+1})R_2$ we have $R' \sim R$ $(\mathbf{C})$, by (5.3). If L'' is not on X_i it meets $X_i \cup X_{i+1}$ in a point W_2 distinct from X_i, X_{i+1} and Z. If L'' is then on X_{i-1} we write $R' = R_1(X_{i-1}, W_2, X_{i+1})R_2$ and have

$$R' \sim R_1(X_{i-1}, W_2, X_i, X_{i+1})R_2 \sim R \ (\mathbf{C}),$$

by (5.3). If instead L'' is not on X_{i-1} it meets the lines L, $X_i \cup X_{i+1}$ and $X_{i-1} \cup X_i$ in distinct points W_1, W_2 and W_3 respectively of $\boldsymbol{M} - \mathbf{C}$. We then write $R' = R_1(X_{i-1}, W_1, W_2, X_{i+1})R_2$ and have $R' \sim R_1(X_{i-1}, W_1, W_3, W_2, X_i, X_{i+1})R_2$ $\sim R_1(X_{i-1}, W_3, X_i, X_{i+1})R_2 \sim R$ $(\mathbf{C})$, by (5.3). For each of these three possibilities we have $R' \sim R$ $(\mathbf{C})$, $u(R') = u(R)$ and $v(R') = v(R) = 2$. Hence we may replace R by R' in the preceding argument. This reduces the problem to the case $U \in \boldsymbol{M} - \boldsymbol{C}$, which we have found to lead to a contradiction.

We now consider the case $v(R) > 2$. By (3.3) there is a connected plane K on $X_{i-1} \cup X_i$ and the connected $v(R)$-flat F. This plane meets E in a line L. Choose a point T distinct from X_{i-1} on L and if possible in $\boldsymbol{C}$. By (3.5) there is a connected $(v(R)-1)$-flat F' on $X_i \cup X_{i+1}$ and F but not on T. Now F' is not on X_{i-1}, for otherwise we would have $F \subseteq F'$, contrary to (2.2). Hence F' meets L in a point T' distinct from X_{i-1} and T. It follows that L is connected, by (3.2), and that $T' \in \boldsymbol{M} - \boldsymbol{C}$. The flats K and F' intersect in a line L' on X_i and T'. If L' is connected we write

$$R' = R_1(X_{i-1}, T', X_i, X_{i+1})R_2$$

and have $R' \sim R$ ($\boldsymbol{C}$), by (5.3). If L' is not connected it is on a connected plane K' on F', by (4.2). K' meets E in a connected line L'' on T', by (4.4). We can find a point U on L'' distinct from T' and in $\boldsymbol{M} - \boldsymbol{C}$. The flat $U \cup X_i$ is a connected line, by (4.3). Using the lemma and the definition of n we find $(T', U, X_i, X_{i-1}, T') \sim 0$ ($\boldsymbol{C}$). In this case we write

$$R' = R_1(X_{i-1}, T', U, X_i, X_{i+1})R_2.$$

Then by (5.2) and (5.3) we have

$$R' \sim R_1(X_{i-1}, T', U, X_i, X_{i-1}, T', X_{i-1}, X_i, X_{i+1})R_2$$
$$\sim R_1(X_{i-1}, T', X_{i-1}, X_i, X_{i+1})R_2 \sim R \ (\boldsymbol{C}).$$

So whether L' is connected or not we have $R' \sim R$ ($\boldsymbol{C}$), $u(R') = u(R)$ and $v(R') < v(R)$, which is contrary to the definition of R.

From the above analysis we deduce that $u(R) = 0$. Hence R is on E and has dimension $< n$. Hence $P \sim R \sim 0$ ($\boldsymbol{C}$), contrary to assumption. The theorem follows.

7. Special cases. With $\boldsymbol{C}$ null in (6.1) we find that every re-entrant path in a matroid $\boldsymbol{M}$ is null-homotopic, as stated in the Introduction.

In applying this result to the circuit-matroid $\boldsymbol{M}$ of a graph G we must remember that a path in $\boldsymbol{M}$ corresponds to a sequence of circuits of G such that any two consecutive circuits form a nonseparable subgraph of rank 2. It can be shown that such a subgraph is made up of three arcs such that any two have both ends but no other edge or vertex in common. Each of the elementary deformations by which a re-entrant sequence of circuits can be transformed into a sequence with only one member operates within some nonseparable subgraph of rank ≤ 3.

Reference

1. Hassler Whitney, *The abstract properties of linear dependence*, Amer. J. Math. vol. 57 (1935) pp. 507–533.

University of Toronto,
Toronto, Ont.

Reprinted from *American Journal of Mathematics*,
volume 57, pages 800 through 804,
by permission of the Johns Hopkins University Press.

A HOMOTOPY THEOREM FOR MATROIDS, II

BY
W. T. TUTTE

1. Introduction. In this paper R denotes a ring, which is either the ring of integers or the ring of residues modulo a prime.

If M is any finite set we define a *chain* on M over R as a mapping f of M into R. If $a \in M$ then $f(a)$ is the *coefficient* of a in f. The set of all $a \in M$ such that $f(a) \neq 0$ is the *domain* $|f|$ of f. If $f(a) = 0$ for all a then f is the *zero* chain on M over R. We denote a zero chain by the symbol 0.

The sum $f+g$ of two chains f and g on M over R is a chain on M over R defined by the following rule:

$$(f+g)(a) = f(a) + g(a), \qquad a \in M. \tag{1.1}$$

With this definition of addition the chains on M over R are the elements of an additive Abelian group $A(M, R)$. Its zero element is the zero chain on M over R and the negative of a chain f is obtained by multiplying each coefficient in f by -1. We define a *chain-group* on M over R as any subgroup of $A(M, R)$.

Let N be any chain-group on M over R. Let N^* be the class of all chains g on M over R such that

$$\sum_{a \in M} f(a)g(a) = 0 \tag{1.2}$$

for all $f \in N$. Then N^* is a chain-group on M over R. We call it the *dual* chain-group of N.

A chain f of N is *elementary* if it is nonzero and there is no nonzero $g \in N$ such that $|g|$ is a proper subset of $|f|$. If R is the ring of integers we define a *primitive* chain of N as an elementary chain f of N in which the coefficients $f(a)$ are restricted to the values 0, 1 and -1.

We call N *binary* if R is the ring of residues mod 2. We call it *regular* if R is the ring of integers and to each elementary chain there corresponds a primitive chain with the same domain. Regular chain-groups are studied in [2]. In that paper it is shown that the groups of integral cycles and coboundaries on an oriented graph are regular chain-groups.

We write $\boldsymbol{M}(N)$ for the class of domains of elementary chains of N. We shall show that $\boldsymbol{M}(N)$ is a matroid on M. We call it the matroid *of* N. We call a matroid *binary* or *regular* if it is the matroid of a binary or regular chain-group respectively. This definition of a binary matroid is clearly equivalent to that given in the Introduction to Paper I [3].

Received by the editors October 26, 1956.

In this paper we investigate the following problem. What are the necessary and sufficient conditions, in terms of geometrical structure, that a given matroid shall be (a) binary or (b) regular? The problem of binary matroids is found to present little difficulty; the condition is simply that there shall be just three points on each connected line. For regular matroids we need the homotopy theorem proved in Paper I. The conditions are that the given matroid shall be binary and that two particular figures shall not occur in its geometrical structure. One of these excluded figures is the Fano configuration of seven lines and seven points in a plane. The second is the *matroid dual* of the Fano configuration. This means that the two figures represent the matroids of two binary chain-groups each of which is the dual of the other.

The two theorems just quoted are members of an interesting sequence. The third member, with which I hope to deal in another paper, gives the conditions for a given regular chain-group to be the cycle-group or coboundary-group of a graph. The fourth, a famous theorem of Kuratowski [1] gives the conditions for a given graph to be a planar graph. In each case the condition imposed is equivalent to the exclusion of certain geometrical figures from the corresponding matroid.

2. **Chain-groups and matroids.** Let N be any chain-group on a set M over R. If $r \in R$ and $f \in N$ we define a chain rf on M over R by the following rule:

$$(rf)(a) = r \times f(a), \qquad a \in M. \tag{2.1}$$

By the definition of R the chain rf can always be represented as a sum of chains, each equal to f or $-f$. Hence rf is always a chain of N.

(2.2) *If $f \in N$ and $a \in |f|$ then there exists an elementary chain g of N such that $a \in |g| \subseteq |f|$.*

Proof. If possible choose f and a so that the theorem fails and $|f|$ has the least number of elements consistent with this. Since f is nonzero there is an elementary chain g of N such that $|g| \subseteq |f|$. We must have $a \notin |g|$. Choose $b \in |g|$ and write $k = g(b)f - f(b)g$. Then $a \in |k| \subset |f|$. Hence, by the definition of f, there exists an elementary chain g' of N such that $a \in |g'| \subseteq |k|$. But then $a \in |g'| \subset |f|$, which is impossible.

(2.3) *The class $\boldsymbol{M}(N)$ of domains of elementary chains of N is a matroid on M.*

Proof. We must show that $\boldsymbol{M}(N)$ satisfies Axioms I and II of the Introduction to Paper I. It satisfies Axiom I by the definition of an elementary chain. To verify Axiom II suppose $(X, Y) \in \boldsymbol{M}(N)$, $a \in X \cap Y$ and $b \in X - (X \cap Y)$. There are chains f and g of N such that $|f| = X$ and $|g| = Y$. We have $b \in |g(a)f - f(a)g| \subseteq (X \cup Y) - \{a\}$. Hence, by (2.2), there exists $Z \in \boldsymbol{M}(N)$ such that $b \in Z \subseteq (X \cup Y) - \{a\}$. Thus Axiom II holds for $\boldsymbol{M}(N)$.

As stated in the Introduction we call $\boldsymbol{M}(N)$ the matroid of N. It follows from (2.2) that the domain of any chain of N is a flat of $\boldsymbol{M}(N)$.

Chains $f_1, \cdots, f_k$ of N are ***linearly dependent*** if they satisfy a relation $r_1f_1+\cdots+r_kf_k=0$ in which the coefficients r_i are elements of R not all zero. If no such relation holds the chains are ***linearly independent***. The ***rank*** $r(N)$ of N is the maximum number of linearly independent chains of N.

(2.4) *Let S be a flat of $\boldsymbol{M}(N)$ of dimension dS. Then the maximum number of linearly independent chains of N with domains in S is $dS+1$.*

Proof. If possible choose S so that the theorem fails and dS has the least value consistent with this. Then $dS\geqq 0$ for otherwise S is null and the theorem is trivially true.

Choose $a\in S$. Since S is a flat of $\boldsymbol{M}(N)$ there exists a chain f of N such that $a\in|f|\subseteq S$. But $d\langle S-\{a\}\rangle=dS-1$, by I(2.3). Hence, by the choice of S, there exist dS linearly independent chains of N with domains in $\langle S-\{a\}\rangle$. Adjoining f to these we obtain $dS+1$ linearly independent chains of N with domains in S.

We deduce that there exist $k=dS+2$ linearly independent chains $f_1, \cdots, f_k$ of N with domains in S. At most $k-2$ of them have domains in $\langle S-\{a\}\rangle$. We can adjust the notation so that $a\in|f_i|$ if $i\leqq s$ and $a\notin|f_i|$ if $i>s$, where s is some integer satisfying $2\leqq s\leqq k$. For each $i\leqq s$ we write $g_i=f_1(a)f_i-f_i(a)f_1$. Then $|g_i|\subseteq\langle S-\{a\}\rangle$. By the choice of S we can find elements $r_2, \cdots, r_k$ of R, not all zero, such that

$$\sum_{i=2}^{s} r_ig_i+\sum_{i=s+1}^{k} r_if_i=0.$$

Writing the left hand side of this equation in terms of the k chains f_i we find that these chains are not linearly independent. This contradiction establishes the theorem.

(2.5) *Let $\boldsymbol{M}$ be a matroid on a set M. With each $X\in\boldsymbol{M}$ let there be associated a chain $f(X)$ on M over R so that the following conditions are satisfied*:

(i) $|f(X)|=X$,

(ii) *If X, Y and Z are distinct points on the same connected line of $\boldsymbol{M}$ there exist nonzero elements r, s and t of R such that $rf(X)+sf(Y)+tf(Z)=0$.*

Let N be the chain-group on M generated by the chains $f(X)$.

Then $\boldsymbol{M}=\boldsymbol{M}(N)$.

Proof. Let T be any member of $\boldsymbol{M}(N)$. There exists $f\in N$ such that $|f|=T$. By the definition of N we can find a flat S of $\boldsymbol{M}$ with the following properties:

(i) There is a nonzero $r\in R$ such that rf is a sum of chains $\pm f(X)$ with $X\subseteq S$,

(ii) dS has the least value consistent with (i).

If possible choose $a\in S-T$. Write $S'=\langle S-\{a\}\rangle$. By I(2.2) any line of $\boldsymbol{M}$ which is not a subset of S' has at most one point in common with S'. Hence the class $\boldsymbol{C}$ of all points of $\boldsymbol{M}$ which are subsets of S' is a convex subclass of $\boldsymbol{M}$.

If g and h are chains of N we write $g \equiv h \bmod S'$ to denote that $g-h$ is a sum of chains $\pm f(X)$ such that $X \subseteq S'$. In particular we may write

(2.5a) $$rf \equiv \sum_i \pm f(Y_i) \bmod S',$$

where the Y_i are points of $\boldsymbol{M}$ satisfying $a \in Y_i \subseteq S$.

Choose a point Y of $\boldsymbol{M}$ such that $a \in Y \subseteq S$. Then for each Y_i the flat $Y \cup Y_i$ is connected. Hence, by I(5.1), there is a path P_i from Y_i to Y on S whose terms are all points of $\boldsymbol{M}-\mathbf{C}$. If U and V are consecutive terms of P_i then the line $U \cup V$ meets S' in a point W of $\mathbf{C}$, by I(2.5), and by hypothesis we have

(2.5b) $$x_i f(U) \equiv y_i f(V) \bmod S',$$

where x_i and y_i are nonzero members of R. If $Y_i \neq Y$ we obtain (2.5b) for each pair of consecutive terms of P_i and combine the results. We thus obtain

(2.5c) $$r_i f(Y_i) \equiv s_i f(Y) \bmod S',$$

where r_i and s_i are nonzero elements of R. Multiplying (2.5a) by the product of the elements r_i we obtain

(2.5d) $$r' f \equiv s' f(Y) \bmod S',$$

where r' and s' are elements of R, r' being nonzero.

Since $a \notin |f|$ we must have $s'=0$. Then (2.5d) contradicts the definition of S. We deduce that $S=T$. Hence there exists $T' \in \boldsymbol{M}$, on S, such that $T' \subseteq T$.

Conversely let T' be any member of $\boldsymbol{M}$. Then $f(T') \in N$. Hence there exists $T \in \boldsymbol{M}(N)$ such that $T \subseteq T'$.

Applying Axiom I to these results we find that the classes $\boldsymbol{M}$ and $\boldsymbol{M}(N)$ are identical.

(2.6) *Let $\boldsymbol{M}$ be a matroid, and L a line of $\boldsymbol{M}$ on which there are just three points, X, Y and Z. Then each cell of L belongs to just two of the points X, Y and Z.*

Proof. Let a be any cell of L. By I(3.1) we have $L = X \cup Y = Y \cup Z = Z \cup X$. Hence a belongs to at least two of the points on L. But if a belongs to all three then, by I(2.3), the flat $\langle L - \{a\} \rangle$ is a point on L distinct from X, Y and Z.

(2.7) *A matroid $\boldsymbol{M}$ is binary if and only if there are just three points on each connected line of $\boldsymbol{M}$.*

Proof. $\boldsymbol{M}$ is a matroid on a set M. If $S \subseteq M$ we define $f(S)$ as the chain on M over the ring of residues mod 2 such that $f(a)=1$ if $a \in S$ and $f(a)=0$ if $a \in M-S$.

Suppose $\boldsymbol{M}$ is binary. Let L be any connected line of $\boldsymbol{M}$. There are three distinct points X, Y and Z on L, by I(3.2). If T is a fourth point on L we have $f(X)+f(Y)+f(Z)=0$ and $f(X)+f(Y)+f(T)=0$, by (2.4). But then

$f(Z) = f(T)$ and $Z = T$, contrary to the definition of T.

Conversely suppose there are just three points on each connected line of $\boldsymbol{M}$. Then if X, Y and Z are the three points on any connected line L we have $f(X)+f(Y)+f(Z)=0$, by (2.6). Hence $\boldsymbol{M}=\boldsymbol{M}(N)$, by (2.5), where N is the binary chain-group on M generated by those chains $f(S)$ for which $S \in \boldsymbol{M}$. Thus $\boldsymbol{M}$ is a binary matroid.

3. Some operations on matroids. Let $\boldsymbol{M}$ be any matroid on a set M, and let S be any subset of M.

Let $\boldsymbol{M} \times S$ be the class of all points of $\boldsymbol{M}$ which are subsets of S. Then $\boldsymbol{M} \times S$ satisfies Axioms I and II, and is thus a matroid on S. The flats of $\boldsymbol{M} \times S$ are those flats of $\boldsymbol{M}$ which are subsets of S. By the definition of dimension a flat T of $\boldsymbol{M} \times S$ has the same dimension in $\boldsymbol{M} \times S$ as in $\boldsymbol{M}$. It is also clear that T is connected in $\boldsymbol{M} \times S$ if and only if it is connected in $\boldsymbol{M}$.

Let $\boldsymbol{M}_S$ be the class of intersections with S of members of $\boldsymbol{M}$, and let $\boldsymbol{M} \cdot S$ be the class of all minimal non-null members of $\boldsymbol{M}_S$.

(3.1) *If $X \in \boldsymbol{M}$ and $a \in S \cap X$, then there exists $Z \in \boldsymbol{M} \cdot S$ such that $a \in Z \subseteq S \cap X$.*

Proof. We can find $Y \in \boldsymbol{M}$ such that $a \in S \cap Y \subseteq S \cap X$ and $S \cap Y$ has the least number of elements consistent with this property. If $S \cap Y \in \boldsymbol{M} \cdot S$ the theorem holds with $Z = S \cap Y$. In the remaining case there exists $Y' \in \boldsymbol{M}$ such that $\emptyset \subset S \cap Y' \subset S \cap Y$ and, by the choice of Y, $a \notin Y'$. Choose $c \in S \cap Y'$. By Axiom II there exists $Y'' \in \boldsymbol{M}$ such that $a \in Y'' \subseteq (Y \cup Y') - \{c\}$. But then $a \in S \cap Y'' \subset S \cap Y \subseteq S \cap X$, contrary to the definition of Y. The theorem follows.

(3.2) *$\boldsymbol{M} \cdot S$ is a matroid on S.*

Proof. By its definition $\boldsymbol{M} \cdot S$ satisfies Axiom I. Suppose $(X, Y) \in \boldsymbol{M} \cdot S$, $a \in X \cap Y$ and $b \in X - (X \cap Y)$. There exist elements X' and Y' of $\boldsymbol{M}$ such that $X = S \cap X'$ and $Y = S \cap Y'$. Since $\boldsymbol{M}$ satisfies Axiom II there exists $U \in \boldsymbol{M}$ such that $b \in U \subseteq (X' \cup Y') - \{a\}$. By (3.1) there exists $Z \in \boldsymbol{M} \cdot S$ such that $b \in Z \subseteq S \cap U \subseteq (X \cup Y) - \{a\}$. Thus $\boldsymbol{M} \cdot S$ satisfies Axiom II.

(3.3) *If T is a flat of $\boldsymbol{M}$ then $S \cap T$ is a flat of $\boldsymbol{M} \cdot S$.*

This follows from (3.1).

We define a *carrier* of $\boldsymbol{M} \cdot S$ in $\boldsymbol{M}$ as a subset Z of M such that $(\boldsymbol{M} \times Z) \cdot S = \boldsymbol{M} \cdot S$, and such that $d\langle Z \rangle$ has the least value consistent with this property. Such a carrier exists, since $(\boldsymbol{M} \times M) \cdot S = \boldsymbol{M} \cdot S$. We must have $S \subseteq Z$.

(3.4) *If Z is a carrier of $\boldsymbol{M} \cdot S$ in $\boldsymbol{M}$, then $d\langle Z - S \rangle = -1$ in $\boldsymbol{M}$.*

Proof. Assume the theorem false. Then there exists $X \in \boldsymbol{M}$ such that $X \subseteq Z - S$. Choose $a \in X$ and write $\boldsymbol{M}' = (\boldsymbol{M} \times (Z - \{a\})) \cdot S$.

If $Y \in \boldsymbol{M}'$ there exists $U \in \boldsymbol{M} \times Z$ such that $Y = S \cap U$. Hence there exists $Y' \in (\boldsymbol{M} \times Z) \cdot S = \boldsymbol{M} \cdot S$ such that $Y' \subseteq Y$. Conversely if $Y' \in \boldsymbol{M} \cdot S = (\boldsymbol{M} \times Z) \cdot S$ there exists $V \in \boldsymbol{M} \times Z$ such that $Y' = S \cap V$. If $a \in V$ then by Axiom II there exists $U \in \boldsymbol{M} \times Z$ such that $\emptyset \subset S \cap U \subseteq Y'$ and $a \notin U$. If $a \notin V$ we write $U = V$. In each case $U \in \boldsymbol{M} \times (Z - \{a\})$ and $S \cap U$ is non-null. Hence there exists $Y \in \boldsymbol{M}'$ such that $Y \subseteq S \cap U \subseteq Y'$.

Applying Axiom I to these results we find that $\boldsymbol{M}'$ and $\boldsymbol{M}\cdot S$ are identical. But $d\langle Z-\{a\}\rangle=d\langle\langle Z\rangle-\{a\}\rangle<d\langle Z\rangle$, by I(2.3). Thus the definition of Z as a carrier of $\boldsymbol{M}\cdot S$ is contradicted.

(3.5) *Let Z be a carrier of $\boldsymbol{M}\cdot S$ in $\boldsymbol{M}$. Then if $X\in\boldsymbol{M}\cdot S$ there is just one $Y\in\boldsymbol{M}\times Z$ such that $S\cap Y=X$.*

Proof. Such a point Y exists, since $\boldsymbol{M}\cdot S=(\boldsymbol{M}\times Z)\cdot S$. Suppose Y' is another such point of $\boldsymbol{M}\times Z$. Choose $a\in S\cap Y=S\cap Y'$. By Axiom I we can find $b\in Y'-(Y'\cap Y)$. By Axiom II there exists $U\in\boldsymbol{M}\times Z$ such that $b\in U\subseteq(Y\cup Y')-\{a\}$. Now $S\cap U\subset S\cap Y=X$. Hence $S\cap U$ is null, since $\boldsymbol{M}\cdot S$ satisfies Axiom I. Hence $U\subseteq Z-S$. This is impossible, by (3.4).

Given a carrier Z of $\boldsymbol{M}\cdot S$ in $\boldsymbol{M}$ we define a mapping θ of the flats of $\boldsymbol{M}\cdot S$ into flats of $\boldsymbol{M}\times Z$ in the following way:

(i) If T is the null flat of $\boldsymbol{M}\cdot S$ then θT is the null flat of $\boldsymbol{M}\times Z$,

(ii) If T is a point of $\boldsymbol{M}\cdot S$ then θT is the unique point Y of $\boldsymbol{M}\times Z$ such that $S\cap Y=T$,

(iii) If T is a d-flat of $\boldsymbol{M}\cdot S$, where $d>1$, then θT is the union of all points θX of $\boldsymbol{M}\times Z$ such that $X\subseteq T$. We refer to θ as the (S, Z)-*mapping* of $\boldsymbol{M}$. We note that if T is any flat of $\boldsymbol{M}\cdot S$, then $T=\theta(T)\cap S$.

(3.6) *If T and U are flats of $\boldsymbol{M}\cdot S$ then $\theta T\subseteq\theta U$ if and only if $T\subseteq U$.*

This follows at once from the definition of θ. It implies that θT and θU are distinct if T and U are distinct, and that $\theta T\subset\theta U$ if and only if $T\subset U$.

(3.7) *If V is a flat of $\boldsymbol{M}\times Z$ then $V\subseteq\theta(S\cap V)$.*

Proof. By (3.3) and the definition of θ we have $S\cap V\subseteq\theta(S\cap V)$.

If b is any given cell of $V-(S\cap V)$ we can find $X_b\in\boldsymbol{M}\times Z$ such that $b\in X_b$ and $S\cap X_b\subseteq S\cap V$. Choose such an X_b so that $S\cap X_b$ has the least possible number of elements. Then $S\cap X_b$ is non-null, by (3.4).

Assume $S\cap X_b\notin\boldsymbol{M}\cdot S$. There exists $Y\in\boldsymbol{M}\cdot S$ such that $Y\subset S\cap X_b$. Choose $a\in Y$. We have $b\notin\theta Y$, by the choice of X_b. By Axiom II we can find $Y'\in\boldsymbol{M}\times Z$ such that $b\in Y'\subseteq(X_b\cup\theta Y)-\{a\}$. But then $S\cap Y'\subset S\cap X_b$ and the definition of X_b is contradicted.

We deduce that $S\cap X_b\in\boldsymbol{M}\cdot S$. Hence $b\in\theta(S\cap X_b)\subseteq\theta(S\cap V)$. Since this is true for all $b\in V-(S\cap V)$ we have $V-(S\cap V)\subseteq\theta(S\cap V)$. But $S\cap V\subseteq\theta(S\cap V)$. The theorem follows.

(3.8) *If T is a flat of $\boldsymbol{M}\cdot S$ then $dT=d\theta T$.*

Proof. If possible choose T so that the theorem fails and dT has the least value consistent with this. Clearly T is non-null. Choose $a\in T$. Let U be the union of all points of $\boldsymbol{M}\cdot S$ which are subsets of $T-\{a\}$.

Now θU is a subset of the flat $\langle\theta T-\{a\}\rangle$ of $\boldsymbol{M}\times Z$, and $S\cap\langle\theta T-\{a\}\rangle\subseteq U$ by (3.3). Hence $\theta U\subseteq\langle\theta T-\{a\}\rangle\subseteq\theta(S\cap\langle\theta T-\{a\}\rangle)\subseteq\theta U$, by (3.6) and (3.7), and therefore $\theta U=\langle\theta T-\{a\}\rangle$. Now $dU=dT-1$ in $\boldsymbol{M}\cdot S$ and $d\langle\theta T-\{a\}\rangle=d\theta T-1$ in $\boldsymbol{M}\times Z$, by I(2.3). But $dU=d\theta U=d\langle\theta T-\{a\}\rangle$, by the choice of T. Hence $dT=d\theta T$, contrary to the choice of T.

4. Binary and regular matroids. Let N be a chain-group on a set M over the ring R. If $f\in N$ we define the *restriction* $f\cdot S$ of f to a subset S of M as that chain on S in which each $a \in S$ has the same coefficient as in f.

We write $N \cdot S$ for the class of restrictions to S of the chains of N, and $N \times S$ for the class of restrictions to S of those chains f of N for which $|f| \subseteq S$. It is clear that $N \cdot S$ and $N \times S$ are chain-groups on M over R which satisfy the two following identities.

(4.1) $$\boldsymbol{M}(N \times S) = \boldsymbol{M}(N) \times S,$$

(4.2) $$\boldsymbol{M}(N \cdot S) = \boldsymbol{M}(N) \cdot S.$$

It is shown in [2, (3.5)], that $N \cdot S$ and $N \times S$ are regular if N is regular. Hence (4.1) and (4.2) imply

(4.3) *Let $\boldsymbol{M}$ be a binary (regular) matroid on a set M, and suppose $S \subseteq M$. Then $\boldsymbol{M} \times S$ and $\boldsymbol{M} \cdot S$ are binary (regular) matroids.*

(4.4) *Every regular matroid is binary.*

Proof. Let $\boldsymbol{M}$ be a regular matroid on a set M. There is a regular chain-group N on M such that $\boldsymbol{M} = \boldsymbol{M}(N)$. If $f \in N$ let ϕf be the chain on M obtained by replacing each coefficient in f by its residue mod 2. The chains ϕf such that $f \in N$ constitute a binary chain-group N' on M.

Suppose $X \in \boldsymbol{M}$. There is a primitive chain g of N such that $|g| = X$. Then $|\phi g| = X$. Hence there exists $X' \in \boldsymbol{M}(N')$ such that $X' \subseteq X$. Conversely suppose $X' \in \boldsymbol{M}(N')$. There is a chain $f \in N$ such that $|\phi f| = X'$. There is a chain $h \in N$ which is a 2-representative of f. (See [2, §6].) Then $|h| = |\phi f| = X'$. Hence there exists $X \in \boldsymbol{M}$ such that $X \subseteq X'$. Using Axiom I we deduce from these results that $\boldsymbol{M} = \boldsymbol{M}(N')$. Hence $\boldsymbol{M}$ is binary.

By a *figure of Type* BI in a binary matroid $\boldsymbol{M}$ we mean an assemblage of flats of $\boldsymbol{M}$ consisting of a plane P, seven distinct lines on P, and the points of intersection of these lines. If X is any point on P there is a line L on P which is not on X, by I(2.4). There are at most three points on L, by (2.7). Hence at most three lines on P are on X. It follows that each line of the figure is on three distinct points, each of which is on two other lines of the figure, and there can be no eighth line on P. By I(3.2) the seven lines are connected and therefore P is connected. The seven lines intersect in threes at seven points and, by I(3.4), there can be no eighth point on P.

We enumerate the seven lines as $L_1, \cdots, L_7$, and denote the point of intersection, if any, of three lines L_i, L_j and L_k by X_{ijk}. We may assume a point X_{135} on L_1, L_3 and L_5. The line L_7, which is not on X_{135}, meets L_1, L_3 and L_5 in points X_{147}, X_{367} and X_{257}. The lines L_2 and L_4 each meet L_3 in its point other than X_{135} and X_{367}, which is therefore X_{234}. Similarly the remainings two points of the figure are X_{456} and X_{126}.

We can construct a matroid $\boldsymbol{M}$ having a figure of Type BI on any set $S = \{a_1, \cdots, a_7\}$ of seven elements. The rule is that X_{ijk} is the set of all elements of S other than a_i, a_j and a_k, and the X_{ijk} are the only points of $\boldsymbol{M}$. Axioms I and II are easily verified. We can verify also that the chains mod 2 on S whose domains are the X_{ijk} generate a binary chain-group on S whose

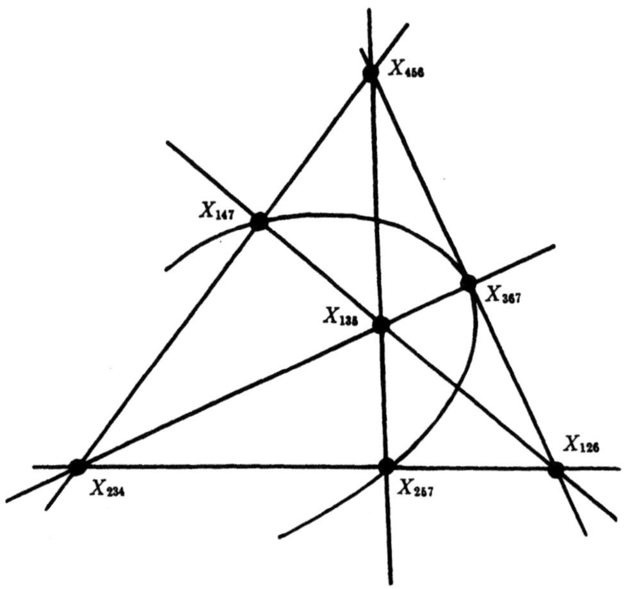

FIG. I

matroid is $\boldsymbol{M}$. The lines of $\boldsymbol{M}$ are the sets $L_i = S - \{a_i\}$ and S is the only plane of $\boldsymbol{M}$.

By a figure of Type BII in a binary matroid $\boldsymbol{M}$ we mean an assemblage of flats of $\boldsymbol{M}$ consisting of a 3-flat E, seven distinct planes of $\boldsymbol{M}$ on E such that no three are on a common line, the 21 lines of intersection of these planes, and those points of E which are common to three or more of the seven planes.

We enumerate the seven planes as $P_1, \cdots, P_7$ and write L_{ij} for the line of intersection of P_i and P_j $(0 \leqq i < j \leqq 7)$.

By the argument already used for Type BI a point on the line L_{ij} can be on at most three lines on the plane P_i. It follows that each of the lines L_{ij} is on just three points of the figure, of which one is on just three of the seven planes and each of the others on just four of them. Hence each of the lines L_{ij} is connected. It follows that the planes P_i and the flat E are all connected.

We write X_{ijk} for a point which is on just three planes P_i, P_j and P_k of the figure, and X_{ijkl} for a point on just four such planes P_i, P_j, P_k and P_l. Each point of the first kind will be on just three lines of the figure, and each point of the second kind on just six. It follows that the figure has just seven points of each kind.

We obtain a detailed notation as follows. The line L_{17} may be supposed on X_{147}, X_{1237} and X_{1567}. The line L_{47} is on two points Y and Z distinct from X_{147}. On P_7 each of these is on three lines of the figure meeting L_{17} in three distinct points. Hence we may suppose Y to be on L_{27} and L_{67} while Z is on L_{37} and L_{57}. Then Y is X_{2467} and Z is X_{3457}. The remaining points of the figure on P_7 are $X_{367}=\langle L_{37}\cap L_{67}\rangle$ and $X_{257}=\langle L_{27}\cap L_{57}\rangle$. The lines L_{13} and L_{14} on P_1 meet in a point X_{134x}, since L_{14} is on X_{147}. Since no three planes of the figure are on a common line this point cannot have just three suffices in common with any of the points of the figure on P_7. It is therefore X_{1346}. But X_{1346} is on the lines L_{13}, L_{14}, L_{16}, L_{34}, L_{36} and L_{46}. These meet P_7 in X_{1237}, X_{147}, X_{1567}, X_{3457}, X_{367} and X_{2467} respectively. The third points on these lines are therefore X_{135}, X_{1245}, X_{126}, X_{234}, X_{2356} and X_{456}. We have now accounted for all 14 points of the figure.

We may illustrate this structure in a modification of Figure I of Paper I. The points represented by four-pointed stars are considered to lie on P_7 and are accordingly given the extra suffix 7. Z_1, Z_2 and Z_3 are X_{2356}, X_{1346} and X_{1245} respectively. The disconnected lines $Z_1\cup Z_2$, $Z_2\cup Z_3$ and $Z_3\cup Z_1$ become the connected lines L_{36}, L_{14} and L_{25} respectively, meeting P_7 in new points X_{367}, X_{147} and X_{257}.

We can construct a matroid $\boldsymbol{M}'$ having a figure of Type BII on the set $S=\{a_1, \cdots, a_7\}$. The rule is that a point Z of the figure is represented by the set of all elements of S with suffices other than those of Z. It may be verified that the resulting 14 sets constitute a binary matroid $\boldsymbol{M}'$ of the required kind on S. Moreover each of the two binary chain-groups on S corresponding to $\boldsymbol{M}$ and $\boldsymbol{M}'$ is the dual of the other.

(4.5) *A matroid is regular if and only if it is binary and includes no figure of Type* BI *or* BII.

Proof. Let $\boldsymbol{M}$ be a regular matroid on a set M. There is a regular chain-group N on M such that $\boldsymbol{M}=\boldsymbol{M}(N)$. $\boldsymbol{M}$ is binary, by (4.4).

Suppose $\boldsymbol{M}$ includes a figure of Type BI. For each line L_i of the figure we choose $a_i\in P-L_i$. Then $L_i=\langle P-\{a_i\}\rangle$ for each L_i, and the seven cells a_i are distinct. There are primitive chains g_1, g_2 and g_3 of N such that $|g_1|=X_{234}$, $|g_2|=X_{135}$ and $|g_3|=X_{126}$. Since each of these chains may be replaced by its negative we may suppose $g_1(a_1)=g_2(a_2)=g_3(a_3)=1$.

We note that $g_i(a_j)=0$ if $1\leqq(i,j)\leqq 3$ and $i\neq j$. By (2.4) the set $\{a_1, a_2, a_3\}=D$ say, meets the domain of every nonzero chain of $N\times P$. Clearly no proper subset of D has this property.

In the terminology of [2, §4], D is a *dendroid* of $N\times P$. This chain-group is regular, by [2, (3.5)]. The *representative matrix* of $N\times P$ associated with D has the *representative vectors* of the chains $g_i\cdot P$ as its rows. If its first seven columns are taken to correspond to the seven cells a_i, in the order of the suffices, we may write them as

$$\begin{Bmatrix} 1 & 0 & 0 & 0 & e_{15} & e_{16} & e_{17} \\ 0 & 1 & 0 & e_{24} & 0 & e_{26} & e_{27} \\ 0 & 0 & 1 & e_{34} & e_{35} & 0 & e_{37} \end{Bmatrix},$$

where each of the numbers e_{ij} is 1 or -1. By [2, (4.6)] we have

$$e_{16}e_{27} = e_{17}e_{26},$$

$$e_{24}e_{37} = e_{27}e_{34},$$

$$e_{35}e_{17} = e_{37}e_{15},$$

$$\begin{vmatrix} 0 & e_{15} & e_{16} \\ e_{24} & 0 & e_{26} \\ e_{34} & e_{35} & 0 \end{vmatrix} = e_{16}e_{24}e_{35} + e_{15}e_{26}e_{34} = 0.$$

But these equations are contradictory, for they imply $-1 = (e_{16}e_{24}e_{35}) \cdot (e_{15}e_{26}e_{34})e_{17}^2e_{27}^2e_{37}^2 = (e_{16}e_{27}e_{17}e_{26})(e_{24}e_{37}e_{27}e_{34})(e_{35}e_{17}e_{37}e_{15}) = 1$.

Suppose next that $\boldsymbol{M}$ includes a figure of Type BII. We proceed as before, choosing $a_i \in E - P_i$ and taking four primitive chains g_i of N satisfying $|g_1| = X_{1567}$, $|g_2| = X_{2467}$, $|g_3| = X_{3457}$ and $|g_4| = X_{456}$. As the first seven columns of a representative matrix of the regular chain-group $N \times E$ associated with the dendroid $\{a_4, a_5, a_6, a_7\}$ we then obtain

$$\begin{Bmatrix} 0 & e_{24} & e_{34} & 1 & 0 & 0 & 0 \\ e_{15} & 0 & e_{35} & 0 & 1 & 0 & 0 \\ e_{16} & e_{26} & 0 & 0 & 0 & 1 & 0 \\ e_{17} & e_{27} & e_{37} & 0 & 0 & 0 & 1 \end{Bmatrix},$$

where each of the numbers e_{ij} is 1 or -1. Applying [2, (4.6)] we obtain the same contradictory equations as before.

We have now proved that if a matroid is regular then it is binary and includes no figure of Type BI or BII.

Assume the theorem false. Then there exists a matroid $\boldsymbol{M}$ on a set M which is binary and which includes no figure of Type BI or BII, but which is not regular. Choose such a matroid so that the number $\alpha(\boldsymbol{M})$ of its cells is as small as possible. Clearly $\alpha(\boldsymbol{M}) \geq 1$. Choose $a \in M$ and write $\boldsymbol{M}' = \boldsymbol{M} \cdot (M - \{a\})$. Then $\boldsymbol{M}'$ has a carrier Z in $\boldsymbol{M}$, and Z is either $M - \{a\}$ or M. Let θ be the $(M - \{a\}, Z)$-mapping of $\boldsymbol{M}$.

$\boldsymbol{M}'$ is binary, by (4.3). If it includes a figure of Type BI or Type BII then the mapping θ induces another such figure in $\boldsymbol{M}$, which is impossible by the choice of $\boldsymbol{M}$. Since $\alpha(\boldsymbol{M}') < \alpha(\boldsymbol{M})$ we deduce that $\boldsymbol{M}'$ is regular. Hence there is a regular chain-group N' on $M - \{a\}$ such that $\boldsymbol{M}' = \boldsymbol{M}(N')$. For each $X \in \boldsymbol{M}'$ we select a primitive chain $g(X)$ of N' such that $|g(X)| = X$. For each $b \in M - \{a\}$ we denote the coefficient of b in $g(X)$ by $s(X, b)$. Then $s(X, b)$ is 1 or -1 if $b \in X$ and zero otherwise.

Let $\mathbf{C}$ denote the class of all points T of $\boldsymbol{M}'$ such that $a \notin \theta T$. Suppose X, Y and Z are distinct points of $\boldsymbol{M}'$ on a line L, X and Y being members of $\mathbf{C}$. Then θX, θY and θZ are distinct points of $\boldsymbol{M}$ on θL, and θL is a connected line of $\boldsymbol{M}$, by (3.8) and I(3.2). Hence $a \notin \theta X \cup \theta Y = \theta L$, by I(3.1). Hence $a \notin \theta Z$, that is $Z \in \boldsymbol{C}$. It follows that $\boldsymbol{C}$ is a convex subclass of $\boldsymbol{M}'$.

Let X and Y be distinct points of $\boldsymbol{M}' - \boldsymbol{C}$ on the same connected line L. By I(3.1) there exists $d \in X \cap Y$. We write

$$t(X, Y) = s(X, d)s(Y, d).$$

The number $t(X, Y)$ thus defined is unique. For suppose e is a second member of $X \cap Y$. By Axiom I we can find $b \in X - (X \cap Y)$ and $c \in Y - (X \cap Y)$. Using (2.4) we find that the representative vectors of $g(X)$ and $g(Y)$ are the rows of a representative matrix K of $N' \times L$ associated with the dendroid $\{b, c\}$. If the first four columns of K are taken to correspond, in order, to b, c, d and e we may write them as

$$\begin{Bmatrix} 1 & 0 & s(X, d) & s(X, e) \\ 0 & 1 & s(Y, d) & s(Y, e) \end{Bmatrix}.$$

Now $N' \times L$ is regular, by [2, (3.5)]. Hence, by [2, (4.6)] we have $s(X, d)s(Y, e) - s(X, e)s(Y, d) = 0$. Since each of the numbers concerned is 1 or -1 we may rewrite this as $s(X, d)s(Y, d) = s(X, e)s(Y, e)$.

If $R = (X_1, \cdots, X_k)$ is any nondegenerate path off $\boldsymbol{C}$ in $\boldsymbol{M}'$ we write

$$u(R) = \prod_{i=1}^{k-1} t(X_i, X_{i+1}).$$

We proceed to show that $u(R) = 1$ whenever R is re-entrant.

We first note that for any re-entrant path R off $\boldsymbol{C}$ in $\boldsymbol{M}'$ there is a finite integer $m(R)$ which is the least number of elementary deformations with respect to $\boldsymbol{C}$ by which R can be converted into a degenerate path. This follows from the homotopy theorem of Paper I.

If possible choose a nondegenerate re-entrant path R off $\boldsymbol{C}$ in $\boldsymbol{M}'$ such that $u(R) \neq 1$ and $m(R)$ has the least value consistent with this. If $m(R) > 1$ we can write either $R = R_1QR_2$ and $R' = R_1R_2$ or $R' = R_1QR_2$ and $R = R_1R_2$, where Q is an elementary re-entrant path with respect to $\boldsymbol{C}$ and $m(R') = m(R) - 1$. In each case $u(R) = u(Q)u(R')$. This is impossible since the choice of R implies $u(Q)u(R') = 1$. We deduce that $m(R) = 1$, that is R is an elementary re-entrant path with respect to $\boldsymbol{C}$.

If R is of the first kind we may write $R = (X, Y, X)$. Then $u(R) = (t(X, Y))^2 = 1$.

If R is of the second kind we may write $R = (X, Y, Z, X)$, where X, Y and Z are distinct points on the same line or plane.

If X, Y and Z are on the same line L then θX, θY and θZ are distinct points of $\boldsymbol{M}$ on the same line θL of $\boldsymbol{M}$. But then $a \in \theta X \cap \theta Y \cap \theta Z$ and therefore θL has a fourth point $\langle \theta L - \{a\} \rangle$. This is impossible, by (2.7).

We may now suppose that X, Y and Z are on a common plane P of $\boldsymbol{M}'$ but not on a common line. Hence there exist cells b, c and d of P, each belonging to just one of the points X, Y and Z, such that $b \in X$, $c \in Y$ and $d \in Z$.

If there is a cell h of P such that $h \in X \cap Y \cap Z$ we have $u(R) = (s(X, h)s(Y, h))(s(Y, h)s(Z, h))(s(Z, h)s(X, h)) = 1$, contrary to the choice of R. Hence there are cells e, f and g of P, each belonging to just two of the points X, Y and Z, such that $e \notin X$, $f \notin Y$ and $g \notin Z$.

If x is any one of b, c, d, e, f and g we write L_x for the line $\langle P - \{x\} \rangle$. Clearly the six lines L_x of $\boldsymbol{M}'$ on P are all distinct. Hence the six lines θL_x on the plane θP of $\boldsymbol{M}$ are all distinct. But each is on one of the points θX, θY and θZ, and therefore includes the cell a. Hence there is a seventh distinct line $\langle \theta P - \{a\} \rangle$ on θP. Thus $\boldsymbol{M}$ includes a figure of Type BI, contrary to its definition.

If R is of the third kind we have $R = (X, Y, Z, T, X)$, where X, Y, Z and T are distinct points on the same plane P, of which no three are on a common line. We can find cells b, c, d and e of P such that $b \in P - (X \cup Y)$, $c \in P - (Y \cup Z)$, $d \in P - (Z \cup T)$ and $e \in P - (T \cup X)$. Then b belongs to Z and T but not to X or Y, c belongs to T and X but not to Y or Z, and so on. The four chains $g(X)$, $g(Y)$, $g(Z)$ and $g(T)$ are linearly dependent, by (2.4). Hence

$$\begin{vmatrix} 0 & 0 & s(Z, b) & s(T, b) \\ s(X, c) & 0 & 0 & s(T, c) \\ s(X, d) & s(Y, d) & 0 & 0 \\ 0 & s(Y, e) & s(Z, e) & 0 \end{vmatrix} = 0,$$

that is $s(Z, b)s(T, c)s(X, d)s(Y, e) - s(T, b)s(X, c)s(Y, d)s(Z, e) = 0$. It follows that $u(R) = (s(X, d)s(Y, d))(s(Y, e)s(Z, e))(s(Z, b)s(T, b))(s(T, c)s(X, c)) = 1$.

If R is of the fourth kind we write $R = (A, X, B, Y, A)$ and then use the notation set out in I, §5 for this case. We observe that no three of the planes P_i on E have a common line, that each is on just two points of $\mathbf{C}$, and that no two of the planes P_i have more than one point of $\mathbf{C}$ in common. It follows that on the 3-flat θE of $\boldsymbol{M}$ the seven planes $\theta P_1, \cdots, \theta P_6, \langle \theta E - \{a\} \rangle$ are all distinct and that no three of them are on a common line. But then $\boldsymbol{M}$ includes a figure of Type BII, contrary to its definition.

For each kind of elementary re-entrant path we have obtained a contradiction. We deduce that $u(R) = 1$ whenever R is a nondegenerate re-entrant path off $\mathbf{C}$ in $\boldsymbol{M}'$.

We can partition the points of $\boldsymbol{M}' - \mathbf{C}$ into disjoint non-null classes $K_1, \cdots, K_q$ such that for any two points X and Y of $\boldsymbol{M}' - \mathbf{C}$ there is a path from X to Y off $\mathbf{C}$ in $\boldsymbol{M}'$ if and only if X and Y belong to the same class K_i. For each class K_i we select an arbitrary representative T_i and write $v(T_i) = 1$. For any other point X in K_i we write

$$v(X) = u(Q)v(T_i),$$

where Q is any path from T_i to X off $\mathbf{C}$ in $\boldsymbol{M}'$. If Q' is any other such path we have $u(Q)u(Q') = u(Q^{-1})u(Q') = u(Q^{-1}Q') = 1$, since $Q^{-1}Q'$ is re-entrant. Hence the number $v(X)$ is uniquely defined for each $X \in \boldsymbol{M}' - \mathbf{C}$ when the representatives T_i have been chosen.

For each $U \in \boldsymbol{M}$ we define a chain f_U on M over the ring of integers as follows.

Suppose $a \in U$. If $U = \{a\}$ we write $f_U(a) = 1$ and $f_U(b) = 0$ if $b \neq a$. If $U \neq \{a\}$ then $U - \{a\}$ is a non-null flat of $\boldsymbol{M}'$, by (3.3). The flat $\theta(U - \{a\})$ of $\boldsymbol{M} \times Z$ is $U - \{a\}$ or U, and the former alternative is ruled out by Axiom I. Hence $U - \{a\}$ is a point X of $\boldsymbol{M}' - \mathbf{C}$, by (3.8). We choose f_U to agree with $g(X)$ in $M - \{a\}$ and to satisfy $f_U(a) = v(X)$.

Suppose $a \notin U$. Then U is a flat of $\boldsymbol{M}'$, by (3.3), and $\theta U = U$ or $U \cup \{a\}$. In the former case U is a point of $\boldsymbol{M}'$, by (3.8). We then choose f_U to agree with $g(U)$ in $M - \{a\}$ and to satisfy $f_U(a) = 0$. In the latter case $U \cup \{a\}$ is a line of $\boldsymbol{M}$, by I(2.3), and U is a line of $\boldsymbol{M}'$, by (3.8). Now each point of $\boldsymbol{M}'$ on U is mapped by θ onto a point of $\boldsymbol{M}$ on $U \cup \{a\}$. But there are at most three points of $\boldsymbol{M}$ on $U \cup \{a\}$, by (2.7), and one of these is U. Hence U is a disconnected line of $\boldsymbol{M}'$, having only two points X and Y. Write $V = \theta X$ and $W = \theta Y$. Then $V = X \cup \{a\}$ and $W = Y \cup \{a\}$, by Axiom I. Thus f_V and f_W are already defined. We take f_U to be one of the differences $f_V(a)f_W - f_W(a)f_V$ and $f_W(a)f_V - f_V(a)f_W$.

We observe that for each $U \in \boldsymbol{M}$ we have $|f_U| = U$. Moreover $f_U \cdot (M - \{a\})$ is a chain of N'.

Suppose U, V and W are the three points on a connected line L of $\boldsymbol{M}$. We proceed to prove that f_U, f_V and f_W are linearly dependent.

We discuss first the case $a \in L$. We may suppose $a \in V \cap W$ and $a \notin U$, by (2.6). Then there are points $X = V - \{a\}$ and $Y = W - \{a\}$ of $\boldsymbol{M}'$ such that $\theta X = V$ and $\theta Y = W$. Write $L' = L \cap (M - \{a\}) = L - \{a\}$. Then L' is a flat of $\boldsymbol{M}'$, by (3.3). It clearly satisfies $\theta L' = L$, and is therefore a line of M'. The set U is also a flat of $\boldsymbol{M}'$, by (3.3). It must be either the line L' or a point on it.

If U is a point of $\boldsymbol{M}'$ on L' then $\theta U = U$, and U is distinct from X and Y. The line L' is then connected, by I(3.2). We can define $v(X)$ by a path Q off $\mathbf{C}$ from a representative point T_i to X, and then define $v(Y)$ by the path $Q(X, Y)$. We then have

$$v(X)v(Y) = u(Q)v(T_i) \times u(Q)t(X, Y)v(T_i) = t(X, Y).$$

Hence for each $b \in X \cap Y$ we have $v(X)v(Y) = s(X, b)s(Y, b)$. By (2.6) this implies

$$|\, v(X)g(Y) - v(Y)g(X) \,| \subseteq U.$$

Since $g(U)$ is an elementary chain of N' it follows that

$$g(U) = \pm\,(v(X)g(Y) - v(Y)g(X)).$$

By the definitions of f_U, f_V and f_W we now have

(4.5a) $$f_U = \pm (f_V(a) f_W - f_W(a) f_V).$$

If U is the line L' of $\boldsymbol{M}'$ we have $\theta U = U \cup \{a\}$, by (3.8). Then (4.5a) follows at once from the definition of f_U. Thus (4.5a) holds whenever $a \in L$, and the chains f_U, f_V and f_W are linearly dependent in this case.

Now suppose $a \notin L$. If U, V and W are points of $\boldsymbol{M}'$ and L is a line of $\boldsymbol{M}'$ there is a linear relation between $g(U)$, $g(V)$ and $g(W)$, by (2.4). A corresponding relation must hold between f_U, f_V and f_W. In the remaining case $\theta L = L \cup \{a\}$. Then θL is a connected plane of $\boldsymbol{M}$, by I(2.3) and the fact that some point on θL must meet both L and $\{a\}$. By I(3.4) there are connected lines L_U, L_V and L_W on θL, distinct from L, whch are on U, V and W respectively. We can choose these lines to have no common point. For suppose T is such a common point. There is a point T', distinct from T and U, on L_U. By I(3.4) there is a connected line L', distinct from L_U, on L and T'. Without loss of generality we may suppose $\langle L \cap L' \rangle = V$. Replacing L_V by L' we obtain three lines of the kind required. Write $U' = \langle L_V \cap L_W \rangle$, $V' = \langle L_W \cap L_U \rangle$ and $W' = \langle L_U \cap L_V \rangle$. We note that $a \in U' \cap V' \cap W'$. Hence we can apply (4.5a) to the lines L_U, L_V and L_W and obtain

$$f_U = \pm (f_{V'}(a) f_{W'} - f_{W'}(a) f_{V'}),$$
$$f_V = \pm (f_{W'}(a) f_{U'} - f_{U'}(a) f_{W'}),$$
$$f_W = \pm (f_{U'}(a) f_{V'} - f_{V'}(a) f_{U'}).$$

Thus f_U, f_V and f_W are linearly dependent in this case also.

The preceding analysis shows that the chains f_U, $U \in \boldsymbol{M}$ satisfy the conditions of (2.5). They therefore generate a chain-group N on M such that $\boldsymbol{M} = \boldsymbol{M}(N)$. This implies that the f_U are primitive chains of N, and hence that N is regular. But then $\boldsymbol{M}$ is regular, contrary to its definition. This contradiction establishes the theorem.

References

1. C. Kuratowski, *Sur le probleme des courbes gauches en Topologie*, Fund. Math. vol. 15 (1930) pp. 271–283.

2. W. T. Tutte, *A class of Abelian groups*, Canadian J. Math. vol. 8 (1956) pp. 13–28.

3. ———, *A homotopy theorem for matroids*, I. Trans. Amer. Math. Soc. vol. 86 (1958) pp. 144–160.

University of Toronto,
Toronto, Ont.

Reprinted from *American Journal of Mathematics*,
volume 57, pages 800 through 804,
by permission of the Johns Hopkins University Press.

Matroid Representation over *GF*(3)

P. D. SEYMOUR*

Department of Pure Mathematics, University College, Swansea SA2 8PP, Wales

Communicated by the Editors

Received July 31, 1975

We show that a matroid is representable over *GF*(3) if and only if no minor is the five-point line or the Fano matroid, or their duals. Tutte's famous characterization of the regular matroids is a corollary. A key lemma states that two representations of the same matroid in the same vector space over *GF*(3) may be transformed one into the other by inverting some points through the origin and taking a linear transformation; no result of this kind holds in larger fields.

1. INTRODUCTION

If T is a finite subset of the points of a vector space $V_n(F)$ over a field F, the minimal linearly dependent subsets of T are the circuits of a matroid. Such matroids are said to be *vectorial*, or *representable* (over F). Vectorial matroids were discussed by Whitney in the first paper on matroid theory [12], but they have not yet been satisfactorily characterized. It is known that any minor of a vectorial matroid is vectorial (and representable over the same field F), so one might hope for a forbidden minor theorem—see [4] for disillusionment.

An easier problem is the determination of the minor-minimal matroids not representable over some particular field F—easier insofar as it has been solved in one case, when F is $GF(2)$. Tutte [7] showed that a matroid is binary if and only if it has no U_4^2 minor. In this paper we solve the problem when F is $GF(3)$.

The main theorem of this paper has also been obtained (independently) by Bixby [2], and he attributes it to Ralph Reid; however, Reid's proof was never published.

The proof given in this paper is of interest, because we find that it is possible to abstract most of it out of the context of representable matroids into general matroid theory—the most complicated part of our proof consists of lemmas saying that certain connectivity/isomorphism properties imply the existence of certain minors. One might hope that these lemmas

* Present address: Department of Combinatorics and Optimization, Faculty of Mathematics, University of Waterloo, Waterloo, Ontario, Canada.

would also give Tutte's forbidden minor characterization of the regular matroids [7]. But here we derive Tutte's result as a corollary of the main theorem.

We shall assume a basic knowledge of matroid theory; [3, 9, 10, 11] give good introductions to the subject, and our terminology is that of [3], which is universal enough to be clear, extended as follows. $E(\mathbf{M})$ is the set of elements of a matroid $\mathbf{M}$, and for $T \subseteq E(\mathbf{M})$, $\mathbf{M}\backslash T = \mathbf{M} \times (E(\mathbf{M}) - T)$ is the matroid with elements $E(\mathbf{M}) - T$ and with circuits just those circuits of $\mathbf{M}$ contained in $E(\mathbf{M}) - T$. $\mathbf{M}/T = \mathbf{M} \circ (E(\mathbf{M}) - T) = (\mathbf{M}^*\backslash T)^*$, where $\mathbf{M}^*$ is the dual matroid of $\mathbf{M}$. The prefix "co-" dualizes a term (for example, coloop, coparallel, cocircuit). $U_n{}^k$ denotes the matroid with n elements, every k-subset of which is a base. F_7 denotes the Fano matroid.

2. Connectivity Lemmas

The binary relation "$x = y$ or there is a circuit of $\mathbf{M}$ containing both x and y" is an equivalence relation [12], and divides $E(\mathbf{M})$ into equivalence classes called the *elementary separators* of $\mathbf{M}$. Any union of elementary separators is a *separator* of $\mathbf{M}$. $T \subseteq E(\mathbf{M})$ is a separator of $\mathbf{M}$ if and only if $\mathbf{M} \circ T = \mathbf{M} \times T$. If T is an elementary separator of $\mathbf{M}$, then $\mathbf{M} \times T$ $(= \mathbf{M} \circ T)$ is called a *component* of $\mathbf{M}$. $\mathbf{M}$ is *connected* if it has at most one component (that is, exactly one, unless $E(\mathbf{M}) = \varnothing$) and *disconnected* otherwise.

Lemma 2.1 (Tutte [8]). *If* $\mathbf{M}$ *is connected and* $x \in E(\mathbf{M})$, *then either* $\mathbf{M}\backslash\{x\}$ *or* $\mathbf{M}/\{x\}$ *is connected.*

Proof. Suppose that both $\mathbf{M}\backslash\{x\}$ and $\mathbf{M}/\{x\}$ are disconnected. Let Y, Z be elementary separators of $\mathbf{M}\backslash\{x\}$, $\mathbf{M}/\{x\}$, respectively. Then

$$\varnothing \neq Y \neq E(\mathbf{M}) - \{x\} \neq Z \neq \varnothing.$$

Choose distinct $y, z \in E(\mathbf{M}) - \{x\}$ such that

$$|Y \cap \{y, z\}| = |Z \cap \{y, z\}| = 1$$

(it is easy to see that this is possible). $\mathbf{M}$ is connected, so there is a circuit C of $\mathbf{M}$ containing y and z. If $x \notin C$ then C is a circuit of $\mathbf{M}\backslash\{x\}$; and if $x \in C$ then $C - \{x\}$ is a circuit of $\mathbf{M}/\{x\}$. Each of these is contrary to our choice of y, z, as required.

Lemma 2.2. *If* $\mathbf{M}$ *is connected, but* $\mathbf{M}\backslash\{x\}$ *is disconnected for each element* x, *then*

$$X(\mathbf{M}) = \{x : \textit{there exists } y \in E(\mathbf{M}) \textit{ coparallel to } x\}$$

contains a base of $\mathbf{M}^*$.

Proof. We proceed by induction on $|E(\mathbf{M}) - X(\mathbf{M})|$. If $X(\mathbf{M}) = E(\mathbf{M})$ the result is true, so we suppose that $X(\mathbf{M}) \subset E(\mathbf{M})$. Choose $x \in E(\mathbf{M}) - X(\mathbf{M})$. $\mathbf{M}\backslash\{x\}$ is not connected, so by Lemma 2.1, $\mathbf{M}/\{x\}$ is connected. If $y \in E(\mathbf{M}/\{x\})$, then $\mathbf{M}\backslash\{y\}$ is disconnected, so there is a partition Y_1, Y_2 of $E(\mathbf{M}) - \{y\}$ with $Y_1 \neq \varnothing \neq Y_2$ such that any circuit of $\mathbf{M}$ intersecting both Y_1 and Y_2 contains y; $x \in Y_1$, say. If $Y_1 = \{x\}$, then any circuit of $\mathbf{M}$ containing x contains y, because x is not a loop. But then x, y are coparallel, contrary to our choice of x. Thus $Y_1 - \{x\} \neq \varnothing$, and no circuit of $\mathbf{M}/\{x\}\backslash\{y\}$ intersects both $Y_1 - \{x\}$ and Y_2, so $\mathbf{M}/\{x\}\backslash\{y\}$ is disconnected.

Now if $u, v \in E(\mathbf{M}/\{x\})$, u, v are coparallel in $\mathbf{M}$ if and only if u, v are coparallel in $\mathbf{M}/\{x\}$, so $X(\mathbf{M}/\{x\}) = X(\mathbf{M})$. Thus $X(\mathbf{M})$ contains a base of $(\mathbf{M}/\{x\})^*$, by induction, and so contains a base of $\mathbf{M}^*$ as required.

LEMMA 2.3. *If* $\mathbf{M}$ *is connected and* $|E(\mathbf{M})| \geqslant 2$, *and for each element* x *both* $\mathbf{M}\backslash\{x\}$ *and* $\mathbf{M}/\{x\}$ *are connected, and for every pair of distinct elements* x, y *both* $\mathbf{M}\backslash\{x, y\}$ *and* $\mathbf{M}/\{x, y\}$ *are disconnected, then* $\mathbf{M} = U_4^2$.

Proof. $\mathbf{M}$ has no loops or coloops, because $|E(\mathbf{M})| \geqslant 2$ and $\mathbf{M}$ is connected. $|E(\mathbf{M})| \neq 2$, for $\mathbf{M}\backslash E(\mathbf{M})$ is connected; thus $|E(\mathbf{M})| \geqslant 3$. No pair of elements are parallel or coparallel, because $\mathbf{M}\backslash\{x\}$ and $\mathbf{M}/\{x\}$ are connected for each $x \in E(\mathbf{M})$. Let $\mathscr{F}$ be the set of 3-element circuits of $\mathbf{M}$, and let $\mathscr{G}$ be the set of 3-element cocircuits. For any element x, $\mathbf{M}\backslash\{x\}$ satisfies the hypotheses of Lemma 2.2, so $\cup\{D \in \mathscr{G}: x \in D\}$ contains a base of $\mathbf{M}^*$. Similarly (working with $\mathbf{M}^*$), $\cup\{C \in \mathscr{F}: x \in C\}$ contains a base of $\mathbf{M}$.

No pair of elements is parallel in $\mathbf{M}$, so $rk(\mathbf{M}) \geqslant 2$; and similarly $rk(\mathbf{M}^*) \geqslant 2$. Thus for each $x \in E(\mathbf{M})$, there exists $C \in \mathscr{F}$, $D \in \mathscr{G}$ with $x \in C \cap D$; and $|C \cap D| \neq 1$ because C is a circuit and D is a cocircuit.

Suppose that $|C_1 \cap C_2| = 2$ for some $C_1, C_2 \in \mathscr{F}$, and $C_1 = \{1, 2, 3\}$, $C_2 = \{1, 2, 4\}$ say. 3, 4 are not parallel, so $\mathbf{M} \times \{1, 2, 3, 4\} = U_4^2$, and $\{1, 3, 4\} \in \mathscr{F}$, $\{2, 3, 4\} \in \mathscr{F}$. If $D \in \mathscr{G}$ and $1 \in D$, then $|D \cap \{1, 2, 3\}| \neq 1$, so $2 \in D$ say. But $|D \cap \{2, 3, 4\}| \neq 1$, so $D \subsetneq \{1, 2, 3, 4\}$; $D = \{1, 2, 3\}$ say. There exists $D' \in \mathscr{G}$ with $4 \in D'$, so similarly $D' \subsetneq \{1, 2, 3, 4\}$, and

$$\mathbf{M} \circ \{1, 2, 3, 4\} = U_4^2 = \mathbf{M} \times \{1, 2, 3, 4\};$$

hence $\{1, 2, 3, 4\}$ is a separator of $\mathbf{M}$. But $\mathbf{M}$ is connected, and so $\mathbf{M} = U_4^2$.

Thus we may suppose that $|C_1 \cap C_2| \leqslant 1$ for distinct $C_1, C_2 \in \mathscr{F}$, and similarly that $|D_1 \cap D_2| \leqslant 1$ for distinct $D_1, D_2 \in \mathscr{G}$. If there is a unique $C \in \mathscr{F}$ containing x, then C contains a base of $\mathbf{M}$, and so $rk(\mathbf{M}) \leqslant 2$, and $\mathbf{M}$ is a line U_k^2 for some k; $k \leqslant 3$ since otherwise there exist $C_1, C_2 \in \mathscr{F}$ with $|C_1 \cap C_2| = 2$. But U_2^2 and $U_3^2\backslash\{x\}$ are disconnected, contrary to our hypotheses. Thus for $x \in E(\mathbf{M})$ there exist distinct $C_1, C_2 \in \mathscr{F}$ with $x \in C_1$, C_2; and hence $C_1 \cap C_2 = \{x\}$. Choose $D \in \mathscr{G}$ with $x \in D$. Then

$$|D \cap C_1| \neq 1 \neq |D \cap C_2|,$$

so $D \subseteq C_1 \cup C_2$, and if $x \in C \in \mathscr{F}$ then $C = C_1$ or C_2, because $C \cap D \neq \{x\}$. So $C_1 \cup C_2$ contains a base of **M** and hence $Y = C_1 \cup C_2 - \{x\}$ contains a base of **M**. Y contains a circuit of **M**, but no member of $\mathscr{F}$, and so Y is a circuit of **M**. If $z \in E(\mathbf{M}) - (C_1 \cup C_2)$, there is a circuit C_z with $z \in C_z$, and $C_z - \{z\} \subseteq Y$; and hence for any $z \in E(\mathbf{M}) - (C_1 \cup C_2)$, $\mathbf{M}\backslash\{x, z\}$ is connected. So $C_1 \cup C_2 = E(\mathbf{M})$; but then for $y \in Y$ there is a unique member of $\mathscr{F}$ containing y, a contradiction, which completes the proof.

LEMMA 2.4. *If a, b are distinct elements of* **M**, *such that*

(i) *for any $X \subseteq E(\mathbf{M}) - \{a, b\}$, $X \cup \{a\}$ is a circuit of* **M** *if and only if $X \cup \{b\}$ is,*

(ii) *a, b are not both loops and not both coloops; and are not parallel or coparallel,*

(iii) $\mathbf{M}\backslash\{a, b\}$ *is connected,*

then **M** *has a minor* U_5^2.

Proof. We proceed by induction on $|E(\mathbf{M})|$. The result is vacuously true when $|E(\mathbf{M})| \leqslant 2$, and so we suppose that $|E(\mathbf{M})| \geqslant 3$. From (i) and (ii), neither of a, b is a loop or a coloop.

Choose $z \in E(\mathbf{M}) - \{a, b\}$. We may suppose that $\mathbf{M}\backslash\{z\}$ has no U_5^2 minor, so that, by induction, one of the conditions (i), (ii), (iii) fails for $\mathbf{M}\backslash\{z\}$. But (i) certainly holds for $\mathbf{M}\backslash\{z\}$; and if (ii) fails, then either a, b are both coloops of $\mathbf{M}\backslash\{z\}$, or a, b are coparallel in $\mathbf{M}\backslash\{z\}$—thus $\{a, b, z\}$ is a cocircuit of **M** (any other possibility would contradict the truth of (ii) for **M**). Then z is a coloop of $\mathbf{M}\backslash\{a, b\}$, which is connected, and so $E(\mathbf{M}) = \{a, b, z\}$—but this is contrary to (ii). So (iii) must fail for $\mathbf{M}\backslash\{z\}$, and $\mathbf{M}\backslash\{a, b, z\}$ is not connected.

By Lemma 2.1, $\mathbf{M}\backslash\{a, b\}/\{z\}$ is connected, and so condition (iii) holds for $\mathbf{M}/\{z\}$. Certainly, condition (i) holds for $\mathbf{M}/\{z\}$, and so we may assume that (ii) fails, and hence that $\{a, b, z\}$ is a circuit of **M**. This conclusion follows for any $z \in E(\mathbf{M}) - \{a, b\}$, and so **M** may be obtained from a matroid U_t^2 (for some $t \geqslant 2$) which has a, b amongst its elements, by adding new elements parallel to some of the other $t - 2$ old elements. $t \neq 2$, because a, b are not both coloops. $t \neq 3$, because a, b are not coparallel. $t \neq 4$, because $\mathbf{M}\backslash\{a, b\}$ is connected; so $t \geqslant 5$, and **M** has a U_5^2 minor which completes the proof.

[The purpose of the lemma is to give us suitable conditions under which it follows from (i) that a, b are parallel. It is clearly necessary to exclude the other possibilities in (ii), because in each of these cases (i) is trivially true. Possibly condition (iii) could be weakened, but we must exclude U_4^2 somehow; and U_5^2 satisfies all three conditions, so must be mentioned.]

LEMMA 2.5. *Suppose that Z is a circuit and a hyperplane of the matroid* **M**, *and that $a, b \in Z$ are distinct and not coparallel, and that* $\mathbf{M}\backslash\{a, b\}$ *is connected.*

Suppose further that for any two circuits C_1, C_2, *if* $|C_1 - Z| = |C_2 - Z| = 2$ *and* $|C_1 \cup C_2 - Z| = 3$ *then either* $a, b \in C_1 \cup C_2$ *or* $C_1 \triangle C_2$ *is a circuit. Then* $\mathbf{M}$ *has a minor* F_7. [$\triangle$ *denotes* symmetric difference, *that is, sum modulo* 2.]

Proof. We may assume that no pair of elements of $E(\mathbf{M}) - Z$ are parallel; for if $x, y \in E(\mathbf{M}) - Z$ are parallel, then $\mathbf{M}\backslash\{x\}$ satisfies all the hypotheses of the theorem, and if the conclusion holds for $\mathbf{M}\backslash\{x\}$ then it holds for $\mathbf{M}$. Z is a hyperplane, so there is no circuit C with $|C - Z| = 1$; and for any distinct $x, y \in E(\mathbf{M}) - Z$ there is a circuit C with

$$x, y \in C \subseteq Z \cup \{x, y\}.$$

Moreover, since Z is a circuit, for any $z \in Z$ there exists such a C with $z \notin C$. Then C is uniquely defined—for if C' is another circuit with

$$x, y \in C' \subseteq (Z \cup \{x, y\}) - \{z\},$$

then there is a circuit contained in $C \cup C' - \{x\}$, which is impossible. So if $x, y \in C_1, C_2$ and $C_1, C_2 \subseteq Z \cup \{x, y\}$ and C_1, C_2 are distinct, then $Z \subseteq C_1 \cup C_2$.

For distinct $x, y \in E(\mathbf{M}) - Z$, let $A(x, y)$, $B(x, y)$ be the subsets of Z such that $\{x, y\} \cup A(x, y)$, $\{x, y\} \cup B(x, y)$ are circuits and $a \notin A(x, y)$, $b \notin B(x, y)$. If $x, y, z \in E(\mathbf{M}) - Z$ are distinct,

$$A(x, y) \triangle A(y, z) \triangle A(z, x) = \varnothing$$

by hypothesis; and so we may partition $E(\mathbf{M}) - Z$ into two disjoint (possibly empty) sets Y_1, Y_2, such that for distinct $x, y \in E(\mathbf{M}) - Z$, $b \notin A(x, y)$ if and only if x, y are in the same Y_i. But $b \in A(x, y)$ if and only if $a \in B(x, y)$, so $a \notin B(x, y)$ if and only if x, y are in the same Y_i.

Suppose that C is a circuit with $a \notin C$, $b \in C$, and that $C \cap Y_2 = \varnothing$. If $|C \cap Y_1| \geqslant 2$, choose $x, y \in C \cap Y_1$, distinct—then there is a circuit

$$C' \subseteq C \cup A(x, y) - \{x\},$$

with $b \in C'$. So $C' \cap Y_2 = \varnothing$, $a \notin C'$, $b \in C'$, and $C' \cap Y_1 \subset C \cap Y_1$. We continue this process to find a circuit C^* with $a \notin C^*$, $b \in C^*$ and $C^* \cap Y_2 = \varnothing$, $|C^* \cap Y_1| \leqslant 1$. But Z is a hyperplane, yet $|C^* - Z| \leqslant 1$; this is impossible, because $C^* \neq Z$ since $a \notin C^*$. Thus any circuit which contains b and not a intersects both Y_1 and Y_2.

Now a, b are not coparallel, so there is a circuit containing b and not a. This intersects both Y_1 and Y_2, so Y_1 and Y_2 are nonempty.

Suppose that $|Y_1| = 1$, and $Y_1 = \{y_1\}$ say. Any circuit of $\mathbf{M}\backslash\{a\}$ containing b intersects Y_1 and so contains y_1. Thus either $\{b\}$ is a coloop or $\{b, y_1\}$ is a cocircuit of $\mathbf{M}\backslash\{a\}$, and so one of $\{a, b\}$, $\{a, b, y_1\}$, $\{b\}$, $\{b, y_1\}$ is a cocircuit of $\mathbf{M}$. Z is a circuit and $b \in Z$, $y_1 \notin Z$, so the last two cases cannot occur. By hypothesis, the first case does not occur. If $\{a, b, y_1\}$ is a cocircuit, then $\{y_1\}$ is a coloop of $\mathbf{M}\backslash\{a, b\}$, which is connected; and so $E(\mathbf{M}) = \{a, b, y_1\}$, which is impossible. Thus $|Y_1| \geqslant 2$, and similarly $|Y_2| \geqslant 2$.

For $i = 1, 2$, put $Z_i = \cup(A(x, y)\colon x, y \in Y_i$, distinct). $|Y_i| \geqslant 2$, and for distinct $x, y \in Y_i$, $A(x, y) \neq \varnothing$ since by assumption no two elements of $E(\mathbf{M}) - Z$ are parallel. Thus $Z_1, Z_2 \neq \varnothing$.

Suppose that $Z_1 \cap Z_2 = \varnothing$. $\mathbf{M}\backslash\{a, b\}$ is connected, so there is a circuit C with $a, b \notin C$, intersecting both Z_1 and Z_2. Choose C with $C - Z$ minimal. If $|C \cap Y_1| \geqslant 2$, choose distinct $x, y \in C \cap Y_1$; there is a circuit

$$C' \subseteq C \cup A(x, y) - \{x\}$$

with $C' \cap Z_2 \neq \varnothing$ (because

$$\varnothing \neq C \cap Z_2 \subseteq C - A(x, y)),$$

and then $C' \cap Z_1 \neq \varnothing$, because $C' \cap A(x, y) \neq \varnothing$ (since $C' \not\subseteq C$). This contradicts the minimality of $C - Z$. On the other hand, $|C - Z| \geqslant 2$; so $|C \cap Y_1| = |C \cap Y_2| = 1$, and $C \cap Y_i = \{y_i\}$ say $(i = 1, 2)$. But then $C = A(y_1, y_2)$, and $b \notin A(y_1, y_2)$—yet $y_1 \in Y_1$, $y_2 \in Y_2$ contrary to the definition of Y_1, Y_2. So $Z_1 \cap Z_2 \neq \varnothing$.

Choose $z \in Z_1 \cap Z_2$; and choose $x_i, y_i \in Y_i$ such that $z \in A(x_i, y_i)$ $(i = 1, 2)$.

$$A(x_1, y_1) \triangle A(y_1, x_2) = A(x_1, x_2)$$

by hypothesis, so we may exchange x_1, y_1 if necessary so that $z \in A(x_1, x_2)$ and $z \notin A(y_1, x_2)$. It follows that $z \notin A(x_1, y_2)$, and $z \in A(y_1, y_2)$. Now $A(x_1, y_2) \neq B(x_1, y_2)$, because $b \in A(x_1, y_2)$; so $A(x_1, y_2) \cup B(x_1, y_2) \supseteq Z$. Thus $z \in B(x_1, y_2)$; and $z \notin B(x_1, x_2)$, $z \notin B(y_1, y_2)$, and $z \in B(y_1, x_2)$.

Put $\mathbf{M}_0 = \mathbf{M} \times (Z \cup \{x_1, x_2, y_1, y_2\}) \circ \{x_1, x_2, y_1, y_2, a, b, z\}$. We claim that $\mathbf{M}_0$ is F_7. For $\{a, b, z\}$ is a circuit of $\mathbf{M}_0$, so it suffices to show that for distinct $u, v \in \{x_1, x_2, y_1, y_2\}$, $A(u, v) \cap E(\mathbf{M}_0)$ and $B(u, v) \cap E(\mathbf{M}_0)$ are circuits of $\mathbf{M}_0$. But certainly they are dependent in $\mathbf{M}_0$—and if C is a circuit of $\mathbf{M}$ with

$$C - (Z - \{a, b, z\}) \subseteq A(u, v),$$

then $b \notin C$, so $C \neq Z$ and $u, v \in C$; thus $C = A(u, v)$. So $A(u, v) \cap E(\mathbf{M}_0)$ is a circuit of $\mathbf{M}_0$, and $B(u, v)$ is treated similarly—the conclusion follows.

LEMMA 2.6. *If Z is a circuit and a hyperplane of a matroid $\mathbf{M}_1$, with elements S, and if the matroid $\mathbf{M}_2$ is obtained from $\mathbf{M}_1$ by replacing the old circuit Z by new circuits $Z \cup \{z\}$ for each $z \in S - Z$; and if further there exist $a, b \in Z$ which are distinct and not coparallel in $\mathbf{M}_1$ such that $\mathbf{M}_1\backslash\{a, b\}$ is connected—then either $\mathbf{M}_1$ has a minor F_7 or $\mathbf{M}_2$ has a minor U_5^2.*

Proof. From Lemma 2.5 applied to $\mathbf{M}_1$, we may suppose that there are circuits C_1, C_2 of $\mathbf{M}_1$, with $C_1 - Z = \{x_1, y\}$, $C_2 - Z = \{x_2, y\}$ say, with x_1, x_2, y distinct, such that $\{a, b\} \not\subseteq C_1 \cup C_2$ and $C_1 \triangle C_2$ is not a circuit of $\mathbf{M}_1$. Without loss of generality, we assume that $a \notin C_1 \cup C_2$.

For distinct $u, v \in S - Z$, there is a circuit $C(u, v)$ of $\mathbf{M}_1$ with

$$u, v \in C(u, v) \subseteq Z \cup \{u, v\},$$

because Z is a hyperplane; and we may choose $C(u, v)$ such that $a \notin C(u, v)$, because Z is a circuit. Then $C(u, v)$ is uniquely defined.

Now $C_1 = C(x_1, y)$, and $C_2 = C(x_2, y)$. For any $z \in C_1 \triangle C_2 - \{x_1, x_2\}$, there is a circuit C of $\mathbf{M}_1$ with

$$z \in C \subseteq C_1 \cup C_2 - \{y\};$$

then $a \notin C$, so $x_1, x_2 \in C$ (because Z is a hyperplane), and $C = C(x_1, x_2)$. Thus $C_1 \triangle C_2 \subseteq C(x_1, x_2)$. But $C_1 \triangle C_2$ is not a circuit, so there exists $z \in C(x_1, x_2) \cap C_1 \cap C_2$.

We claim that $\mathbf{M}_2 \times (Z \cup \{x_1, x_2, y\}) \circ \{a, z, x_1, x_2, y\}$ is U_5^2. For C_1, C_2, $C(x_1, x_2)$, $Z \cup \{x_1\}$, $Z \cup \{x_2\}$, $Z \cup \{y\}$ are circuits of $\mathbf{M}_2$, so it suffices to check that there is no circuit C_0 of $\mathbf{M}_2$ with

$$C_0 \subseteq Z \cup \{x_1, x_2, y\}$$

and

$$| C_0 \cap \{a, z, x_1, x_2, y\}| \leqslant 2.$$

But if C_0 is not a circuit of $\mathbf{M}_1$, then $C_0 = Z \cup \{w\}$ for some $w \in \{x_1, x_2, y\}$ and $a, z \in C_0$; and if C_0 is a circuit of $\mathbf{M}_1$, then $C_0 \neq Z$ (because Z is not a circuit of $\mathbf{M}_2$) so $| C_0 - Z | \geqslant 2$ (because Z is a hyperplane of $\mathbf{M}_1$) and either $a \in C_0$ or C_0 is one of C_1, C_2, $C(x_1, x_2)$ each of which contains z. The conclusion follows.

LEMMA 2.7. *If $\mathbf{M}_1$, $\mathbf{M}_2$ are distinct matroids, both with elements S, and $a, b \in S$ are distinct elements such that*

(i) $\mathbf{M}_1 \backslash \{a\} = \mathbf{M}_2 \backslash \{a\}$,

(ii) $\mathbf{M}_1 \backslash \{b\} = \mathbf{M}_2 \backslash \{b\}$,

(iii) $\mathbf{M}_1 \backslash \{a, b\}$ $(= \mathbf{M}_2 \backslash \{a, b\})$ *is connected,*

(iv) $\{a, b\}$ *is not a cocircuit of* $\mathbf{M}_1$ *or of* $\mathbf{M}_2$,

then one of $\mathbf{M}_1$, $\mathbf{M}_2$ *has a minor* U_5^2 *or* F_7.

Proof. We proceed by induction on $| S |$. The result is vacuously true when $| S | \leqslant 1$.

$\mathbf{M}_1 \neq \mathbf{M}_2$, so there is a subset of S independent in one of $\mathbf{M}_1$, $\mathbf{M}_2$ and dependent in the other. Choose a minimal such subset Z. Then Z is a circuit of one of $\mathbf{M}_1$, $\mathbf{M}_2$ ($\mathbf{M}_1$ say) and independent in the other ($\mathbf{M}_2$). $\mathbf{M}_1\backslash\{a\} = \mathbf{M}_2\backslash\{a\}$, and $\mathbf{M}_1\backslash\{b\} = \mathbf{M}_2\backslash\{b\}$, so $a, b \in Z$.

For $z \in E(\mathbf{M}) - \{a, b\}$, we may assume that either $\mathbf{M}_1\backslash\{z\}$, $\mathbf{M}_2\backslash\{z\}$ are not distinct, or that one of conditions (i)$\cdots$(iv) fails for this pair of matroids—for otherwise, by induction, one of these matroids (and hence one of $\mathbf{M}_1$, $\mathbf{M}_2$) has a minor U_5^2 or F_7. But conditions (i) and (ii) cannot fail, and if (iv) fails, then $\{a, b, z\}$ is a cocircuit of $\mathbf{M}_1$ or of $\mathbf{M}_2$—but $\mathbf{M}_1\backslash\{a, b\}$ is connected, so $S = \{a, b, z\}$, and then a, b are coparallel in $\mathbf{M}_2$, contrary to (iv). Thus either $\mathbf{M}_1\backslash\{z\} = \mathbf{M}_2\backslash\{z\}$ or $\mathbf{M}_1\backslash\{a, b, z\}$ is not connected. Similarly, either $\mathbf{M}_1/\{z\} = \mathbf{M}_2/\{z\}$ or $\mathbf{M}_1\backslash\{a, b\}/\{z\}$ is not connected.

Now if $z \in S - Z$, then Z is a circuit of $\mathbf{M}_1\backslash\{z\}$ and is independent in $\mathbf{M}_2\backslash\{z\}$; so $\mathbf{M}_1\backslash\{z\} \neq \mathbf{M}_2\backslash\{z\}$. Hence $\mathbf{M}_1\backslash\{a, b, z\}$ is not connected.

On the other hand, if $z \in Z - \{a, b\}$, then $Z - \{z\}$ is a circuit of $\mathbf{M}_1/\{z\}$ and is independent in $\mathbf{M}_2/\{z\}$, so $\mathbf{M}_1/\{z\} \neq \mathbf{M}_2/\{z\}$. Thus $\mathbf{M}_1\backslash\{a, b\}/\{z\}$ is not connected.

But from Lemma 2.1, for any $z \in S - \{a, b\}$, either $\mathbf{M}_1\backslash\{a, b, z\}$ is connected or $\mathbf{M}_1\backslash\{a, b\}/\{z\}$ is connected; so $Z - \{a, b\}$ is the set of all elements $z \in S - \{a, b\}$ such that $\mathbf{M}_1\backslash\{a, b, z\}$ is connected. Thus Z is uniquely defined, and Z is the only subset of S which is a circuit of one of $\mathbf{M}_1$, $\mathbf{M}_2$ and independent in the other.

If $z \in S - Z$, then $\mathbf{M}_1\backslash\{a, b, z\}$ is not connected, as before, so $\mathbf{M}_1\backslash\{a, b\}/\{z\}$ is connected, by Lemma 2.1. Thus $\mathbf{M}_1/\{z\} = \mathbf{M}_2/\{z\}$. Z is a circuit of $\mathbf{M}_1$, and so is a union of circuits of $\mathbf{M}_1/\{z\}$; but Z is independent in $\mathbf{M}_2$, so contains at most one circuit of $\mathbf{M}_2/\{z\}$. Thus Z is a circuit of $\mathbf{M}_1/\{z\} = \mathbf{M}_2/\{z\}$, and so $Z \cup \{z\}$ is a circuit of $\mathbf{M}_2$, for each $z \in S - Z$.

If C is a circuit of one of $\mathbf{M}_1$, $\mathbf{M}_2$ and not of the other, then by the uniqueness of Z, $Z \subseteq C$; so $C = Z$ or $Z \cup \{z\}$ for some $z \in S - Z$. In these circumstances Lemma 2.6 applies, and either $\mathbf{M}_1$ has a minor F_7 or $\mathbf{M}_2$ has a minor U_5^2, as required.

3. Chain-Groups

Let R be a ring which is either the integers ($\mathbb{Z}$) or a field. Let S be a finite set. A *chain* g on S over R is a map from S to R. The *domain* $|g|$ of g is $\{x \in S: g(x) \neq 0\}$. If f, g are chains, $f + g$ is defined by

$$(f + g)(x) = f(x) + g(x) \qquad (x \in S),$$

and if $\gamma \in R$, γf is defined by

$$(\gamma f)(x) = \gamma(f(x)) \qquad (x \in S).$$

A *chain-group* G on S over R is a class G of chains on S over R closed under the above two operations. $g \in G$ is an *elementary* chain of G if $g \neq 0$ but for any $f \in G$, if $|f| \subset |g|$ then $f = 0$. The domains of the elementary chains of a chain-group G are the circuits of a matroid $\mathbf{M}(G)$. (For proofs of this and some subsequent elementary results about chain-groups, see [9, 10].)

If g is a chain on S over R and $T \subseteq S$, $g \mid T$ is the chain on T over R such that $(g \mid T)(x) = g(x)$ $(x \in T)$. When G is a chain-group on S over R,

$$G \circ T = \{g \mid T : g \in G\},$$

and

$$G \times T = \{g \mid T : g \in G \text{ and } g \mid (S - T) = 0\}.$$

Both $G \circ T$ and $G \times T$ are chain-groups on T over R, and $\mathbf{M}(G) \times T = \mathbf{M}(G \times T)$, $\mathbf{M}(G) \circ T = \mathbf{M}(G \circ T)$. Let G^* be the set of chains g' on S over R such that $\sum_{x \in S} g(x)\, g'(x) = 0$ for each $g \in G$; then $(\mathbf{M}(G))^* = \mathbf{M}(G^*)$, and $(G \times T)^* = G^* \circ T$ for $T \subseteq S$, and $(G^*)^* = G$.

A *representation* of a matroid $\mathbf{M}$ over a field F is a map φ: $E(\mathbf{M}) \to V_n(F)$ for some positive integer n (where $V_n(F)$ is a vector space of dimension n over F) such that for $X \subseteq E(\mathbf{M})$, X is independent in $\mathbf{M}$ if and only if the restriction of φ to X is one-to-one and $\{\varphi(x): x \in X\}$ is linearly independent in $V_n(F)$. If λ is a nonsingular linear transformation of $V_n(F)$, let $\lambda(\varphi)$: $E(\mathbf{M}) \to V_n(F)$ be defined by

$$\lambda(\varphi)(x) = \lambda(\varphi(x)) \qquad (x \in E(\mathbf{M})).$$

Then $\lambda(\varphi)$ is a representation of $\mathbf{M}$ if φ is.

If φ is a representation of $\mathbf{M}$ over F, we associate a chain-group $G(\varphi)$ containing just those chains g on $E(\mathbf{M})$ over F such that $\sum_{x \in E(\mathbf{M})} g(x)\varphi(x) = 0$. Then $\mathbf{M}(G(\varphi)) = \mathbf{M}$, and it is easily seen that $\mathbf{M}$ is representable over a field F if and only if there is a chain-group G over F with $\mathbf{M} = \mathbf{M}(G)$. Moreover, two representations φ, ψ of $\mathbf{M}$ over F (mapping into the same vector space $V_n(F)$) satisfy $G(\varphi) = G(\psi)$ if and only if there is a nonsingular linear transformation λ of $V_n(F)$ such that $\lambda(\varphi) = \psi$. It is also easy to check that $\mathbf{M}$ is representable over the rationals if and only if $\mathbf{M} = \mathbf{M}(G)$ for some chain-group G over the integers.

Thus it is possible to work exclusively with the "chain-group" rather than the "representation" formulation without in theory any significant loss, and to do so gives a marked increase in clarity and simplicity—however, some results have an intuitive geometric significance which is lost in the chain-group formulation, and also the best way to specify a particular chain-group is often to give some corresponding representation; so both formulations should be borne in mind.

If R is either $\mathbb{Z}$ or a field as before, we say that $\mathbf{M}$ is *representable* over R if $\mathbf{M} = \mathbf{M}(G)$ for some chain-group G over R. This is consistent with our previous definition if R is a field.

We observe that if G is a chain-group on S over R and α is a map from S to the invertible elements of R, then the chain-group αG, containing just those chains f on S over R such that for some $g \in G$, $f(x) = \alpha(x)g(x)$ $(x \in S)$, satisfies $\mathbf{M}(\alpha G) = \mathbf{M}(G)$. $\alpha^{-1}(\alpha G) = G$, where $\alpha^{-1}(x) = (\alpha(x))^{-1}$ $(x \in S)$, so this provides an equivalence relation on those chain-groups G such that $\mathbf{M}(G) = \mathbf{M}$.

There are several characterizations of *binary* matroids [that is, the matroids representable over $GF(2)$]:

THEOREM 3.1. *The following are equivalent:*

(i) $\mathbf{M}$ *is binary;*

(ii) $\mathbf{M}$ *has no minor* U_4^2 (*Tutte* [7]);

(iii) *any symmetric difference* (*that is, sum modulo* 2) $C_1 \triangle \cdots \triangle C_k$ *of circuits of* $\mathbf{M}$ *is a disjoint union of circuits of* $\mathbf{M}$ (*Whitney* [12]);

(iv) *for any circuit* C *and cocircuit* D, $| C \cap D |$ *is even;*

(v) *for any circuit* C *and cocircuit* D, $| C \cap D | \neq 3$ (*this follows from* (ii)).

There are also characterizations for connected matroids in terms of the circuits containing some fixed element; see [1, 5, 6].

If $\mathbf{M}$ is binary, there is a unique chain-group G over $GF(2)$ such that $\mathbf{M} = \mathbf{M}(G)$, because $GF(2)$ has only one nonzero element. Thus any representation of $\mathbf{M}$ in a vector space over $GF(2)$ can be taken by means of a linear transformation into any other. This is not true in any other field; because multiplying a point (used by a representation) of the vector space by any nonzero scalar gives another representation, and this cannot in general be obtained from the first by a linear transformation. However, in $GF(3)$ the next best thing is true:

LEMMA 3.2: *If* G_1, G_2 *are chain-groups on* S *over* $GF(3)$ *and* $\mathbf{M}(G_1) = \mathbf{M}(G_2) = \mathbf{M}$, *then there is a map* $\alpha: S \to \{+1, -1\}$ *such that* $\alpha G_1 = G_2$. [*We write the elements of* $GF(3)$ *as* $+1, 0, -1$.]

Proof. We proceed by induction on $| S |$; the result is trivial when $| S | \leqslant 1$, so we suppose that $| S | \geqslant 2$. We may assume that $\mathbf{M}$ is connected, for other-otherwise $\mathbf{M}$ satisfies the theorem, by our inductive hypothesis applied to its components.

Now $\alpha G_1 = G_2$ if and only if $\alpha G_1^* = G_2^*$; choose $z \in S$, and then taking duals if necessary we may assume that $\mathbf{M}\backslash\{z\}$ is connected, by Lemma 2.1. Choose $\alpha': S \to \{+1, -1\}$ such that

$$(\alpha' G_1) \times (S - \{z\}) = G_2 \times (S - \{z\});$$

this is possible by application of our inductive hypothesis to $G_1 \times (S - \{z\})$, $G_2 \times (S - \{z\})$ and arbitrary extension of the resulting map to domain S. If there exists $\alpha'': S \to \{+1, -1\}$ such that $\alpha''(\alpha' G_1) = G_2$ then the theorem is satisfied (defining $\alpha(x) = \alpha'(x)\, \alpha''(x)$ $(x \in S)$), and so it suffices to deal with the case when

$$G_1 \times (S - \{z\}) = G_2 \times (S - \{z\});$$

we henceforth assume this.

Let $z' \notin S$ be a new element, and let G_2' be an isomorphic copy of G_2 on the set $(S - \{z\}) \cup \{z'\}$, obtained by replacing z by z'. Let G^+ be the chain-group on $S \cup \{z'\}$ over $GF(3)$ with chains $g_1 + g_2'$ $(g_1 \in G_1, g_2' \in G_2')$. Then $G^+ \times S = G_1$, and $G^+ \times ((S - \{z\}) \cup \{z'\}) = G_2'$, because $G_1 \times (S - \{z\}) = G_2' \times (S - \{z\})$. So for $X \subseteq S - \{z\}$, $X \cup \{z\}$ is a circuit of $\mathbf{M}(G^+)$ if and only if $X \cup \{z'\}$ is. z, z' are not coparallel in $\mathbf{M}(G^+)$, and z is not a loop or coloop of $\mathbf{M}(G^+)$, because z is not a loop or coloop of $\mathbf{M}(G^+)\backslash\{z'\} = \mathbf{M}$ (for $\mathbf{M}$ is connected and $|S| \geqslant 2$). Moreover, $\mathbf{M}(G^+)\backslash\{z, z'\} = \mathbf{M}\backslash\{z\}$ and hence is connected. Thus by Lemma 2.4, z, z' are parallel in $\mathbf{M}(G^+)$, because U_5^2 is not representable over $GF(3)$. Let f be a chain of G^+ with $|f| = \{z, z'\}$; we may assume that $f(z) = 1$, and $f(z') = \gamma = \pm 1$. If $g \in G_1$ and $z \in |g|$, then the chain $g' = g - g(z)f$ is a chain of G_2', because $g'(z) = 0$. But $g'(x) = g(x)$ $(x \in S - \{z\})$, and $g'(z') = -\gamma g(z)$; and hence defining $\alpha: S \to \{+1, -1\}$ by $\alpha(x) = 1$ $(x \neq z)$, $\alpha(z) = -\gamma$, we have $\alpha G_1 = G_2$, as required.

Thus any representation of $\mathbf{M}$ over $GF(3)$ can be taken to any other by negating some points and applying a linear transformation.

COROLLARY 3.3. *If* $\mathbf{M}$ *is representable over* $GF(3)$ *in* $V_n(GF(3))$, *and* φ^- *is a representation of* $\mathbf{M}\backslash\{x\}$ *for some element* $x \in E(\mathbf{M})$, *then there is a representation* φ *of* $\mathbf{M}$ *in* $V_n(GF(3))$ *such that* $\varphi(y) = \varphi^-(y)$ $(y \in E(\mathbf{M}) - \{x\})$.

Proof. Let ψ be some representation of $\mathbf{M}$ in $V_n(GF(3))$. By Lemma 3.2 there exists a map $\alpha: E(\mathbf{M}) - \{x\} \to \{+1, -1\}$, such that $G(\varphi^-) = G(\alpha\psi^-)$, where ψ^- is the restriction of ψ to $E(\mathbf{M}) - \{x\}$. There is a nonsingular linear transformation λ of $V_n(GF(3))$ such that

$$\lambda(\alpha\psi^-(y)) = \varphi^-(y) \qquad (y \in E(\mathbf{M}) - \{x\}).$$

Define $\varphi(x) = \lambda\psi(x)$, and $\varphi(y) = \varphi^-(y)$ $(y \in E(\mathbf{M}) - \{x\})$. Then φ is the required representation.

No such result holds in larger fields, because if F is a field with an element γ such that $\gamma \neq 0$, 1 and $\gamma^2 \neq 1$ (and every field except $GF(2)$ and $GF(3)$ has such an element), then

$$\{(1, 0, 0), (0, 1, 0), (1, 1, 0), (\gamma, 1, 1), (\gamma, 1, \gamma), (0, 0, 1)\}$$

and

$$\{(1, 0, 0), (0, 1, 0), (1, \gamma, 0), (1, 1, 1), (\gamma, \gamma, 1), (0, 0, 1)\}$$

are both representations of the rank 3 matroid with two disjoint lines of length 3; adding (1, 1, 1) to the first representation produces a matroid not obtainable by an extension of the second representation. It follows that no analog of Lemma 3.2 holds in larger fields either; and in fact that scalar multiplication together with automorphisms of the associated projective geometry will not transform one representation into the other.

If Γ is a set of chains on S over $\mathbb{Z}$, the chain-group *generated* by Γ is the set of all finite sums of members of Γ and their negatives.

If G is a chain-group on S over $\mathbb{Z}$, G is said to be *regular* if every elementary chain of G is a multiple of a *primitive* chain, that is, an elementary chain $g \in G$ such that $-1 \leqslant g(x) \leqslant 1$ $(x \in S)$. $\mathbf{M}(G)$ is then said to be a *regular* matroid. Such matroids are of great theoretical interest—see [7, 9, 10].

If g is a chain on S over $GF(3)$, let g^+ be the chain on S over $\mathbb{Z}$ such that $-1 \leqslant g^+(x) \leqslant 1$ $(x \in S)$ and for each $x \in S$, $g(x)$ is the residue modulo 3 of $g^+(x)$. [That is, label the elements of $GF(3)$ as $+1, 0, -1$; then g^+ is the chain g with its values regarded as integers.]

If g, h are chains on S, we define $g \cdot h$ to be $\sum_{x \in S} g(x)h(x)$.

LEMMA 3.4. $\mathbf{M}$ *is regular if and only if* $\mathbf{M}$ *is representable over both* $GF(2)$ *and* $GF(3)$.

Proof. If $\mathbf{M}$ is regular, then $\mathbf{M}$ is representable over every field (essentially, a result of Tutte [9]). For the converse, suppose that $\mathbf{M} = \mathbf{M}(G)$, where G is a chain-group over $GF(3)$, and that $\mathbf{M}$ is binary. Let Γ be the set of elementary chains of G, and let $\Gamma^+ = \{g^+ : g \in \Gamma\}$. Let G^+ be the chain-group over $\mathbb{Z}$ generated by Γ^+. We claim that $\mathbf{M}(G^+) = \mathbf{M}$. If we can show this, the result follows, because G^+ must then be a regular chain-group; for each elementary chain f of G^+ has domain $|g|$ for some $g \in \Gamma$, and so f is a multiple of g^+, which is a primitive chain of G^+.

Each circuit of $\mathbf{M}$ is the domain of a chain in $\Gamma^+ \subseteq G^+$, and so includes a circuit of $\mathbf{M}(G^+)$. It remains to show the converse inclusion. But if g_0 is an elementary chain of G^+ and $|g_0|$ is independent in $\mathbf{M}$, then there exists $x \in |g_0|$, and there exists $h \in G^*$ such that $|g_0| \cap |h| = \{x\}$ — and so $g_0 \cdot h^+ \neq 0$. g_0 is a sum of members of Γ^+; so it is sufficient to show that for each $g \in \Gamma$ and for each elementary chain h of G^*, $g^+ \cdot h^+ = 0$.

Put $|g| = C$, $|h| = D$. We proceed by induction on $|C \cap D|$. The result is true if $|C \cap D| \leqslant 2$, because certainly $g \cdot h = 0$, so $g^+ \cdot h^+ \equiv 0$ mod 3, and for each x, $g^+(x)\, h^+(x)$ is $+1$, 0, or -1. We assume that $|C \cap D| \geqslant 3$. D is a cocircuit of $\mathbf{M}$, so each pair of elements of D is the intersection of D with some circuit. Thus we can choose a circuit C' of $\mathbf{M}$ with $C' - (C \cup D)$ minimal subject to the conditions $\varnothing \neq C' \cap D \subset C \cap D$. Choose $x \in (C \cap D) - C'$, and $y \in C' \cap D$. Now $\mathbf{M}$ is binary, so $C \triangle C'$ is a disjoint union of circuits $C_1, \ldots, C_k$ say, with $x \in C_1$. Now $C_i \cap D \subseteq C \cap D -$

$\{y\}$ for each i, and $C_1 \cap D \neq \varnothing$; so by the choice of C', $C' - C = C_1 - C$. Since $C_1, \ldots, C_k$ are disjoint, each C_i with $i \neq 1$ is a subset of $C - \{y\}$, which is impossible; thus $k = 1$, and $C \triangle C'$ is a circuit.

Let $C' = |g_1|$ and $C \triangle C' = |g_2|$, where g_1, g_2 are elementary chains of G and $g_2(x) = g(x)$. Let

$$f_1 = g - g_2 - g_1,$$
$$f_2 = g - g_2,$$
$$f_3 = g - g_2 + g_1;$$

f_1, f_2, f_3 are all chains of G whose domains do not contain x. Now $C' \nsubseteq C$; choose $z \in C' - C$. $g_1(z) \neq 0$, so $f_1(z), f_2(z), f_3(z)$ are distinct and one of them is zero; but $f_2(z) = -g_2(z) \neq 0$, so (replacing g_1 by $-g_1$ if necessary) we may assume that $f_1(z) = 0$. Thus, by choice of C', $|f_1| \cap D = \varnothing$ (because $|f_1|$ is a union of circuits of $\mathbf{M}$) and so for each $w \in D$, $g(w) = g_1(w) + g_2(w)$. It follows that for $w \in D$, $g^+(w) = g_1^+(w) + g_2^+(w)$; because certainly this is true modulo 3, and $|g| \cap |g_1| \cap |g_2| = \varnothing$. By choice of C', both $|g_1| \cap D$ and $|g_2| \cap D$ are proper subsets of $C \cap D$, and so $g_1^+ \cdot h^+ = g_2^+ \cdot h^+ = 0$ by induction; it follows that $g^+ \cdot h^+ = 0$, as required.

4. The Principal Result

Theorem 4.1. $\mathbf{M}$ *is representable over* $GF(3)$ *if and only if* $\mathbf{M}$ *has no minor* U_5^2, U_5^3, F_7, *or* F_7^*.

Proof. These four matroids are not representable over $GF(3)$ (this is well known and easily verified). For the converse, we proceed by induction on $|S|$, where $S = E(\mathbf{M})$. The result is trivial when $|S| \leqslant 1$, and so we assume that $|S| \geqslant 2$. We may assume that $\mathbf{M}$ is connected, for otherwise we can treat its components separately.

Suppose that $z \in S$ and $\mathbf{M} \backslash \{z\}$ is disconnected; let Z be an elementary separator. Let $\mathbf{M}_1 = \mathbf{M} \circ (Z \cup \{z\})$, and $\mathbf{M}_2 = \mathbf{M} \circ (S - Z)$. Let G_1, G_2 be chain-groups on $E(\mathbf{M}_1)$, $E(\mathbf{M}_2)$ over $GF(3)$ such that $\mathbf{M}_1 = \mathbf{M}(G_1)$, $\mathbf{M}_2 = \mathbf{M}(G_2)$ (we may assume that these exist, for otherwise, by induction, one of $\mathbf{M}_1$, $\mathbf{M}_2$—and hence $\mathbf{M}$— has one of the proscribed minors). Let G be the chain-group on S over $GF(3)$ with chains g defined by

$$g(x) = g_1(x) + g_2(x) \qquad (x \neq z),$$
$$g(z) = g_1(z) = g_2(z),$$

for any pair of chains $g_1 \in G_1$, $g_2 \in G_2$ such that $g_1(z) = g_2(z)$. Then $G \circ (Z \cup \{z\}) = G_1$, and $G \circ (S - Z) = G_2$. If $C \subseteq S$, and $z \notin C$, then C is a circuit of $\mathbf{M}$ if and only if C is a circuit of one of $\mathbf{M}_1$, $\mathbf{M}_2$; and if $z \in C$, then

C is a circuit of $\mathbf{M}$ if and only if $C \cap (Z \cup \{z\})$ is a circuit of $\mathbf{M}_1$, and $C \cap (S - Z)$ is a circuit of $\mathbf{M}_2$. Thus $\mathbf{M} = \mathbf{M}(G)$, and $\mathbf{M}$ is representable over $GF(3)$ as required.

We may thus assume that $\mathbf{M}\backslash\{z\}$ is connected for any $z \in S$. Now $\mathbf{M}$ is representable over $GF(3)$ if and only if $\mathbf{M}^*$ is, and $\mathbf{M}$ has one of the four proscribed minors if and only if $\mathbf{M}^*$ does; it follows that we may assume that $\mathbf{M}/\{z\}$ is connected for each $z \in S$.

If $\mathbf{M} = U_4^2$, then $\mathbf{M}$ is representable over $GF(3)$, so we may assume that $\mathbf{M} \neq U_4^2$; and by Lemma 2.3, there exist $a, b \in S$, $a \neq b$, such that $\mathbf{M}\backslash\{a, b\}$ is connected (taking duals if necessary).

Let G_1, G_2 be chain-groups on $S - \{a\}$, $S - \{b\}$, respectively, over $GF(3)$, such that $\mathbf{M}\backslash\{a\} = \mathbf{M}(G_1)$ and $\mathbf{M}\backslash\{b\} = \mathbf{M}(G_2)$ (we may assume that these exist, by induction). By Lemma 3.2 (or Corollary 3.3) we may choose G_2 such that

$$G_2 \times (S - \{a, b\}) = G_1 \times (S - \{a, b\}).$$

Let G have chains $g_1 + g_2$ for $g_1 \in G_1$, $g_2 \in G_2$. Then G is a chain-group on S over $GF(3)$, and $G \times (S - \{a\}) = G_1$, and $G \times (S - \{b\}) = G_2$. $\mathbf{M}\backslash\{a, b\}$ is connected, and $\mathbf{M}(G)\backslash\{a\} = \mathbf{M}\backslash\{a\}$, and $\mathbf{M}(G)\backslash\{b\} = \mathbf{M}\backslash\{b\}$, so by Lemma 2.7 either $\mathbf{M} = \mathbf{M}(G)$ or one of $\mathbf{M}$, $\mathbf{M}(G)$ has a U_5^2 or F_7 minor. $\mathbf{M}(G)$ has none of these minors, because they are not representable over $GF(3)$. So either $\mathbf{M} = \mathbf{M}(G)$, or $\mathbf{M}$ has a minor U_5^2 or F_7, completing the proof.

COROLLARY 4.2 (Tutte [7]). $\mathbf{M}$ *is regular if and only if* $\mathbf{M}$ *has no minor* U_4^2, F_7, *or* F_7^*.

Proof. $\mathbf{M}$ is not representable over $GF(2)$ if and only if $\mathbf{M}$ has a U_4^2 minor (Theorem 3.1). $\mathbf{M}$ is not representable over $GF(3)$ if and only if $\mathbf{M}$ has a minor U_5^2, U_5^3, F_7, or F_7^* (Theorem 4.1), and U_5^2, U_5^3 have U_4^2 minors. $\mathbf{M}$ is regular if and only if $\mathbf{M}$ is representable over $GF(2)$ and $GF(3)$ (Lemma 3.4), and the result follows.

It is clear that the crux of the proof of our main theorem is Lemma 2.7; and it may seem remarkable (and lucky) that this lemma should be true. But in fact it can be shown quite easily, from Lemma 3.2, that if $\mathbf{M}_1$, $\mathbf{M}_2$ are related as in Lemma 2.7, then one of $\mathbf{M}_1$, $\mathbf{M}_2$ is not representable over $GF(3)$; and so a weak form of Lemma 2.7 is implied by the truth of our main result.

Note added in proof. Lemma 3.2 is proved in [13].

REFERENCES

1. R. E. BIXBY, l-matrices and a characterization of binary matroids, *Discrete Math.* **8** (1974), 139–145.
2. R. E. BIXBY, On Reid's characterization of the matroids, *J. Combinatorial Theory Ser. B* **26** (1979), 174–204.

3. F. Harary and D. J. A. Welsh, "Matroids versus Graphs," Lecture Notes in Mathematics, Vol. 110, pp. 155–170, Springer, Berlin, 1969.
4. A. W. Ingleton, Representation of matroids, *in* "Combinatorial Mathematics and Its Applications," pp. 149–167, Academic Press, New York, 1971.
5. A. Lehman, A solution of the Shannon switching game, *J. SIAM* **12** (1964), 687–725.
6. P. D. Seymour, The forbidden minors of binary clutters, *J. London Math. Soc.* (2), **12** (1976), 356–360.
7. W. T. Tutte, A homotopy theorem for matroids, I, II, *Trans. Amer. Math. Soc.* **88** (1958), 144–174.
8. W. T. Tutte, Connectivity in matroids, *Canad. J. Math.* **18** (1966), 1301–1324.
9. W. T. Tutte, "Introduction to the Theory of Matroids," RAND Corp. Report R-448-PR, 1966.
10. W. T. Tutte, Lectures on matroids, *J. Res. Nat. Bur. Standards Sect. B* **69** (1965), 1–47.
11. D. J. A. Welsh, "Matroid Theory," Academic Press, London, 1976.
12. H. Whitney, On the abstract properties of linear dependence, *Amer. J. Math.* **57** (1935), 509–533.
13. T. H. Brylawski and T. D. Lucas, Uniquely representable combinatorial geometries, *in* "Proc. Internat. Colloq. Combinatorial Theory, Rome, Italy, 1975," pp. 83–104.

Chapter III
Enumeration in geometric lattices

1. Möbius functions

The papers reprinted in this chapter are concerned with enumeration within a geometric lattice. They are closely related to papers on the Tutte decomposition reprinted in the next chapter.

A central concept in enumeration in partially ordered sets is the Möbius function. The importance of the Möbius function was first pointed out and established in the classic paper of Rota, "On the foundations of combinatorial theory I. Theory of Möbius functions," the first paper reprinted in this chapter. This paper also initiated a theory-oriented (rather than problem-solving) approach to combinatorics and is a significant influence on recent developments in combinatorics. There is a thorough discussion of the background to Möbius functions and a detailed summary of the paper in §1.

A very restricted list of related papers is given in the references.

References

Baclawski, K.: The Möbius algebra as a Grothendieck ring, *J. Algebra* *57*(1979), 167-179.

Edelman, P.H.: Zeta polynomials and the Möbius function, *Europ. J. Combin.* *1*(1980), 335-340.

Greene, C.: On the Möbius algebra of a partially ordered set, *Advances in Math.* *10*(1973), 177–187.

Rota, G.-C. and Smith, D.: Enumeration under group action, *Ann. Scuola Norm. Sup. Pisa Cl. Sci. (4)* *4*(1977), 637-646.

Solomon, L.: The Burnside algebra of a finite group, *J. Combin. Theory* *2*(1967), 603-615.

2. Homology

In §6 of [III. 1], Rota defined a homology theory for a lattice relative to a cross-cut and conjectured that it is independent of the cross-cut. This conjecture was first proved in Folkman's paper "The homology groups of a lattice," the second paper reprinted in this chapter.

The idea of the proof is to show that the homology groups $H_n(C)$ relative to the cross-cut C are isomorphic to the homology groups $H_n(L)$ of the chain complex $K(L)$ of the lattice L (Theorem 3.1). This is done by showing that the simplicial complex $K(C)$ defined by a cross-cut C is the nerve of a suitable covering of the chain complex of L (§3). The proof is then completed by using the technical result (Theorem 2.1) proved in §2.

In §4, by taking the cross-cut in a geometric lattice L consisting of all the points, the homology groups $H_n(L)$ are computed (cf. [III. 1], Proposition 2, §7). It turns out that if L is of rank at least three, all of its homology groups are trivial, except for $H_0(L)$, which is the free abelian group on one generator, and $H_{n-2}(L)$, which is the free abelian group on $|\mu(0, 1)|$ generators.

This paper is one of the first of many papers on the homology of partially ordered sets and the application of homology to combinatorics. Some of these papers are listed below.

Euler characteristic: Rota [71].
Homology of geometric lattices: Baclawski [75], Björner [82], Orlik and Solomon [80].
Homology of partially ordered sets and related structures: Farmer [75, 75′, 77, 78/79], Griffiths [72].
Galois connections: Baclawski [77], Griffiths [76].
Homotopy of partially ordered sets: Björner [81], Björner and Walker [83], Brini [82], Lakser [71], Mather [66], Quillen [78], Walker [81].
Fixed points: Baclawski and Björner [77, 81].
Cohen-Macaulay partially ordered sets: Baclawski [80, 82, pre], Björner [80], Björner, Garsia and Stanley [82], Stanley [79].

References

Baclawski, K.: Whitney numbers of geometric lattices, *Advances in Math.* *16*(1975), 125-138.

Baclawski, K.: Galois connections and the Leray spectral sequence, *Advances in Math.* *25*(1977), 191-215.

Baclawski, K.: Cohen-Macaulay ordered sets, *J. Algebra 63*(1980), 226-258.

Baclawski, K.: Cohen-Macaulay connectivity and geometric lattices, *Europ. J. Combin.* *3*(1982), 293-305.

Baclawski, K.: Nonpositive Cohen-Macaulay connectivity, preprint.

Baclawski, K. and Björner, A.: Fixed points in partially ordered sets, *Advances in Math. 31*(1979), 263-287.

Baclawski, K. and Björner, A.: Fixed points and complements in finite lattices, *J. Combin. Theory Ser. A 30*(1981), 335-338.

Björner, A.: Shellable and Cohen-Macaulay partially ordered sets, *Trans. Amer. Math. Soc. 260*(1980), 159-183.

Björner, A.: Homotopy type of posets and lattice complementation, *J. Combin. Theory Ser. A 30*(1981), 90-100.

Björner, A.: On the homology of geometric lattices, *Algebra Universalis 14*(1982), 107-128.

Björner, A., Garsia, A.M. and Stanley, R.P.: An introduction to Cohen-Macaulay partially ordered sets, *Ordered Sets* (I. Rival, ed.), pp. 583-615, Reidel, Dordrecht and Boston, 1982.

Björner, A. and Walker, J.W.: A homotopy complementation formula for partially ordered sets, *Europ. J. Combin. 4*(1983), 11-19.

Brini, A.: Some homological properties of partially ordered sets, *Advances in Math. 43*(1982), 197-201.

Farmer, F.D.: Homology of reflexive relations, *Math. Japon. 20*(1975), 303-310.

Farmer, F.D.: Homology of products and joins of reflexive relations, *Discrete Math. 11*(1975), 23-27.

Farmer, F.D.: Characteristic for reflexive relations, *Aequationes Math. 15*(1977), 195-199.

Farmer, F.D.: Cellular homology for posets, *Math. Japon. 23*(1978/79), 607-613.

Griffiths, H.B.: The homology groups of some ordered systems, *Acta Math. 129*(1972), 195-235.

Griffiths, H.B.: An exact homology sequence induced by a Galois connection, *Proc. London Math. Soc. (3) 32*(1976), 101-116.

Lakser, H.: The homology of a lattice, *Discrete Math. 1*(1971), 187-192.

Mather, J.: Invariance of the homology of a lattice, *Proc. Amer. Math. Soc. 17*(1966), 1120-1124.

Orlik, P. and Solomon, L.: Combinatorics and topology of complements of hyperplanes, *Invent. Math. 56*(1980), 167-189.

Quillen, D.: Homotopy properties of the poset of non-trivial *p*-subgroups of a group, *Advances in Math. 28*(1978), 101-128.

Rota, G.-C.: On the combinatorics of the Euler characteristic, *Studies in Pure Mathematics (Presented to Richard Rado)*, pp. 221-233, Academic Press, London, 1971.

Stanley, R.P.: Balanced Cohen-Macaulay complexes, *Trans. Amer. Math. Soc. 249*(1979), 139-157.
Walker, J.W.: Homotopy type and Euler characteristic, *Europ. J. Combin. 2*(1981), 373-384.

3. Modular factorizations

As shown in [III. 1] and [IV. 2], the zeros of the characteristic polynomial have important combinatorial interpretations. Thus, any information on how it factors is of great interest. The only general result known in this direction is the modular factorization theorem of Stanley. This appeared in his paper "Modular elements in geometric lattices," the third paper reprinted in this chapter.

Briefly, the content of this paper is as follows. §1 gives a characterization of modular elements: an element is modular if and only if no two of its complements are comparable. This result, together with Crapo's complementation theorem, is used in §2 to prove the main result. This asserts that if x is a modular element in a geometric lattice L, then the characteristic polynomial of the lower interval $[0, x]$ is a factor of the characteristic polynomial of L. In §3, geometric lattices which contain a maximal chain of modular elements (or *supersolvable* geometric lattices) are discussed. Finally, a characterization of modular elements in the lattice of flats of the cycle matroid of a graph is given. (The following "footnote" to the last paragraph of §1 may be helpful to the reader. An element s of a lattice L is said to be *standard* if

$$x \wedge (s \vee y) = (x \wedge s) \vee (x \wedge y) \quad \text{for all} \quad x \text{ and } y,$$

or, equivalently,

(i) the function $x \mapsto x \vee s$ defined from L to itself is a lattice endomorphism (i.e. it preserves both meets and joins), and

(ii) $x \vee s = y \vee s$ and $x \wedge s = y \wedge s$ imply $x = y$.

The second condition implies trivially that if a standard element has a complement, then it is unique.)

We end with a list of related papers. Supersolvable lattices are further studied in Stanley [72, 74]; interesting examples can be found in Dowling [73]. A proof and generalization of the modular factorization theorem using Möbius algebras can be found in Greene [73]. Another proof, which yields an interpretation of the quotient $p_L(\lambda)/p_x(\lambda)$ as the characteristic polynomial of a geometric lattice, is in Brylawski [75]. Some extensions of this are in Brylawski and Oxley [80]. The relation of modular factorization to the join-decomposition of the broken-circuit complex is studied in Brylawski and Oxley [81].

References

Brylawski, T.: Modular constructions for combinatorial geometries, *Trans. Amer. Math. Soc. 203*(1975), 1-44.

Brylawski, T. and Oxley, J.G.: Several identities for the characteristic polynomial of a combinatorial geometry, *Discrete Math. 31*(1980), 161-170.

Brylawski, T. and Oxley, J.G.: The broken-circuit complex: its structure and factorization, *Europ. J. Combin. 2*(1981), 107-121.

Dowling, T.A.: A class of geometric lattices based on finite groups, *J. Combin. Theory Ser. B 14*(1973), 61-86.

Greene, C.: On the Möbius algebra of a partially ordered set, *Advances in Math. 10*(1973), 177-187.

Stanley, R.P.: Supersolvable lattices, *Algebra Universalis 2*(1972), 197-217.

Stanley, R.P.: Finite lattices and Jordan-Hölder sets, *Algebra Universalis 4*(1974), 361-371.

4. Whitney numbers of the second kind

The last two papers reprinted in this chapter are concerned, directly or indirectly, with Whitney numbers of the second kind. Let L be a geometric lattice. The *Whitney numbers* $W_k(L)$ (*of the second kind*) are defined by

$$W_k(L) = \text{number of elements of rank } k \text{ in } L.$$

There are three classical examples of Whitney numbers.

1. Binomial coefficients. Let B_n be the Boolean algebra of all subsets of a set of size n. Then

$$W_k(B_n) = \binom{n}{k} = \frac{n!}{k!\,(n-k)!}.$$

2. Gaussian coefficients. Let $L_n(q)$ be the lattice of subspaces of a vector space of dimension n over the finite field with q elements. Then

$$W_k(L_n(q)) = \begin{bmatrix} n \\ k \end{bmatrix}_q = \frac{(q^n - 1)(q^{n-1} - 1) \cdots (q^{n-k+1} - 1)}{(q^k - 1)(q^{k-1} - 1) \cdots (q - 1)}.$$

3. Stirling numbers of the second kind. Let Π_n be the lattice of partitions of a set of size n. Then

$$W_k(\Pi_n) = S(n, n-k)$$

where $S(n, n-k)$ is the number of partitions of a set of size n into $n-k$ (non-empty) blocks.

From the properties of these classical examples, it is natural to make several conjectures. One of them is the *unimodality conjecture.* This states: For a geometric lattice L (of rank n), the sequence $W_k(L)$ of Whitney numbers is unimodal: that is, there exists an integer m such that

$$W_1(L) \leqslant W_2(L) \leqslant \cdots \leqslant W_m(L)$$

and

$$W_m(L) \geqslant W_{m+1}(L) \geqslant \cdots \geqslant W_{n-1}(L).$$

A stronger conjecture is that the sequence $W_k(L)$ is logarithmically concave: that is,

$$[W_k(L)]^2 \geqslant W_{k-1}(L)\, W_{k+1}(L).$$

These conjectures have proved to be intractable and progress so far has been confined to the first three ranks. In [75], Stonesifer proved that $W_2^2 \geqslant W_1 W_3$ for lattices of flats of the cycle matroid of a graph and in [82], Seymour proved that this is true under the more general hypothesis that the geometric lattice contains no five-point lines as minors. (Other, more special, results can be found in Harper [67], Kurtz [72], Lieb [68], Stonesifer [75′]. A useful survey is Mason [72].) There is an analogous unimodality conjecture for the Whitney numbers of the first kind (and as a special case, the coefficients of the chromatic polynomial of a graph). However, almost nothing is known here. We should also remark that the unimodality conjecture for convex polytopes (which states that the sequence f_k, where f_k is the number of k-dimensional faces of a convex polytope, is unimodal) is false. See Björner [81].

Another conjecture states that in a geometric lattice L of rank n,

$$W_k(L) \leqslant W_{n-k}(L)$$

for $k \leqslant n/2$. This conjecture is true when $k=1$. This was essentially shown in Motzkin [51]. Other proofs can be found in Basterfield and Kelly [68], Greene [70, 75], Heron [73], Woodall [76], and Kung [79]. In the last paper reprinted in this chapter, "Whitney number inequalities for geometric lattices" by Dowling and Wilson, the following weaker result is proved:

$$(*) \quad W_1(L)+W_2(L)+\cdots+W_k(L) \\ \leqslant W_{n-k}(L)+W_{n-k+1}(L)+\cdots+W_{n-1}(L)$$

for $k \leqslant n/2$. The proof consists of showing that the incidence matrix with rows indexed by the elements in L of rank 1 to k and columns indexed by the elements of rank $n-k$ to $n-1$ in which the xy-entry equals 1 if $x \leqslant y$ and 0 otherwise is invertible. This is done by computing the right inverse of the incidence matrix explicitly using ingenious Möbius function identities. (Another application of these identities can be found in Dowling [76].) In addition, the geometric lattices for which the inequality $(*)$ is an equality for some k, $k \leqslant n-2$, are characterized: they are the modular geometric lattices.

An extension of the inequality $(*)$ to the sequence $W_k(G; L)$, where $W_k(G; L)$ is the number of orbits of a group G of automorphisms of L acting on the elements of rank k in L, can be found in Dowling [81]. Lower bounds for the Whitney numbers (of both kinds) and the geometric lattices attaining them are studied in Dowling and Wilson [74]. See also Björner [80], Brylawski [77], and Zaslavsky [81, 83].

The independent set numbers $I_k(M)$ of a matroid M are defined by

$$I_k(M) = \text{number of independent sets of size } k \text{ in } M.$$

Many of the questions asked about Whitney numbers can also be asked about the independent set numbers (see Mason [72]), but not much is known. Some partial results on the unimodality conjecture for the independent set numbers can be found in Dowling [80], Mahoney [85] and Stanley [81].

References

Basterfield, J.G. and Kelly, L.M.: A characterization of sets of n points which determine n hyperplanes, *Proc. Cambridge Philos. Soc. 64*(1968), 585-588.

Björner, A.: Some matroid inequalities, *Discrete Math. 31*(1980), 101-103.

Björner, A.: The unimodality conjecture for convex polytopes, *Bull. Amer. Math. Soc. (New Series) 4*(1981), 187-188.

Brylawski, T.: Connected matroids with the smallest Whitney numbers, *Discrete Math. 18*(1977), 243-252.

Dowling, T.A.: Complementing permutations in finite lattices, *J. Combin. Theory Ser. B 23*(1977), 223-226.

Dowling, T.A.: On the independent set numbers of a finite matroid, *Combinatorics 79, Part I, Ann. Discrete Math. 8*(1980), 21-28.

Dowling, T.A.: On the orbit numbers of a finite geometric lattice, preprint, 1981.

Dowling, T.A. and Wilson, R.M.: The slimmest geometric lattices, *Trans. Amer. Math. Soc. 196*(1974), 203-215.

Greene, C.: A rank inequality for finite geometric lattices, *J. Combin. Theory 9*(1970), 357-364.

Greene, C.: A inequality for the Möbius function of a geometric lattice, *Stud. Appl. Math. 54*(1975), 71-74.

Harper, L.H.: Stirling behavior is asymptotically normal, *Ann. Math. Statist. 38*(1967), 410-414.

Heron, A.P.: A property of the hyperplanes of a matroid and an extension of Dilworth's theorem, *J. Math. Anal. Appl. 42*(1973), 119-131.

Kung, J.P.S.: The Radon transforms of a combinatorial geometry, I, *J. Combin. Theory Ser. A 26*(1979), 97-102.

Kurtz, D.C.: A note on concavity properties of triangular arrays of numbers, *J. Combin. Theory Ser. A 13*(1972), 135-139.

Lieb, E.H.: Concavity properties and a generating function for Stirling numbers, *J. Combin. Theory 5*(1968), 203-206.

Mahoney, C.R.: On unimodality of the independent set numbers of a class of matroids, *J. Combin. Theory Ser. B 39*(1985), 77-85.

Mason, J.H.: Matroids: unimodal conjectures and Motzkin's theorem, *Combinatorics (Proc. Conf. Combinatorial Math., Math. Inst., Oxford, 1972)*, pp. 207-220, Inst. Math. Appl., Southend-on-Sea, 1972.

Motzkin, T.S.: The lines and planes connecting the points of a finite set, *Trans. Amer. Math. Soc. 70*(1951), 451-464.

SEYMOUR, P.D.: On the points-lines-planes conjecture, *J. Combin. Theory Ser. B 33*(1982), 17-26.
STANLEY, R.P.: Two combinatorial applications of the Aleksandrov-Fenchel inequalities, *J. Combin. Theory Ser. A 31*(1981), 56-65.
STONESIFER, J.R.: Logarithmic concavity for edge lattices of graphs, *J. Combin. Theory Ser. A 18*(1975), 36-46.
STONESIFER, J.R.: Logarithmic concavity for a class of geometric lattices, *J. Combin. Theory Ser. A 18*(1975), 216-218.
WOODALL, D.R.: The inequality $b \geqslant v$, *Proc. Fifth British Combinatorial Conf. (Univ. Aberdeen, Aberdeen,* 1975), pp. 661-664, *Congressus Numerantium, No. 15,* Utilitas Math., Winnipeg, Man,. 1976.
ZASLAVSKY, T.: The slimmest arrangements of hyperplanes I: Geometric lattices and projective arrangements, *Geometriae Dedicata 14*(1983), 243-259.
ZASLAVSKY, T.: The slimmest arrangements of hyperplanes. II: Basepointed geometric lattices and Euclidean arrangements, *Mathematika 28*(1981), 169-190.

5. The Sperner property

One of the classic results in extremal set theory (for an introduction and further details, see Greene and Kleitman [78]) is the following theorem.

Sperner's theorem. Let $\mathcal{A}$ be a collection of pairwise incomparable subsets of a set of size n. Then,

$$|\mathcal{A}| \leqslant \max_k \binom{n}{k} = \binom{n}{[n/2]}.$$

Here, $[n/2]$ is the greatest integer less than or equal to $n/2$.

Generalizing from this, we say that a geometric lattice L has the *Sperner property* if every antichain (i.e. collection of pairwise incomparable elements) has at most $\max_k W_k(L)$ elements.

Does the Sperner property hold in all geometric lattices? The answer to this question, posed in Harper and Rota [71], is no and the first examples of non-Sperner geometric lattices were given in "A counterexample to the generalization of Sperner's theorem" by Dilworth and Greene, the fourth paper reprinted in this chapter. The examples given are lattices of flats of the cycle matroid of a graph and show that even "nice" geometric lattices may not satisfy the Sperner property. Since then, other examples of non-Sperner geometric lattices have been found. In [80], using design theory, Kahn gives examples of paving matroids whose lattice of flats does not satisfy Sperner's property. (Lattices of flats of paving matroids have the property that every proper lower interval is a Boolean algebra.) More surprising examples are given by a result of Canfield (see Canfield [78], and also Peck [pre] and Shearer [79]) which shows that the lattice of partitions of a set of size n does not satisfy the Sperner property if n is sufficiently large.

On the positive side, the lattice $L_n(q)$ of subspaces of a vector space of dimension n over the finite field of order q satisfies the Sperner property. (See Baker [69] and Harper [74]). More generally, the Sperner property holds in a geometric lattice in which every element of rank k is covered by the same number a_k of elements of rank $k+1$ and covers the same number b_k of elements of rank $k-1$. (See Baker [69].) Both these results are proved using the LYM-inequality.

We end with a research problem posed by Greene: for which k are geometric lattices k-Sperner? For example, since there exists a matching from the points to the copoints such that if p is matched to x then $p \leqslant x$, a geometric lattice of rank n is $(n-2)$-Sperner. (See Greene and Kleitman [76] for a definition of the k-Sperner property and Greene [70] for the matching result.)

References

BAKER, K.A.: A generalization of Sperner's lemma, *J. Combin. Theory* *6*(1969), 224-225.

CANFIELD, E.R.: On a problem of Rota, *Advances in Math.* *29*(1978), 1-10.

GREENE, C.: A rank inequality for finite geometric lattices, *J. Combin. Theory* *9*(1970), 357-364.

GREENE, C. and KLEITMAN, D.J.: The structure of Sperner k-families, *J. Combin. Theory Ser. A* *20*(1976), 41-68.

GREENE, C. and KLEITMAN, D.J.: Proof techniques in the theory of finite sets, *Studies in Combinatorics* (G.-C. Rota, ed.) pp. 22-79, Math. Assoc. Amer., Washington, D.C., 1978.

HARPER, L.H.: The morphology of partially ordered sets, *J. Combin. Theory Ser. A* *17*(1974), 44-58.

HARPER, L.H. and ROTA, G.-C.: Matching theory, an introduction, *Advances in Probability, Vol. 1*(P. Ney, ed.), pp. 169-215, Marcel Dekker, New York, 1971.

KAHN, J.: Some non-Sperner paving matroids, *Bull. London Math. Soc. 12*(1980), 268.

PECK, G.W.: On Canfield type antichains of partitions, preprint.

SHEARER, J.B.: A simple counterexample to a conjecture of Rota, *Discrete Math. 28*(1979), 327-330.

Reprinted from *American Journal of Mathematics*,
volume 57, pages 800 through 804,
by permission of the Johns Hopkins University Press.

On the Foundations of Combinatorial Theory
I. Theory of Möbius Functions

By

GIAN-CARLO ROTA

This work was begun under contract NSF-GP-149, continued under contract with the Office of Naval Research, and concluded while the author was a Fellow of the Sloan Foundation.

* Author's names refer to the bibliography at the end.

Contents

1. Introduction

One of the most useful principles of enumeration in discrete probability and combinatorial theory is the celebrated *principle of inclusion-exclusion* (cf. FELLER*, FRÉCHET, RIORDAN, RYSER). When skillfully applied, this principle has yielded the solution to many a combinatorial problem. Its mathematical foundations were thoroughly investigated not long ago in a monograph by FRÉCHET, and it might at first appear that, after such exhaustive work, little else could be said on the subject.

One frequently notices, however, a wide gap between the bare statement of the principle and the skill required in recognizing that it applies to a particular combinatorial problem. It has often taken the combined efforts of many a combinatorial analyst over long periods to recognize an inclusion-exclusion pattern. For example, for the ménage problem it took fifty-five years, since CAYLEY's attempts, before JACQUES TOUCHARD in 1934 could recognize a pattern, and thence readily obtain the solution as an explicit binomial formula. The situation becomes bewildering in problems requiring an enumeration of any of the numerous collections of combinatorial objects which are nowadays coming to the fore. The counting of trees, graphs, partially ordered sets, complexes, finite sets on which groups act, not to mention more difficult problems relating to permutations with restricted position, such as Latin squares and the coloring of maps, seem to lie beyond present-day methods of enumeration. The lack of a systematic theory is hardly matched by the consummate skill of a few individuals with a natural gift for enumeration.

This work begins the study of a very general principle of enumeration, of which the inclusion-exclusion principle is the simplest, but also the typical case. It often happens that a set of objects to be counted possesses a natural ordering, in general only a partial order. It may be unnatural to fit the enumeration of such a set into a linear order such as the integers: instead, it turns out in a great many cases that a more effective technique is to work with the natural order of the set. One is led in this way to set up a "difference calculus" relative to an arbitrary partially ordered set.

Looked at in this way, a surprising variety of problems of enumeration reveal themselves to be instances of the general problem of inverting an "indefinite sum" ranging over a partially ordered set. The inversion can be carried out by defining an analog of the "difference operator" relative to a partial ordering. Such an operator is the Möbius function, and the analog of the "fundamental theorem of the calculus" thus obtained is the Möbius inversion formula on a partially ordered set. This formula is here expressed in a language close to that of number theory, where it appears as the well-known inverse relation between the Riemann zeta function and the Dirichlet generating function of the classical Möbius function. In fact, the algebra of formal Dirichlet series turns out to be the simplest non-trivial instance of such a "difference calculus", relative to the order relation of divisibility.

Once the importance of the Möbius function in enumeration problems is realized, interest will naturally center upon relating the properties of this function to the structure of the ordering. This is the subject of the first paper of this series; we hope to have at least begun the systematic study of the remarkable properties of this most natural invariant of an order relation.

We begin in Section 3 with a brief study of the incidence algebra of a locally finite partially ordered set and of the invariants associated with it: the zeta function, Möbius function, incidence function, and Euler characteristic. The language of number theory is kept, rather than that of the calculus of finite differences, and the results here are quite simple.

The next section contains the main theorems: Theorem 1 relates the Möbius functions of two sets related by a Galois connection. By suitably varying one of the sets while keeping the other fixed one can derive much information. Theorem 2 of this section is suggested by a technique that apparently goes back to RAMANUJAN. These two basic results are applied in the next section to a variety of special cases; although a number of applications and special cases have been left out, we hope thereby to have given an idea of the techniques involved.

The results of Section 6 stem from an "Ideenkreis" that can be traced back to Whitney's early work on linear graphs. Theorem 3 relates the Möbius function to certain very simple invariants of "cross-cuts" of a finite lattice, and the analogy with the Euler characteristic of combinatorial topology is inevitable. Pursuing this analogy, we were led to set up a series of homology theories, whose Euler characteristic does indeed coincide with the Euler characteristic which we had introduced by purely combinatorial devices.

Some of the work in lattice theory that was carried out in the thirties is useful in this investigation; it turns out, however, that modular lattices are not combinatorially as interesting as a type of structure first studied by WHITNEY, which we have called geometric lattices following BIRKHOFF and the French school. The remarkable property of such lattices is that their Möbius function alternates in sign (Section 7).

To prevent the length of this paper from growing beyond bounds, we have omitted applications of the theory. Some elementary but typical applications will be found in the author's expository paper in the American Mathematical Monthly. Towards the end, however, the temptation to give some typical examples became irresistible, and Sections 9 and 10 were added. These by no means exhaust the range of applications, it is our conviction that the Möbius inversion formula on a partially ordered set is a fundamental principle of enumeration, and we hope to implement this conviction in the successive papers of this series. One of them will deal with structures in which the Möbius function is multiplicative, —-that is, has the analog of the number-theoretic property $\mu(mn) = \mu(m)\,\mu(n)$ if m and n are coprime -- and another will give a systematic development of the Ideenkreis centering around POLYA's Hauptsatz, which can be significantly extended by a suitable Möbius inversion.

A few words about the history of the subject. The statement of the Möbius inversion formula does not appear here for the first time: the first coherent version—with some redundant assumptions—is due to WEISNER, and was independently rediscovered shortly afterwards by PHILIP HALL. Ward gave the statement in full generality. Strangely enough, however, these authors did not pursue the combinatorial implications of their work; nor was an attempt made to systematically investigate the properties of Möbius functions. Aside from HALL's applications to p-groups, and from some applications to statistical mechanics by M. S. GREEN and NETTLETON, little has been done; we give a hopefully complete bibliography at the end.

It is a pleasure to acknowledge the encouragement of G. BIRKHOFF and A. GLEASON, who spotted an error in the definition of a cross-cut, as well as of SEYMOUR SHERMAN and KAI-LAI CHUNG. My colleagues D. KAN, G. WHITEHEAD, and especially F. PETERSON gave me essential help in setting up the homological interpretation of the cross-cut theorem.

2. Preliminaries

Little knowledge is required to read this work. The two notions we shall not define are those of a *partially ordered set* (whose order relation is denoted by $\leq$) and a *lattice*, which is a partially ordered set where max and min of two elements (we call them join and meet, as usual, and write them $\vee$ and $\wedge$) are defined. We shall use instead the symbols $\cup$ and $\cap$ to denote union and intersection *of sets* only. A *segment* $[x, y]$, for x and y in a partially ordered set P, is the set of all elements z between x and y, that is, such that $x \leq z \leq y$. We shall occasionally use open or half-open segments such as $[x, y)$, where one of the endpoints is to be omitted. A segment is endowed with the induced order structure; thus, a segment of a lattice is again a lattice. A partially ordered set is *locally finite* if every segment is finite. We shall only deal with locally finite partially ordered sets.

The *product* $P \times Q$ of partially ordered sets P and Q is the set of all ordered pairs (p, q), where $p \in P$ and $q \in Q$, endowed with the order $(p, q) \geqq (r, s)$ whenever $p \geqq r$ and $q \geqq s$. The product of any number of partially ordered sets is defined similarly. The *cardinal power* Hom (P, Q) is the set of all monotonic functions from P to Q, endowed with the partial order structure $f \geqq g$ whenever $f(p) \geqq g(p)$ for every p in P.

In a partially ordered set, an element p *covers* an element q when the segment $[q, p]$ contains two elements. An *atom* in P is an element that covers a minimal element, and a *dual atom* is an element that is covered by a maximal element.

If P is a partially ordered set, we shall denote by P^* the partially ordered set obtained from P by inverting the order relation.

A *closure relation* in a partially ordered set P is a function $p \to \bar{p}$ of P into itself with the properties (1) $\bar{p} \geqq p$; (2) $\bar{\bar{p}} = \bar{p}$; (3) $p \geqq q$ implies $\bar{p} \geqq \bar{q}$. An element is *closed* if $p = \bar{p}$. If P is a finite Boolean algebra of sets, then a closure relation on P defines a lattice structure on the closed elements by the rules $p \wedge q = p \cap q$ and $p \vee q = \overline{p \cup q}$, and it is easy to see that every finite lattice is isomorphic to one that is obtained in this way. A *Galois connection* (cf. Ore, p. 182ff.) between two partially ordered sets P and Q is a pair of functions $\zeta : P \to Q$ and $\pi : Q \to P$ with the properties: (1) both ζ and π are order-inverting; (2) for p in P, $\pi(\zeta(p)) \geqq p$, and for q in Q, $\zeta(\pi(q)) \geqq q$. Under these circumstances the mappings $p \to \pi(\zeta(p))$ and $q \to \zeta(\pi(q))$ are closure relations, and the two partially ordered sets formed by the closed sets are isomorphic.

In Section 7, the notion of a closure relation with the *Mac Lane-Steinitz exchange property* will be used. Such a closure relation is defined on the Boolean algebra P of subsets of a finite set E and satisfies the following property: if p and q are points of E, and S a subset of E, and if $p \notin \bar{S}$ but $p \in \overline{S \cup q}$, then $q \in \overline{S \cup p}$. Such a closure relation can be made the basis of Whitney's theory of independence, as well as of the theory of geometric lattices. The closed sets of a closure relation satisfying the Mac Lane-Steinitz exchange property where every point is a closed set form a geometric (= matroid) lattice in the sense of Birkhoff (Lattice Theory, Chapter IX).

A partially ordered set P is said to have a 0 or a I if it has a unique minimal or maximal element. We shall always assume $0 \neq I$. A partially ordered set P having a 0 and a I satisfies the *chain condition* (also called the Jordan-Dedekind chain condition) when all totally ordered subsets of P having a maximal number of elements have the same number of elements. Under these circumstances one introduces the ***rank*** $r(p)$ of an element p of P as the length of a maximal chain in the segment $[o, p]$, minus one. The rank of 0 is 0, and the rank of an atom is 1. The height of P is the rank of any maximal element, plus one.

Let P be a finite partially ordered set satisfying the chain condition and of height $n + 1$. The ***characteristic polynomial*** of P is the polynomial $\sum_{x \in P} \mu(0, x) \lambda^{n-r(x)}$, where r is the rank function (see the def. of μ below).

If A is a finite set, we shall write $n(A)$ for the number of elements of A.

3. The incidence algebra

Let P be a locally finite partially ordered set. The *incidence algebra* of P is defined as follows. Consider the set of all real-valued functions of two variables $f(x, y)$, defined for x and y ranging over P, and with the property that $f(x, y) = 0$ if $x \not\leq y$. The sum of two such functions f and g, as well as multiplication by scalars, are defined as usual. The product $h = fg$ is defined as follows:

$$h(x, y) = \sum_{x \leq z \leq y} f(x, z) g(z, y).$$

In view of the assumption that P is locally finite, the sum on the right is well-defined. It is immediately verified that this is an associative algebra over the real field (any other associative ring could do). The incidence algebra has an identity element which we write $\delta(x, y)$, the Kronecker delta.

The *zeta function* $\zeta(x, y)$ of the partially ordered set P is the element of the incidence algebra of P such that $\zeta(x, y) = 1$ if $x \leq y$ and $\zeta(x, y) = 0$ otherwise. The function $n(x, y) = \zeta(x, y) - \delta(x, y)$ is called the *incidence function*.

The idea of the incidence algebra is not new. The incidence algebra is a special case of a semigroup algebra relative to a semigroup which is easily associated with the partially ordered set. The idea of taking "interval functions" goes back to DEDEKIND and E. T. BELL; see also WARD.

Proposition 1. *The zeta function of a locally finite partially ordered set is invertible in the incidence algebra.*

Proof. We define the inverse $\mu(x, y)$ of the zeta function by induction over the number of elements in the segment $[x, y]$. First, set $\mu(x, x) = 1$ for all x in P. Suppose now that $\mu(x, z)$ has been defined for all z in the open segment $[x, y)$. Then set

$$\mu(x, y) = - \sum_{x \leq z < y} \mu(x, z).$$

Clearly μ is an inverse of ζ.

The function μ, inverse to ζ, is called the *Möbius function* of the partially ordered set P.

The following result, simple though it is, is fundamental:

Proposition 2. (Möbius inversion formula). *Let $f(x)$ be a real-valued function, defined for x ranging in a locally finite partially ordered set P. Let an element p exist with the property that $f(x) = 0$ unless $x \geq p$.*

Suppose that

(*) $$g(x) = \sum_{y \leq x} f(y).$$

Then

(**) $$f(x) = \sum_{y \leq x} g(y) \mu(y, x).$$

Proof. The function g is well-defined. Indeed, the sum on the right can be written as $\sum_{p \leq y \leq x} f(y)$, which is finite for a locally finite ordered set.

Substituting the right side of (*) into the right side of (**) and simplifying, we get

$$\sum_{y \leq x} g(y)\mu(y,x) = \sum_{y \leq x} \sum_{z \leq y} f(z)\mu(y,x) = \sum_{y \leq x} \sum_{z} f(z)\zeta(z,y)\mu(y,x).$$

Interchanging the order of summation, this becomes

$$\sum_{z} f(z) \sum_{y \leq x} \zeta(z,y)\mu(y,x) = \sum_{z} f(z)\delta(z,x) = f(x), \quad \text{q. e. d.}$$

Corollary 1. *Let $r(x)$ be a function defined for x in P. Suppose there is an element q such that $r(x)$ vanishes unless $x \leqq q$. Suppose that*

$$s(x) = \sum_{y \leq x} r(y).$$

Then

$$r(x) = \sum_{y \geqq x} \mu(x,y) s(y).$$

The proof is analogous to the above and is omitted.

Proposition 3. (Duality). *Let P^* be the partially ordered set obtained by inverting the order of a locally finite partially ordered set P, and let μ^* and μ be the Möbius functions of P^* and P. Then $\mu^*(x,y) = \mu(y,x)$.*

Proof. We have, in virtue of Proposition 2 and Corollary 1,

$$\sum_{x \geqq^* y \geqq^* z} \mu^*(x,y) = \delta(x,z).$$

Letting $q(x,y) = \mu^*(y,x)$, it follows that q is an inverse of ζ in the incidence algebra of P. Since the inverse is unique, $q = \mu$, q. e. d.

Proposition 4. *The Möbius function of any segment $[x,y]$ of P equals the restriction to $[x,y]$ of the Möbius function of P.*

The proof is omitted.

Proposition 5. *Let $P \times Q$ be the direct product of locally finite partially ordered sets P and Q. The Möbius function of $P \times Q$ is given by*

$$\mu((x,y),(u,v)) = \mu(x,u)\mu(y,v), \quad x,u \in P;\ y,v \in Q.$$

The proof is immediate and is omitted.

The same letter μ has been used for the Möbius functions of three partially ordered sets, and we shall take this liberty whenever it will not cause confusion.

Corollary (Principle of Inclusion-Exclusion). *Let P be the Boolean algebra of all subsets of a finite set of n elements. Then, for x and y in P,*

$$\mu(x,y) = (-1)^{n(y)-n(x)}, \qquad y \geqq x,$$

where $n(x)$ denotes the number of elements of the set x.

Indeed, a Boolean algebra is isomorphic to the product of n chains of two elements, and every segment $[x,y]$ in a Boolean algebra is isomorphic to a Boolean algebra.

Aside of the simple result of Proposition 5, little can be said in general about how the Möbius function varies by taking subsets and homomorphic images of a partially ordered set. We shall see that more sophisticated notions will be required to relate the Möbius functions of two partially ordered sets.

Let P be a finite partially ordered set with 0 and I. The *Euler characteristic* E of P is defined as

$$E = 1 + \mu(0,1).$$

The simplest result relating to the computation of the Euler characteristic was proved by PHILIP HALL by combinatorial methods. We reprove it below with a very simple proof which shows one of the uses of the incidence algebra:

Proposition 6. *Let P be a finite partially ordered set with 0 and I. For every k, let C_k be the number of chains with k elements stretched between 0 and I. Then*

$$E = 1 - C_2 + C_3 - C_4 + \cdots.$$

Proof. $\mu = \zeta^{-1} = (\delta + n)^{-1} = \delta - n + n^2 \ldots$. It is easily verified that $n^{k-1}(x, y)$ equals the number of chains of k elements stretched between x and y. Letting $x = 0$ and $y = I$, the result follows at once.

It will be seen in section 6 that the Euler characteristic of a partially ordered set can be related to the classical Euler characteristic in suitable homology theories built on the partially ordered set.

Proposition 6 is a typical application of the incidence algebra. Several other results relating the number of chains and subsets with specified properties can often be expressed in terms of identities for functions in the incidence algebra. In this way, one obtains generalizations to an arbitrary partially ordered set of some classical identities for binomial coefficients. We shall not pursue this line here further, since it lies out of the track of the present work.

Example 1. The classical Möbius function $\mu(n)$ is defined as $(-1)^k$ if n is the product of k distinct primes, and 0 otherwise. The classical inversion formula first derived by Möbius in 1832 is:

$$g(m) = \sum_{n|m} f(n); \quad f(m) = \sum_{n|m} g(n)\, \mu\left(\frac{m}{n}\right).$$

It is easy to see (and will follow trivally from later results) that $\mu\left(\frac{m}{n}\right)$ is the Möbius function of the set of positive integers, with divisibility as the partial order. In this case the incidence algebra has a distinguished subalgebra, formed by all functions $f(n, m)$ of the form $f(n, m) = G\left(\frac{m}{n}\right)$. The product $H = FG$ of two functions in this subalgebra can be written in the simpler form

$$(*) \qquad H(m) = \sum_{kn=m} F(k)\, G(n).$$

If we associate with the element F of this subalgebra the *formal Dirichlet series* $\hat{F}(s) = \sum_{n=1}^{\infty} F(n)/n^s$, then the product $(*)$ corresponds to the product of two formal Dirichlet series considered as functions of s, $\hat{H}(s) = \hat{F}(s)\, \hat{G}(s)$. Under this representation, the zeta function of the partially ordered set is the classical *Riemann zeta function* $\zeta(s) = \sum_{n=1}^{\infty} 1/n^s$, and the statement that the Möbius function is the inverse of the zeta function reduces to the classical identity $1/\zeta(s) = \sum_{n=1}^{\infty} \mu(n)/n^s$. It is hoped this example justifies much of the terminology introduced above.

Example 2. If P is the set of ordinary integers, then $\mu(m, n) = -1$ if $m = n - 1$, $\mu(m, m) = 1$, and $\mu(m, n) = 0$ otherwise. The Möbius inversion formula reduces to a well known formula of the calculus of finite differences, which is the discrete analog of the fundamental theorem of calculus.

The Möbius function of a partially ordered set can be viewed as the analog of the classical difference operator $\Delta f(n) = f(n + 1) - f(n)$, and the incidence algebra serves as a calculus of finite differences on an arbitrary partially ordered set.

4. Main results

It turns out that the Möbius functions of two partially ordered sets can be compared, when the sets are related by a Galois connection. By keeping one of the sets fixed, and varying the other from among sets with a simpler structure, such as Boolean algebras, subspaces of a finite vector space, partitions, etc., one can derive much information about a Möbius function. This is the program we shall develop. The basic result is the following:

Theorem 1. *Let P and Q be finite partially ordered sets, where P has a 0 and Q has a 0 and a 1. Let μ_P and μ be their Möbius functions. Let*

$$\pi : Q \to P; \quad \varrho : P \to Q$$

be a Galois connection such that

(1) $$\pi(x) = 0 \quad \text{if and only if} \quad x = 1 .$$

(2) $$\varrho(0) = 1 .$$

Then

$$\mu(0, 1) = \sum_{a>0} \mu_P(0, a) \zeta(\varrho(a), 0) = \sum_{[a:\varrho(a)=0]} \mu_P(0, a) .$$

One gets a significant summand on the right for every $a > 0$ in P which is mapped into 0 by ϱ. One therefore expects the right side to contain "few" terms. In general, μ_P is a known function and μ is the function to be determined.

Proof. We shall first establish the identity

(*) $$\sum_{a \geqq b} \delta(\pi(x), a) = \zeta(x, \varrho(b))$$

for every b in P. Here ζ on the right stands for the zeta function of Q. Equation (*) is equivalent to the following statement: $\pi(x) \geqq b$ if and only if $x \leqq \varrho(b)$. But this latter statement is immediate from the properties of a Galois connection. Indeed, if $\pi(x) \geqq b$, then $\varrho(\pi(x)) \leqq \varrho(b)$, but $x \leqq \varrho(\pi(x))$, hence $x \leqq \varrho(b)$, and similarly for the converse implication.

To identity (*) we apply the Möbius inversion formula relative to P, thereby obtaining the identity

(**) $$\delta(\pi(x), 0) = \sum_{a \geqq 0} \mu_P(0, a) \zeta(x, \varrho(a)) .$$

Now, $\delta(\pi(x), 0)$ takes the value 1 if and only if $\pi(x) = 0$, that is, in view of assumption (1), if and only if $x = 1$. For all other values of x, we have $\delta(\pi(x), 0) = 0$. Therefore,

$$\delta(\pi(x), 0) = 1 - n(x, 1).$$

We can now rewrite equation (**) in the form

$$1 - n(x, 1) = \zeta(x, \varrho(0)) + \sum_{a>0} \mu_p(0, a)\zeta(x, \varrho(a))$$

However, in view of assumption (2), $\zeta(x, \varrho(0)) = \zeta(x, 1)$, and this is identically one for all x in Q. Therefore, simplifying,

$$-n(x, 1) = \sum_{a>0} \mu_p(0, a)\zeta(x, \varrho(a)).$$

Now, since $\zeta = \delta + n$, we have $\mu = \delta - \mu n$, hence, recalling that $0 \neq 1$,

$$\mu(0, 1) = -\sum_{0 \leqq x \leqq 1} \mu(0, x) n(x, 1) = \sum_{0 \leqq x \leqq 1} \sum_{a>0} \mu_p(0, a)\mu(0, x)\zeta(x, \varrho(a)).$$

Interchanging the order of summation, we get

$$\mu(0, 1) = \sum_{a>0} \mu_p(0, a) \sum_{0 \leqq x \leqq 1} \mu(0, x)\zeta(x, \varrho(a)).$$

The last sum on the right equals $\delta(0, \varrho(a))$, and this equals $\zeta(\varrho(a), 0)$. The proof is therefore complete.

For simplicity of application, we restate Theorem 1 inverting the order of P.

Corollary. *Let $p: Q \to P$; $q: P \to Q$ be order preserving functions between P and Q such that*

(1) If $p(x) = 1$ then $x = 1$, and conversely.

(2) $q(1) = 1$.

(3) $p(q(x)) \leqq x$ and $q(p(x)) \geqq x$.

Then

$$\mu(0, 1) = \sum_{a<1} \mu_p(a, 1)\zeta(q(a), 0) = \sum_{[a: q(a)=0]} \mu_p(a, 1)$$

where μ is the Möbius function of Q.

The second result is suggested by a technique which apparently goes back to RAMANUJAN (cf. HARDY, RAMANUJAN, page 139).

Theorem 2. *Let Q be a finite partially ordered set with 0, and let P be a partially ordered set with 0. Let $p: Q \to P$ be a monotonic function of Q onto P. Assume that the inverse image of every interval $[0, a]$ in P is an interval $[0, x]$ in Q, and that the inverse image of 0 contains at least two points.*

Then

$$\sum_{[x: p(x)=a]} \mu(0, x) = 0$$

for every a in P.

The *proof* is by induction over the set P. Since $[0, 0]$ is an interval and its inverse image is an interval $[0, q]$ with $q > 0$, we have

$$\sum_{[x: p(x)=0]} \mu(0, x) = \sum_{0 \leqq x \leqq q} \mu(0, x) = 0.$$

Suppose now the statement is true for all b such that $b < a$ in P. Then

$$\sum_{b<a} \sum_{[x:p(x)=b]} \mu(0, x) = 0.$$

It follows that

$$\sum_{[x:p(x)=a]} \mu(0, x) = \sum_{b \leq a} \sum_{[x:p(x)=b]} \mu(0, x).$$

The last sum equals the sum over some interval $[0, r]$ which is the inverse image of the segment $[0, a]$, that is

$$\sum_{b \leq a} \sum_{[x:p(x)=b]} \mu(0, x) = \sum_{0 \leq x \leq r} \mu(0, x) = \delta(0, r).$$

But $r > 0$ because a is strictly greater than 0. Hence $\delta(r, 0) = 0$, and this concludes the proof.

5. Applications

The simplest (and typical) application of Theorem 1 is the following:

Proposition 1. *Let R be a subset of a finite lattice L with the following properties: $1 \notin R$, and for every x of L, except $x = 1$, there is an element y of R such that $y \geqq x$.*

For $k \geqq 2$, let q_k be the number of subsets of R containing k elements whose meet is 0. Then $\mu(0, 1) = q_2 - q_3 + q_4 + \cdots$.

Proof. Let $B(R)$ be the Boolean algebra of subsets of R. We take $P = B(R)$ and $Q = L$ in Theorem 1, and establish a Galois connection as follows. For x in L, let $\pi(x)$ be the set of elements of R which dominate x. In particular, $\pi(1)$ is the empty set. For A in $B(R)$, set $\varrho(A) = \wedge A$, namely, the meet of all elements of A, an empty meet giving as usual the element 1. This is evidently a Galois connection. Conditions (1) and (2) of the Theorem are obviously satisfied.

The function μ_P is given by the Corollary of Proposition 5 of Section 3, and hence the conclusion is immediate.

Two noteworthy special cases are obtained by taking R to be the set of dual atoms of Q, or the set of all elements < 1 (cf. also WEISNER).

Closure relations. A useful application of Theorem 1 is the following:

Proposition 2. *Let $x \to \bar{x}$ be a closure relation on a partially ordered set Q having 1, with the property that $\bar{x} = 1$ only if $x = 1$. Let P be the partially ordered subset of all closed elements of Q. Then:* (a) *If $\bar{x} > x$, then $\mu(x, 1) = 0$;* (b) *If $\bar{x} = x$, then $\mu(x, 1) = \mu_P(x, 1)$, where μ_P is the Möbius function of P.*

Proof. Considering $[x, 1]$, it may be assumed that P has a 0 and $x = 0$. We apply Corollary 1 of Theorem 1, setting $p(x) = \bar{x}$ and letting q be the injection map of P into Q. It is then clear that the assumptions of the Corollary are satisfied, and the set of all a in P such that $q(a) = 0$ is either the empty set or the single element 0, q. e. d.

Corollary (Ph. Hall). *If 0 is not the meet of dual atoms of a finite lattice L, or if 1 is not the join of atoms, then $\mu(0, 1) = 0$.*

Proof. Set $\bar{x} = \wedge A(x)$, where $A(x)$ is the set of dual atoms of Q dominating x, and apply the preceding result. The second assertion is obtained by inverting the order.

Example 1. *Distributive lattices.* Let L be a locally finite distributive lattice. Using Proposition 2, we can easily compute its Möbius function. Taking an interval

$[x, y]$ and applying Proposition 4 of Section 3, we can assume that L is finite. For $a \in L$, define $\bar{a}$ to be the join of all atoms which a dominates. Then $a \to \bar{a}$ is a closure relation in the inverted lattice L^*. Furthermore, the subset of closed elements is easily seen to be isomorphic to a finite Boolean algebra (cf. BIRKHOFF Lattice Theory, Ch. IX) Applying Proposition 5 of Section 3, we find: $\mu(x, y) = 0$ if y is not the join of elements covering x, and $\mu(x, y) = (-1)^n$ if y is the join of n distinct elements covering x.

In the special case of the integers ordered by divisibility, we find the formula for the classical Möbius function (cf. Example 1 of Section 3.).

The Möbius function of cardinal products. Let P and Q be finite partially ordered sets. We shall determine the Möbius function of the partially ordered set $\mathrm{Hom}(P, Q)$ of monotonic functions from P to Q, in terms of the Möbius function of Q. It turns out that very little information is needed about P.

A few preliminaries are required for the statement.

Let R be a subset of a partially ordered set Q with 0, and let $\bar{R}$ be the ideal generated by R, that is, the set of all elements x in Q which are below ($<$) some element of R. We denote by Q/R the partially ordered set obtained by removing off all the elements of $\bar{R}$, and leaving the rest of the order relation unchanged. There is a natural order-preserving transformation of Q onto Q/R which is one-to-one for elements of Q not in $\bar{R}$. We shall call Q/R the *quotient* of Q by the ideal generated by R.

Lemma. *Let $f: P \to Q$ be monotonic with range $R \subset Q$. Then the segment $[f, 1]$ in* Hom (P, Q) *is isomorphic with* Hom $(P, Q/R)$.

Proof. For g in $[f, 1]$, set $g'(x) = g(x)$ to obtain a mapping $g \to g'$ of $[f, 1]$ to Hom $(P, Q/R)$. Since $g \geq f$, the range of g lies above R, so the map is an isomorphism.

Proposition 3. *The Möbius function μ of the cardinal product* Hom (P, Q) *of the finite partially ordered set P with the partially ordered set Q with 0 and 1 is determined as follows:*

(a) *If $f(p) \neq 0$ for some element p of P which is not maximal, then $\mu(0, f) = 0$.*

(b) *In all other cases,*

$$\mu(0, f) = \prod_m \mu(0, f(m)), \qquad f \in P,$$

where the product ranges over all maximal elements of P, and where μ on the right stands for the Möbius function of Q.

(c) *For $f \leqq g$, $\mu(f, g) = \mu(0, g')$, where g' is the image of g under the canomial map of $[f, 1]$ onto Hom $(P, Q/R)$, provided Q/R has a 0.*

Proof. Define a closure relation in $[0, f]^*$, namely the segment $[0, f]$ with the inverted order relation, as follows. Set $\bar{g}(m) = g(m)$ if m is a maximal element of P, and $g(a) = 0$ if a is not a maximal element of P. If $\bar{g} = 0$, then $g(m) = 0$ for all maximal elements m, hence $g(a) = 0$ for all $a <$ some maximal element, since g is monotonic. Hence $g = 0$, and the assumption of Proposition 2 is satisfied. The set of closed elements is isomorphic to Hom (M, P), where M is a set of as many elements as there are maximal elements in P. Conclusion (a) now follows from Proposition 2, and conclusion (b) from Proposition 5 of Section 3. Conclusion (c) follows at once from the Lemma.

We pass now to some applications of Theorem 2.

Proposition 4. *Let $a \to \bar{a}$ be a closure relation on a finite lattice Q, with the property that $\overline{a \vee b} = \bar{a} \vee \bar{b}$ and $\bar{0} > 0$. Then for all $a \in Q$,*

$$\sum_{[x:\bar{x}=a]} \mu(0, x) = 0 .$$

Proof. Let P be a partially ordered set isomorphic to the set of closed elements of L. We define $p(x)$, for x in Q, to be the element of P corresponding to the closed element $\bar{x}$. Since $\bar{0} > 0$, any x between 0 and $\bar{0}$ is mapped into $\bar{0}$. Hence the inverse image of 0 in P under the homomorphism p is the nontrival interval $[0, \bar{0}]$.

Now consider an interval $[0, a]$ in P. Then $p^{-1}([0, a]) = [0, \bar{x}]$, where $\bar{x}$ is the closed element of L corresponding to a. Indeed, if $0 \leqq y \leqq \bar{x}$ then $\bar{y} \leqq \ddot{x} = \bar{x}$, hence $p(y) \leqq a$. Conversely, if $p(y) \leqq a$, then $\bar{y} \leqq \bar{x}$ but $y \leqq \bar{y}$, hence $y \leqq \bar{x}$. Therefore the condition of Theorem 2 is satisfied, and the conclusion follows at once.

Corollary (Weisner).

(a) *Let $a > 0$ in a finite lattice L. Then, for any b in L,*

$$\sum_{x \vee a = b} \mu(0, x) = 0$$

(b) *Let $a < 1$ in L. Then, for any b in L,*

$$\sum_{x \wedge a = b} \mu(x, 1) = 0 .$$

Proof. Take $\bar{x} = x \vee a$. Part (b) is obtained by inverting the order.

Example 2. Let V be a finite-dimensional vector space of dimension n over a finite field with q elements. We denote by $L(V)$ the lattice of subspaces of V. We shall use Proposition 4 to compute the Möbius function of $L(V)$.

In the lattice $L(V)$, every segment $[x, y]$, for $x \leqq y$, is isomorphic to the lattice $L(W)$, where W is the quotient space of the subspace y by the subspace x. If we denote by $\mu_n = \mu_n(q)$ the value of $\mu(0, 1)$ for $L(V)$, it follows that $\mu(x, y) = \mu_j$, when j is the dimension of the quotient space W. Therefore once μ_n is known for for every n, the entire Möbius function is known.

To determine μ_n, consider a subspace a of dimension $n - 1$. In view of the preceding Corollary, we have for all $a < 1$ (where 1 stands for the entire space V):

$$\sum_{x \wedge a = 0} \mu(x, 1) = 0$$

where 0 stands of course for the 0-subspace. Let a be a dual atom of $L(V)$, that is, a subspace of dimension $n - 1$. Which subspaces x have the property that $x \wedge a = 0$? x must be a line in V, and such a line must be disjoint except for 0 from a. A subspace of dimension $n - 1$ contains q^{n-1} distinct points, so there will be $q^n - q^{n-1}$ points outside of a. However, every line contains exactly $q - 1$ points. Therefore, for each subspace a of dimension $n - 1$ there are

$$\frac{q^n - q^{n-1}}{q - 1} = q^{n-1}$$

distinct lines x such that $x \wedge a = 0$. Since each interval $[x, 1]$ is isomorphic to a space of dimension $n - 1$, we obtain

$$\mu_n = \mu(0, 1) = -\sum_{\substack{x \wedge a = 0 \\ x \neq 0}} \mu(x, 1) = - q^{n-1} \mu_{n-1}.$$

This is a difference equation for μ_n which is easily solved by iteration. We obtain the result, first established by PHILIP HALL (see also WEISNER and S. DELSARTE):

$$\mu_n(q) = (-1)^n q^{n(n-1)/2} = (-1)^n q^{\binom{n}{2}}.$$

6. The Euler characteristic

Sharper results relating $\mu(0, 1)$ to combinatorial invariants of a finite lattice can be obtained by application of Theorem 1, when the "comparison set" P remains a Boolean algebra.

A *cross-cut* C of a finite lattice L is a subset of L with the following properties:

(a) C does not contain 0 or 1.

(b) no two elements of C are comparable (that is, if x and y belong to C, then neither $x < y$ nor $x > y$ holds).

(c) Any maximal chain stretched between 0 and 1 meets the set C.

A *spanning subset* S of L is a subset such that $\vee S = 1$ and $\wedge S = 0$.

The main result is the following *Cross-cut Theorem*:

Theorem 3. *Let μ be the Möbius function and E the Euler characteristic of a non-trivial finite lattice L, and let C be a cross-cut of L. For every integer $k \geqq 2$, let q_k denote the number of spanning subsets of C containing k distinct elements. Then*

$$E - 1 = \mu(0, 1) = q_2 - q_3 + q_4 - q_5 + \cdots$$

The *proof* is by induction over the distance of a cross-cut C from the element 1.

Define the distance $d(x)$ of an element x from the element 1 as the maximum length of a chain stretched between x and 1. For example, the distance of a dual atom is two. If C is a cross-cut of L, define the distance $d(C)$ as $\max d(x)$ as x ranges over C. Thus, the distance of the cross-cut consisting of all dual atoms is two, and conversely, this is the only cross-cut having distance two.

It follows from Proposition 1 of Section 5 that the result holds when $d(C) = 2$ (take $R = C$ in the assertion of the Proposition). Thus, we shall assume the truth of the statement for all cross-cuts whose distance is less than n, and prove it for a cross-cut with $d(C) = n$.

If C is a subset of L, we shall write $x > C$ or $x \leqq C$ to mean that there is an element y or C such that $x > y$, or that there is an element y of C such that $x \leqq y$. For a general C, these possibilities may not be mutually exclusive; they are mutually exclusive when C is a cross-cut. We shall repeatedly make use of this remark below.

Define a modified lattice L' as follows. Let L' contain all the elements x such that $x \leqq C$ in the same order. On top of C, add an element 1 covering all the elements of C, but no others; this defines L'.

In L', consider the cross-cut C and apply Proposition 1 of section 5 again. If μ' is the Möbius function of L', then

$$\mu'(0, 1) = p_2 - p_3 + p_4 \cdots,$$

where p_k is the number of all subsets $A \subset C \subset L'$ of k elements, such that $\wedge A = 0$.

Comparing the lattices L and L', we have

$$0 = \sum_{x \leq C} \mu(0, x) + \sum_{x > C} \mu(0, x) = \sum_{x \leq C} \mu'(0, x) + \mu'(0, 1).$$

However, for $x \leq C$, we have $\mu'(0, x) = \mu(0, x)$ by construction of L'. Hence

$$\sum_{x \leq C} \mu(0, x) = -p_2 + p_3 - p_4 + \cdots$$

Since the sets $(x/x \leq C)$ and $(x/x > C)$ are disjoint, we can write

$$\mu(0, 1) = -\sum_{x<1} \mu(0, x) = -[\sum_{x \leq C} \mu(0, x) + \sum_{1>x>C} \mu(0, x)].$$

We now simplify the first summation on the right:

(*) $$\mu(0, 1) = p_2 - p_3 + p_4 \cdots - \sum_{1>x>C} \mu(0, x).$$

Now let $q_k(x)$ be the number of subsets of C having k elements, whose meet is 0 and whose join is x. In particular, $q_k(1) = q_k$. Then clearly

$$p_k = \sum_{x>C} q_k(x), \quad k \geq 2,$$

the summation in (*) can be simplified to

(**) $$\mu(0, 1) = (q_2 - q_3 + q_4 - \cdots) - \sum_{1>x>C} [-q_2(x) + q_3(x) - q_4(x) + \\ + \cdots + \mu(0, x)].$$

For x above C and unequal to 1, consider the segment $[0, x]$. We prove that $C(x) = C \cap [0, x]$ is a cross-cut of the lattice $[0, x]$ such that $d(C(x)) < d(C)$. Once this is done, it follows by the induction hypothesis that every term in brackets on the right of (**) vanishes, and the proof will be complete.

Conditions (a) and (b) in the definition of a cross-cut are trivially satisfied by $C(x)$, and condition (c) is verified as follows. Suppose Q is a maximal chain in $[0, x]$ which does not meet $C(x)$. Choose a maximal chain R in the segment $[x, 1]$; then the chain $Q \cup R$ is maximal in L, and does not intersect C.

It remains to verify that $d(C(x)) < d(C)$, and this is quite simple. There is a chain Q stretched between C and x whose length is $d(C(x))$. Then $d(C)$ exceeds the length of the chain $Q \cup R$, and since $x < 1$, R has length at least 2, hence the length of $Q \cup R$ exceeds that of Q by at least one. The proof is therefore complete.

Theorem 3 gives a relation between the value $\mu(0, 1)$ and the width of narrow cross-cuts or *bottlenecks* of a lattice. The proof of the following statement is immediate.

Corollary 1. (a) *If L has a cross-cut with one element, then* $\mu(0, 1) = 0$.

(b) *If L has a cross-cut with two elements, then the only two possible values of* $\mu(0, 1)$ *are* 0 *and* 1.

(c) *If L has a cross-cut having three elements, then the only possible values of* $\mu(0, 1)$ *are* 2, 1, 0 *and* -1.

In this connection, an interesting combinatorial problem is to determine all possible values of $\mu(0, 1)$, given that L has a cross-cut with n elements.

Reduction of the main formula. In several applications of the cross-cut theorem, the computation of the number q_k of spanning sets may be long, and systematic procedures have to be devised. One such procedure is the following:

Proposition 1. *Let C be a cross-cut of a finite lattice L. For every integer $k \geqq 0$, and for every subset $A \subset C$, let $q(A)$ be the number of spanning sets containing A, and let $S_k = \sum_A q(A)$, where A ranges over all subsets of C having k elements. Set S_0 to be the number of elements of C. Then*

$$\mu(0,1) = S_0 - 2S_1 + 2^2 S_2 - 2^3 S_3 + \cdots.$$

Proof. For every subset $B \subset C$, set $p(B) = 1$ if B is a spanning set, and $p(B) = 0$ otherwise. Then

$$q(A) = \sum_{C \supseteq B \supseteq A} p(B).$$

Applying the Möbius inversion formula on the Boolean algebra of subsets of C, we get

$$p(A) = \sum_{B \supseteq A} q(B)\mu(A, B),$$

where μ is the Möbius function of the Boolean algebra. Summing over all subsets $A \subset C$ having exactly k elements,

$$q_k = \sum_{n(A)=k} p(A) = \sum_{n(A)=k} \quad \sum_{B \supseteq A} q(B)\mu(A, B).$$

Interchanging the order of summation on the right, recalling Proposition 5 of Section 3 and the fact that a set of $k + l$ elements possesses $\binom{k+l}{l}$ subsets of k elements, we obtain

$$q_k = S_k - \binom{k+1}{1} S_{k+1} + \binom{k+2}{2} S_{k+2} \cdots + (-1)^{n-k} \binom{n}{k} S_n.$$

A convenient way of recasting this expression in a form suitable for computation is the following. Let V be the vector space of all polynomials in the variable x, over the real field. The polynomials $1, x, x^2, \ldots$, are linearly independent in V. Hence there exists a linear functional L in V such that

$$L(x^k) = S_k, \quad k = 0, 1, 2, \ldots.$$

Formula (*) can now be rewritten in the concise form

$$q_k = L\left(x^k - (k+1)x^{k+1} + \binom{k+2}{2} x^{k+2} - \cdots\right) = L\left(\frac{x^k}{(1+x)^{k+1}}\right).$$

Upon applying the cross-cut theorem, we find the expression (where q_0 and q_1 are also given by (*), but turn out to be 0)

$$\begin{aligned}\mu(0,1) &= L\left(\frac{1}{1+x} - \frac{x}{(1+x)^2} + \frac{x^2}{(1+x)^3} - \cdots\right) \\ &= L\left(\frac{1}{1+2x}\right) = L(1 - 2x + 4x^2 - 8x^3 + \cdots) \\ &= S_0 - 2S_1 + 4S_2 - \cdots, \quad \text{q.e.d.}\end{aligned}$$

The cross-cut theorem can be applied to study which alterations of the order relation of a lattice preserve the Euler characteristic. Every alteration which preserves meets and joins of the spanning subsets of some cross-cut will preserve the Euler characteristic. There is a great variety of such changes, and we shall not develop a systematic theory here. The following is a simple case.

Following BIRKHOFF and JÓNSSON and TARSKI we define the *ordinal sum* of lattices as follows. Given a lattice L and a function assigning to every element x of L a lattice $L(x)$, (all the $L(x)$ are distinct) the *ordinal sum* $P = \sum_L L(x)$ of the lattices $L(x)$ *over* the lattice L is the partially ordered set P consisting of the set $\bigcup_{x \in L} L(x)$, where $u \leqq v$ if $u \in L(x)$ and $v \in L(x)$ and $u \leqq v$ in $L(x)$, or if $u \in L(x)$ and $v \in L(y)$ and $x < y$. It is clear that P is a lattice if all the $L(x)$ are finite lattices.

Proposition 2. *If the finite lattice P is the ordinal sum of the lattices $L(x)$ over the non-trivial lattice L, and μ_P, μ_x and μ_L are the corresponding Möbius functions, then:*

If $L(0)$ is the one element lattice, then $\mu_P(0, 1) = \mu_L(0, 1)$.

Proof. The atoms of P are in one-to-one correspondence with the atoms of L and the spanning subsets are the same. Hence the result follows by applying the cross-cut theorem to the atoms.

In virtue of a theorem of JÓNSSON and TARSKI, every lattice P has a unique maximal decomposition into an ordinal sum over a "skeleton" L. This can be used in connection with the preceding Corollary to further simplify the computation of $\mu(0, n)$ as n ranges through P.

Homological interpretation. The alternating sums in the Cross-Cut Theorem suggest that the Euler characteristic of a lattice be interpreted as the Euler characteristic in a suitable homology theory. This is indeed the case. We now define* a *homology theory* $H(C)$ relative to an arbitrary cross-cut C of a finite lattice L. For the homological notions, we refer to Eilenberg-Steenrod.

Order the elements of C, say $a_1, a_2, \ldots, a_n$. For $k \geqq 0$, let a *k-simplex* σ be any subset of C of $k+1$ elements which does *not* span. Let C_k be the free abelian group generated by the k-simplices. We let $C_{-1} = 0$; for a given simplex σ, let σ_i be the set obtained by omitting the $(i+1)$-st element of σ, when the elements of σ are ordered according to the given ordering of C. The boundary of a k-simplex is defined as usual as $\partial_k \sigma = \sum_{i=0}^{k} (-1)^i \sigma_i$, and is extended by linearity to all of C_k, giving a linear mapping of C_k into C_{k-1}. The k-th homology group H_k is defined as the abelian group obtained by taking the quotient of the kernel of ∂_k by the image of ∂_{k+1}. The rank b_k of the abelian group H_k, that is, the number of independent generators of infinite cyclic subgroups of H_k, is the k-th *Betti number*.

Let α_k be the rank of C_k, that is, the number of k-simplices. The *Euler characteristic* of the homology $H(C)$ is defined in homology theory as

$$E(C) = \sum_{k=0}^{\infty} (-1)^k \alpha_k .$$

* This definition was obtained jointly with D. KAN, F. PETERSON and G. WHITEHEAD, whom I now wish to thank.

25*

It follows from well-known results in homology theory that

$$E(C) = \sum_{k=0}^{\infty} (-1)^k b_k .$$

Let q_k be the number of spanning subsets with k elements as in Theorem 3. Then $q_{k+1} + \alpha_k$ is the total number of subsets of C having $k + 1$ elements; if C has N elements, then $\alpha_k = \binom{N}{k+1} - q_{k+1}$. It follows from the Cross-Cut Theorem that

$$E(C) = \sum_{k=0}^{\infty} (-1)^k \binom{N}{k+1} - \sum_{k=0}^{\infty} (-1)^k q_{k+1}$$

$$= \sum_{k=0}^{\infty} (-1)^k \binom{N}{k+1} + \mu(0, 1) .$$

We have however

$$\sum_{k=0}^{\infty} (-1)^k \binom{N}{k+1} = - \sum_{i=1}^{\infty} (-1)^i \binom{N}{i} = 1 - \sum_{i=0}^{\infty} (-1)^i \binom{N}{i} = 1 - (1-1)^N = 1 ,$$

and hence

$$E(C) = 1 + \mu(0, 1) = E;$$

in other words:

Proposition 3. *In a finite lattice, the Euler characteristic of the homology of any cross-cut C equals the Euler characteristic of the lattice.*

This result can sometimes be used to compute the Möbius functions of "large" lattices. In general, the numbers q_k are rather redundant, since any spanning subset of k elements gives rise to several spanning subsets with more than k elements. A method for eliminating redundant spanning sets is then called for. One such method consists precisely in the determination of the Betti numbers b_k.

We conjecture that the Betti numbers of $H(C)$ are themselves independent of the cross-cut C, and are also "invariants" of the lattice L, like the Euler characteristic $E(C)$. In the special case of lattices of height 4 satisfying the chain condition, this conjecture has been proved (in a different language) by DOWKER.

Example 1. *The Betti numbers of a Boolean algebra.* We take the cross-cut C of all atoms. If the height of the Boolean algebra is $n + 1$, then every k-cycle, for $k < n - 2$, bounds, so that $b_0 = 1$ and $b_k = 0$ for $0 < k < n - 2$. On the other hand, there is only one cycle in dimension $n - 2$. Hence $b_{n-2} = 1$ and we find $E = 1 + (-1)^{n-2}$, which agrees with Proposition 5 of Section 3.

A notion of Euler characteristic for *distributive* lattices has been recently introduced by HADWIGER and KLEE. For finite distributive lattices, KLEE's Euler characteristic is related to the one introduced in this work. We refer to KLEE's paper for details.

7. Geometric lattices

An ordered structure of very frequent occurrence in combinatorial theory is the one that has been variously called matroid (WHITNEY), matroid lattice (BIRKHOFF), closure relation with the exchange property (MACLANE), geometric lattice

(BIRKHOFF), abstract linear dependence relation (BLEICHER and PRESTON). Roughly speaking, these structures arise in the study of combinatorial objects that are obtained by piecing together smaller objects with a particularly simple structure. The typical such case is a linear graph, which is obtained by piecing together edges. Several counting problems associated with such structures can often be attacked by Möbius inversion, and one finds that the Möbius functions involved have particularly simple properties.

We briefly summarize the needed facts out of the theory of such structures, referring to any of the works of the above authors for the proofs.

A finite lattice L is a *geometric lattice* when every element of L is the join of atoms, and whenever if a and b in L cover $a \wedge b$, then $a \vee b$ covers both a and b. Equivalently, a geometric lattice is characterized by the existence of a rank function satisfying $r(a \wedge b) + r(a \vee b) \leqq r(a) + r(b)$. Notice that this implies the chain condition. In particular if a is an atom, then $r(a \vee c) = r(c)$ or $r(c) + 1$. If M is a semimodular lattice, then the partially ordered subset of all elements which are joins of atoms is a geometric sublattice.

Geometric lattices are most often obtained from a closure relation on a finite set which satisfies the MACLANE-STEINITZ exchange property. The lattice L of closed sets in such a closure relation is a geometric lattice whenever every one-element set is closed. Conversely, every geometric lattice can be obtained in this way by defining one such closure relation on the set of its atoms.

The fundamental property of the Möbius function of geometric lattices is the following:

Theorem 4. *Let μ be the Möbius function of a finite geometric lattice L. Then:*

(a) *$\mu(x, y) \neq 0$ for any pair x, y in L, provided $x \leqq y$.*

(b) *If y covers z, then $\mu(x, y)$ and $\mu(x, z)$ have opposite signs.*

Proof. Any segment $[x, y]$ of a geometric lattice is also a geometric lattice. It will therefore suffice to assume that $x = 0$, $y = 1$ and that z is a dual atom of L.

We proceed by induction. The theorem is certainly true for lattices of height 2, where $\mu(0, 1) = -1$. Assume it is true for all lattices of height $n - 1$, and let L be a lattice of height n. By the Corollary to Proposition 4 of Section 5, with $b = 1$, and a an atom of L, we have

$$\mu(0, 1) = -\sum_{\substack{x \vee a = 1 \\ x \neq 1}} \mu(0, x).$$

Now from the subadditive inequality

$$r(x \wedge a) + r(x \vee a) \leqq r(x) + r(a)$$

we infer that if $x \vee a = 1$, then $n \leqq \dim x + \dim a$, hence $\dim x \geqq n - 1$. The element x must therefore be a dual atom. It follows from the induction assumption and from the fact that L satisfies the chain condition, that all the $\mu(0, x)$ in the sum on the right have the same sign, and none of them is zero. Therefore, $\mu(0, 1)$ is not zero, and its sign is the opposite of that $\mu(0, x)$ for any dual atom x. This concludes the proof.

Corollary. *The coefficients of the characteristic polynomial of a geometric lattice alternate in sign.*

We next derive a combinatorial interpretation of the Euler characteristic of a geometric lattice, which generalizes a technique first used by WHITNEY in the study of linear graphs.

A subset $\{a, b, \ldots, c\}$ of a geometric lattice L is *independent* when

$$r(a \vee b \vee \cdots \vee c) = r(a) + r(b) + \cdots + r(c).$$

Let C_k be the cross-cut of L of all elements of rank $k > 0$. A maximal independent subset $\{a, b, \ldots, c\} \subset C_k$ is a *basis* of C_k. All bases of C_k have the same number of elements, namely, $n - k$ if the lattice has height n. A subset $A \subset C_k$ is a *circuit* (WHITNEY) when it is not independent but every proper subset is independent. A set is independent if and only if it contains no circuits.

Order the elements of L of rank k in a linear order, say $a_1, a_2, \ldots, a_l$. This ordering induces a lexicographic ordering of the circuits of C_k.

If the subset $\{a_{i_1}, a_{i_2}, \ldots, a_{i_j}\}$ $(i_1 < i_2 < \cdots < i_j)$ is a circuit, the subset $a_{i_1}, a_{i_2}, \ldots, a_{i_{j-1}}$ will be called a *broken circuit.*

Proposition 1. *Let L be a geometric lattice of height $n + 1$, and let C_k be the cross-cut of all elements of rank k. Then $\mu(0, 1) = (-1)^n m_k$, where m_k is the number of subsets of C_k whose meet is 0, containing $n - k + 1$ elements each, and not containing all the arcs of any broken circuit.*

Again, the assertion implies that $m_1 = m_2 = m_3 = \cdots$.

Proof. Let the lexicographically ordered broken circuits be $P_1, P_2, \ldots, P_\sigma$, and let S_i be the family of all spanning subsets of C_k containing P_i but not $P_1, P_2, \ldots$, or P_{i-1}. In particular, $S_{\sigma+1}$ is the family of all those spanning subsets not containing all the arcs of any broken circuit. Let q_j^i be the number of spanning subsets of j elements and not belonging to S_i. We shall prove that for each $i \geqq 1$

$$\mu(0, 1) = q_2^i - q_3^i + q_4^i \cdots . \tag{*}$$

First, set $i = 1$. The set S_1 contains all spanning subsets containing the broken circuit P_1. Let $\bar{P}_1$ be the cicuit obtained by completing the broken circuit P_1. — A spanning set contained in S_1 contains either $\bar{P}_1$ or else P_1 but not $\bar{P}_1$; call these two families of spanning subsets A and B, and let q_j^A and q_j^B be defined accordingly. Then $q_j = q_j^1 + q_j^A + q_j^B$, and

$$\begin{aligned}\mu(0, 1) = q_2 - q_3 + q_4 \cdots = q_2^1 - q_3^1 + \cdots + \\ + q_2^A + (q_2^B - q_3^A) - (q_3^B - q_4^A) + \cdots .\end{aligned}$$

Now, $q_2^A = 0$, because no circuit can contain two elements; there is a one-to-one correspondence between the elements of A and those of B, obtained by completing the broken circuit P_1. Thus, all terms in parentheses cancel and the identity (*) holds for $i = 1$.

To prove (*) for $i > 1$, remark that the element c_i of C_k, which is dropped from a circuit to obtain the broken circuit P_i, does not occur in any of the previous circuits, because of the lexicographic ordering of the circuits. Hence the induction can be continued up to $i = \sigma + 1$.

Any set belonging to $S_{\sigma+1}$ does not contain any circuit. Hence, it is an independent set. Since it is a spanning set, it must contain $n - k + 1$ elements. Thus, all the integers $q_{\sigma+1}$ vanish except $q^{\sigma+1}_{n-k+1}$ and the statement follows from (*), q.e.d.

Corollary 1. *Let $q(\lambda) = \lambda^n + m_1\lambda^{n-1} + m_2\lambda^{n-2} + \cdots + m_n$ be the characteristic polynomial of a geometric lattice of height $n + 1$. Then $(-1)^k m_k$ is a positive integer for $1 \leqq k \leqq n$, equal to the number of independent subsets of k atoms not containing any broken circuit.*

The *proof* is immediate: take $k = 1$ in the preceding Proposition.

The homology of a geometric lattice is simpler than that of a general lattice:

Proposition 2. *In the homology relative to the cross-cut C_k of all elements of rank $k = 1$, the Betti numbers $b_1, b_2, \ldots, b_{k-2}$ vanish.*

The *proof* is not difficult.

Example 1. *Partitions of a set.*

Let S be a finite set of n elements. A *partition* π of S is a family of disjoint subsets $B_1, B_2, \ldots, B_k$, called *blocks*, whose union is S. There is a (well-known) natural ordering of partitions, which is defined as follows: $\pi \leqq \sigma$ whenever every block of π is contained in a block of partition σ. In particular, 0 is the partition having n blocks, and I is the partition having one block. In this ordering, the partially ordered set of partitions is a geometric lattice (cf. BIRKHOFF).

The Möbius function for the lattice of partitions was first determined by SCHÜTZENBERGER and independently by ROBERTO FRUCHT and the author. We give a new proof which uses a recursion. If π is a partition, the *class* of π is the (finite) sequence $(k_1, k_2, \ldots)$, where k_i is the number of blocks with i elements.

Lemma. *Let L_n be the lattice of partitions of a set with n elements. If $\pi \in L_n$ is of rank k, then the segment $[\pi, 1]$ is isomorphic to L_{n-k}. If π is of class $(k_1, k_2, \ldots)$, then the segment $[0, \pi]$ is isomorphic to the direct product of k_1 lattices isomorphic to L_1, k_2 lattices isomorphic to L_2,* etc.

The *proof* is immediate.

It follows from the Lemma that if $[x, y]$ is a segment of L_n, then it is isomorphic to a product of k_i lattices isomorphic to L_i, $i = 1, 2, \ldots$. We call the sequence $(k_1, k_2, \ldots)$ the *class* of the *segment* $[x, y]$.

Proposition 3. *Let $\mu_n = \mu(0, 1)$ for the lattice of partitions of a set with n elements. Then $\mu_n = (-1)^{n-1}(n-1)!$.*

Proof. By the Corollary to Proposition 4 of Section 5, $\sum_{x \wedge a = 0} \mu(x, 1) = 0$. Let a be the dual atom consisting of a block C_1 containing $n - 1$ points, and a second block C_2 containing one point. Which non-zero partitions x have the property that $x \wedge a = 0$? Let the blocks of such a partition x be $B_1, \ldots, B_k$. None of the blocks B_i can contain two distinct points of the block C_1, otherwise the two points would still belong to the same block in the intersection. Furthermore, only one of the B_i can contain the block C_2. Hence, all the B_i contain one point, except one, which contains C_2 and an extra point. We conclude that x must be an atom, and there are $n - 1$ such atoms. Hence, $\mu_n = \mu(0, 1) = -\sum_x \mu(x, 1)$, where x ranges over a set of $n - 1$ atoms. By the Lemma, the segment $[x, 1]$ is isomorphic to the lattice of partitions of a set with $n-1$ elements, hence $\mu_n = -(n-1)\mu_{n-1}$. Since $\mu_2 = -1$, the conclusion follows.

Corollary. *If the segment* $[x, y]$ *is of class* $(k_1, k_2, \ldots, k_n)$, *then*

$$\mu(x, y) = \mu_1^{k_1} \mu_2^{k_2} \ldots \mu_n^{k_n} = (-1)^{k_1+k_2+\cdots+k_n-n} (2!)^{k_3} (3!)^{k_4} \ldots ((n-1)!)^{k_n}.$$

The Möbius inversion formula on the partitions of a set has several combinatorial applications; see the author's expository paper on the subject.

8. Representations

There is, as is well known, a close analogy between combinatorial results relating to Boolean algebras and those relating to the lattice of subspaces of a vector space. This analogy is displayed for example in the theory of q-difference equations developed by F. H. JACKSON, and can be noticed in many number-theoretic investigations. In view of it, we are led to surmise that a result analogous to Proposition 1 of Section 5 exists, in which the Boolean algebra of subsets of R is replaced by a lattice of subspaces of a vector space over a finite field. Such a result does indeed exist; in order to establish it a preliminary definition is needed.

Let L be a finite lattice, and let V be a finite-dimensional vector space over a finite field with q elements. A *representation* of L over V is a monotonic map p of L into the lattice M of subspaces of V, having the following properties:

(1) $p(0) = 0$.

(2) $p(a \vee b) = p(a) \vee p(b)$.

(3) Each atom of L is mapped to a line of the vector space V, and the set of lines thus obtained spans the entire space V.

A representation is *faithful* when the mapping p is one-to-one. We shall see in Section 9 that a great many ordered structures arising in combinatorial problems admit faithful representations. Given a representation $p: L \to M$, one defines the *conjugate map* $q: M \to L$ as follows.

Let K be the set of atoms of M (namely, lines of V), and let A be the image under p of the set of atoms of L. For $s \in M$, let $K(s)$ be the set of atoms of M dominated by s, and let $B(s)$ be a minimal subset of A which spans (in the vector space sense) every element of $K(s)$. Let $A(s)$ be the subset of A which is spanned by $B(s)$. A simple vector-space argument, which is here omitted, shows that the set $A(s)$ is well defined, that is, that it does not depend upon the choice of $B(s)$, but only upon the choice of s.

Let $C(s)$ be the set of atoms of L which are mapped by p onto $A(s)$. Set $q(s) = \vee C(s)$ in the lattice L; this defines the map q. It is obviously a monotonic function.

Lemma. *Let* $p: L \to M$ *be a faithful representation and let* $q: M \to L$ *be the conjugate map. Assume that every element of* L *is a join of atoms. Then* $p(q(s)) \geq s$ *and* $q(p(x)) \leq x$.

Proof. By definition, $q(s) = \vee C(s)$, where $C(s)$ is the inverse image of $A(s)$ under p. By property (2) of a representation,

$$p(q(s)) = p(\vee C(s)) = \vee p(C(s)) = \vee A(s).$$

But this join of the set of lines $A(s)$ in the lattice M is the same as their span in the vector space V. Hence $\bigvee A(s) \geqq s$, and we conclude that $p(q(s)) \geqq s$.

To prove that $q(p(x)) \leqq x$, it suffices to show that $A(p(x)) = B$, where B is the set of atoms in A dominated by $p(x)$. Clearly $B \subset A(p(x))$, and it will suffice to establish the converse implication. By (2), and by the fact that x is a join of atoms, we have $p(x) = \bigvee B$. Therefore every line l dominated by $p(x)$ is spanned by a subset of B. If in addition $l \in A$, then $l \leqq \bigvee C$ for some subset $C \subset B$, hence $l \in B$. This shows $B \supset A(p(x))$, q.e.d.

Theorem 5. *Let L be a finite lattice, where every element is a join of atoms, let $p: L \to M$ be a faithful representation of L into the lattice M of subspaces of a vector space V over a finite field with $\underline{q}$ elements, and let $q: M \to L$ be the conjugate map. For every $k \geqq 2$, let m_k be the number of k-dimensional subspaces s of V such that $q(s) = I$. Then*

$$(*) \qquad \mu(0, 1) = \underline{q}^{\binom{2}{2}} m_2 - \underline{q}^{\binom{3}{2}} m_3 + \underline{q}^{\binom{4}{2}} m_4 - \cdots,$$

where μ is the Möbius function of L.

Proof. Let $Q = L^*$, let $c: L \to Q$ and $c^*: Q \to L$ be the canonical isomorphisms between L and Q. Define $\pi: Q \to M$ as $\pi = pc^*$, and $\varrho: M \to Q$ as $\varrho = cq$. We verify that π and ϱ give a Galois connection between Q and M satisfying the hypothesis of Theorem 1. If $\pi(x) = 0$, then there is a $y \in L$ such that $y = c^*(x)$ and $p(y) = 0$. It follows from the definition of a representation that $y = 0$. Hence $x = c(y) = 1$. Furthermore, $\varrho(0) = c(q(0)) = 1$. It follows from the preceding Lemma that π and ϱ are a Galois connection. Applying Theorem 1 and the result of Example 2 of Section 5, formula (*) follows at once.

Remark. It is easy to see that every lattice having a faithful representation is a geometric lattice. The converse is however not true, as an example of T. Lazarson shows.

A reduction similar to that of Proposition 1 of Section 7 can be carried out with Theorem 5 and representations, and another combinatorial property of the Euler characteristic is obtained.

9. The coloring of graphs

By way of illustration of the preceding theory, we give some applications to the classic problem of coloring of graphs, and to the problem of constructing flows in networks with specified properties. Our results extend previous work of G. D. Birkhoff, D. C. Lewis, W. T. Tutte and H. Whitney.

A *linear graph* $G = (V, E)$ is a structure consisting of a finite set V, whose elements are called vertices, together with a family E of two-element subsets of V, called edges. Two vertices a and b are adjacent when the set (a, b) is an edge; the vertices a and b are called the endpoints of (a, b). Alternately, one calls the vertices *regions* and calls the graph a *map*, and we use the two terms interchangeably, considering them as two words for the same object. If S is a set of edges, the *vertex set* $V(S)$ consists of all vertices which are incident to some edge in S.

A set of edges S is *connected* when in any partition $S = A \cup B$ into disjoint non-empty sets A and B, the vertex sets $V(A)$ and $V(B)$ are not disjoint. Every set of edges is the union of disjoint connected *blocks*.

The *bond closure* on a graph $G = (V, E)$ is a closure relation defined on the set E of edges as follows. If $S \subset E$, let $\bar{S}$ be the set of all edges both of whose endpoints belong to one and the same block of S. Every set consisting of a single edge is closed, and these are the only minimal non-empty closed sets.

Lemma 1. *The bond closure $S \to \bar{S}$ has the exchange property.*

Proof. Suppose e and f are edges, $S \subset E$, and $e \in \overline{S \cup f}$ but $e \notin \bar{S}$. Then every endpoint of e which is not in $V(S)$ is an endpoint of f; on the other hand, S and f have at least one point in common, otherwise $e \in \bar{S}$. Thus both e and f either connect the same two blocks of S, or else they have one endpoint in S and one common endpoint; hence $f \in \overline{S \cup e}$, q.e.d.

The lattice $L = L(G)$ of bond-closed subsets of E is called the *bond lattice* of the graph G. Suppose that E has n blocks and $p(\lambda)$ is the characteristic polynomial of L, then the polynomial $\lambda^n p(\lambda)$ is the *chromatic polynomial* of the graph G, first studied by G. D. BIRKHOFF. From Theorem 4 we infer at once the theorem of WHITNEY that the coefficients of the chromatic polynomial alternate in sign.

The chromatic polynomial has the following combinatorial interpretation. Let C be a set of n elements, called colors. A function $f: V \to C$ is a *proper coloring* of the graph, when no two adjacent vertices are assigned the same color. To every coloring f — not necessarily proper — there corresponds a subset of E, the *bond* of f, defined as the set of all edges whose endpoints are assigned the same color by f. The bond of f is a closed set of edges. For every closed set S, let $p(\lambda, S)$ be the number of colorings whose bond is S. Then we shall prove that $p(\lambda, S) = \lambda^n q(\lambda, S)$, where $q(\lambda, S)$ is the characteristic polynomial of the segment $[S, I]$ in the lattice L. Since every coloring has a bond $\sum_{T \geqq S} p(\lambda, T)$ equals the total number of colorings having some bond $T \geqq S$. But this number is evidently $\lambda^{k-r(s)}$, where k is the number of vertices of the graph and $r(S)$ is the rank of S in L. Applying the Möbius inversion formula on the bond-lattice, we get

$$(*) \qquad p(\lambda) = p(\lambda, 0) = \sum_{T \in L} \lambda^{k-r(T)} \mu(0, T).$$

But the number of colorings whose bond is the null set 0 is exactly the number of proper colorings.

WHITNEY's evaluation (cf. A logical expansion in Mathematics) of the chromatic polynomials of a graph in terms of the number of subgraphs of s edges and p connected components is an immediate consequence of the cross-cut theorem applied to the atoms of the bond-lattice of G. This result of WHITNEY's can now be sharpened in two directions: first, a cross-cut other than that of the atoms can be taken; secondly, the computation of the coefficients of the chromatic polynomial can be simplified by Proposition 1 of Section 8. The cross-cut of all elements of rank 2 is particularly suited for computation, and can be programmed. The interested reader may wish to explicitly translate the cross-cut theorem and the results of Section 8 into the geometric language of graphs.

Example 1. For a *complete graph* on n vertices, where every two-element subset is an edge, the bond-lattice is isomorphic to the lattice of partitions of a set with n elements. The chromatic polynomial is evidently $(\lambda)_n = \lambda(\lambda - 1) \ldots (\lambda - n + 1)$, and the coefficients $s(n, k)$ are the *Stirling numbers of the first kind*.

Thus, $\sum_{r(\pi)=k} \mu(0, \pi) = s(n, k)$. This gives a combinatorial interpretation to the Stirling numbers of the first kind.

For a map $\mathfrak{m}$ embedded in the plane, where regions and boundaries have their natural meaning and no region bounds with itself, one obtains an interesting geometric result by applying the cross-cut theorem to the dual atoms of the bond lattice $L(\mathfrak{m})$.

Let $\mathfrak{m}$ be a connected map in the plane; without loss of generality we can assume: (a) that all the regions of $\mathfrak{m}$, except one which is unbounded, lie inside a convex polygon, the outer boundary of $\mathfrak{m}$; (b) that all boundaries are segments of straight lines. The *dual graph* of $\mathfrak{m}$ is the linear graph made up of the boundaries of $\mathfrak{m}$. A *circuit* in a linear graph is defined as a simple closed curve contained in the graph. We give an expression of the polynomial $P(\lambda, \mathfrak{m})$ in terms of the circuits of the dual graph. The outer boundary is always a circuit.

A set of circuits of a map $\mathfrak{m}$ in the plane *spans*, when their union — in the set-theoretic sense — is the entire boundary of $\mathfrak{m}$.

Proposition 1. *For every integer $k \geqq 1$, let C_k be the number of spanning sets of k distinct circuits of a map $\mathfrak{m}$ in the plane. Then*

$$\mu_{\mathfrak{m}}(0, 1) = -C_1 + C_2 - C_3 + C_4 - \cdots$$

Proof. If the map has two regions, then $C_1 = 1$ and all other $C_c = 0$, so the result is trivial. Assume now that $\mathfrak{m}$ has at least 3 regions. Then $C_1 = 0$. All we have to prove is that the integers C_k are the integers q_k of Theorem 3, relative to the cross-cut of $L(\mathfrak{m})$ consisting of all the dual atoms.

By the Jordan curve theorem, every circuit divides the plane into two regions: this gives a one-to-one correspondence of the circuits with the dual atoms of $L(\mathfrak{m})$. Conversely, because we can assume that the map is of the special type described above, every dual atom in $L(\mathfrak{m})$ is a map with two connected regions, and so must have as a boundary a simple closed curve, q.e.d.

It has been shown by Richard Rado (p. 312) that the bond-lattice $L(G)$ of any linear graph G has a faithful representation. Accordingly, Theorem 5 can also be applied to obtain expression for $\mu(0,1)$. These expressions usually give sharper bounds than similar expressions based upon the cross-cut of atoms.

Farther-reaching techniques for the computation of the Möbius function of $L(G)$ are obtained by applying Theorem 1 to situations where P and Q are both bond-lattices of graphs. This we shall now do. A *monomorphism* of a graph G into a graph H is a one-to-one function f of the vertices of G *onto* the vertices of H, which induces a map $\bar{f}$ of the edges of G *into* the edges of H. Every monomorphism $f: G \to H$ induces a monotonic map $p: L(G) \to L(H)$, where $p(S)$ is defined as the closure of the image $\bar{f}(S)$ in H. It also induces a monotonic map $q: L(H) \to L(G)$, where $q(T)$ is defined as the set of edges of G whose image is in T.

Lemma 2. *$q(p(S)) = S$ for S in $L(G)$ and $p(q(T)) \leqq T$ for T in $L(H)$.*

Proof. Intuitively, $p(S)$ is obtained by "adding edges" to S, and $q(p(S))$ simply removes the added edges. Thus, the first statement is graphically clear. The second one can be seen as follows. $q(T)$ is obtained from T by removing a number of edges. Taking $p(q(T))$, some of the edges may be replaced, but in general not all. Thus, $p(q(T)) \leq T$.

Taking $M = L(H)^*$ and $c : L(H) \to M$ to be the canonical order-inverting map, we see that $\pi = cp$ and $\zeta = qc$ give a Galois connection between $L(G)$ and M. Now, $\pi(x) = 0$ is equivalent to $p(x) = 1$ for $x \in L(G)$. This can happen only if x has only one component, that is — since x is closed — only if $x = 1$ in $L(G)$. Thus $\pi(x) = 0$ if and only if $x = 1$. Secondly, $\varrho(0) = q(1) = 1$, evidently. We have verified all the hypotheses of Theorem 1, and we therefore obtain:

Proposition 2. *Let $f : G \to H$ be a monomorphism of a linear graph G into a linear graph H, and let μ_G and μ_H be the Möbius functions of the bond-lattices. Then*

$$\mu_G(0, 1) = \sum_{[a \in L(H);\, q(a)=0]} \mu_H(a, 1),$$

where q is the map of $L(H)$ into $L(G)$ naturally associated with f, as above.

Proposition 1 can be used to derive a great many of the reductions of G. D. BIRKHOFF and D. C. LEWIS, and provides a systematic way of investigating the changes of Möbius functions — and hence of the chromatic polynomial — when edges of a graph are removed. It has a simple geometric interpretation.

An interesting application is obtained by taking H to be the complete lattice on n elements. We then obtain a formula for μ which completes the statements of Theorems 3 and 5. Let G be a linear graph on n vertices. Let C be the family of two-element subsets of G which are not edges of G. Let F be the family of all subsets of C which are closed sets in the bond-lattice of the complete graph on n vertices built on the vertices of G. Then,

Corollary.
$$\mu_G(0, 1) = \sum_{a \in F} \mu(a, 1).$$

where μ is the Möbius function of the lattice of partitions (cf. Example 5) of a set of n elements.

Stronger results can be obtained by considering "epimorphisms" rather than "monomorphisms" of graphs, relating μ_G to the Möbius function obtained from G by "coalescing" points. In this way, one makes contact with G. A. DIRAC's theory of critical graphs. We leave the development of this topic to a later work.

10. Flows in networks

A network $N = (V, E)$ is a finite set V of vertices, together with a set of ordered pairs of vertices, called edges.

We shall adopt for networks the same language as for linear graphs.

A *circuit* is a sequence of edges S such that every vertex in $V(S)$ belongs to exactly two edges of S. Every edge has a positive and a negative endpoint. Given a function Φ from E to the integers from 0 to $\lambda - 1$, let for each vertex v, $\overline{\Phi}(v)$ be defined as

$$\overline{\Phi}(v) = \sum_{e} \eta(e, v)\, \Phi(e),$$

where the sum ranges over all edges incident to v, and the function $\eta(e, v)$ takes the value +1 or −1 according as the positive or negative end of the edge e abuts at the vertex v, and the value zero otherwise. The function Φ is a *flow* (mod. λ) when $\bar{\Phi}(v) = 0$ (mod. λ) for every vertex v. The value $\Phi(e)$ for an edge e is called the *capacity* of the *flow* through e. The mod. λ restriction is inessential, but will be kept throughout.

A *proper flow* is one in which no edge is assigned zero capacity. TUTTE was the first to point out the importance of the problem of counting proper flows (cf. A contribution to the theory of chromatic polynomials) in combinatorial theory.

We shall reduce the solution of the problem to a Möbius inversion on a lattice associated with the network. This will give an expression for the number of proper flows as a polynomial in λ, whose coefficients are the values of a Möbius function.

Every flow through N is a proper flow of a suitable subnetwork of N, obtained by removing those edges which are assigned capacity 0. However, the converse of this assertion is not true: given a subnetwork S of N, it may not be possible to find a flow which is proper on the complement of N. This happens because every flow which assigns capacity zero to each edge of S may assign capacity zero to some further edges. We are therefore led to define a closure relation on the set of all subgraphs as follows: $\bar{S}$ shall be the set of all edges which necessarily are assigned capacity zero, in any flow of N which assigns capacity zero to every edge of S. In other words, if $e \notin \bar{S}$, then there is a flow in N which assigns capacity $\neq 0$ to the edge e, but which assigns capacity zero to all the edges of $\bar{S}$. It is immediately verified that $S \to \bar{S}$ is a closure relation. We call it the *circuit closure* of S. The circuit closure *has the exchange property*: if $e \in \overline{S \cup p}$ but $e \notin \bar{S}$, then $p \in \overline{S \cup e}$. Before verifying it, we first derive a geometric characterization of the circuit closure. A set S is circuit closed ($S = \bar{S}$) if and only if through every edge e not in S there passes a circuit which is disjoint from S. For if S is closed and $e \notin S$, then there is a flow through e and disjoint from S. But this can happen only if there is a circuit through e.

If there is a circuit through the edge p disjoint from $\overline{S \cup e}$, and a circuit through e disjoint from $\bar{S}$ and containing p, then there is — as has been observed by WHITNEY — also a circuit through e not containing $\bar{S} \cup p$. This implies that e is not in the closure of $\bar{S} \cup p$, and verifies the exchange property.

The lattice $C(N)$ of closed subsets of edges of the network N is the *circuit lattice of* N. An atom in this lattice is not necessarily a single edge.

Proposition 1. *The number of proper flows,* (mod. λ) *on a network N with v vertices, e edges and p connected components is a polynomial $p(\lambda)$ of degree $e - v + p$. This polynomial is the characteristic polynomial of the circuit lattice of N. The coefficients alternate in sign.*

Proof. The last statement is an immediate consequence of Theorem 4 of Section 8.

The total number of flows on N (not necessarily proper) is determined as follows. Assume for simplicity that N is connected. Remove a set D of $v - 1$ edges from N, one adjacent to each but one of the vertices.

Every flow on N can be obtained by first assigning to each of the edges not in D an arbitrary capacity, between 0 and $\lambda - 1$, and then filling in capacities for the edges in D to match the requirement of zero capacity through each vertex. There are λ^{e-v+1} ways of doing this, and this is therefore the total number of flows mod. λ. If the network is in p connected components, the same argument gives λ^{e-v+p}. Now, every flow on G is a proper flow on a unique closed subset $\bar{S}$, obtained by removing all edges having capacity zero.

Hence

$$\lambda^{e-v+p} = \sum_{\bar{S} \in \underline{C}(G)} p(\bar{S}, \lambda),$$

where $p(\bar{S}, \lambda)$ is the characteristic polynomial of the closed subgraph $\bar{S}$. Setting $n(s) = e(s) - v(s) + p(s)$, the number of edges, vertices and components of s, and applying the inversion formula, we get

$$p(G, \lambda) = \sum_{S \in \underline{C}(G)} \lambda^{n(s)} \mu(S, G), \quad \text{q.e.d.}$$

In the course of the proof we have also shown that $n(s)$ is the *rank* of S in the circuit lattice of G. The rank of the null subgraph is one.

The four-color problem is equivalent to the statement that every planar network without an isthmus has a proper flow mod 5. (An isthmus is an edge that disconnects a component of the network when removed.)

Most of the results of the preceding section extend to circuit lattices of a network, and give techniques for computation of the flow polynomials of networks. We shall not write down their translation into the geometric language of networks.

References

AUSLANDER, L., and H. M. TRENT: Incidence matrices and linear graphs. J. Math. Mech. 8, 827--835 (1959).

BELL, E. T.: Algebraic Arithmetic. New York: Amer. Math. Soc. (1927).

-- Exponential polynomials. Ann. of Math., II. Ser. **35**, 258--277 (1934).

BERGE, C.: Théorie des graphes et ses applications. Paris: Dounod 1958.

BIRKHOFF, GARRETT: Lattice Theory, third preliminary edition. Harvard University, 1963.

-- Lattice Theory, revised edition. American Mathematical Society, 1948.

BIRKHOFF, G. D.: A determinant formula for the number of ways of coloring a map. Ann. of Math., II. Ser. **14**, 42--46 (1913).

--, and D. C. LEWIS: Chromatic polynomials. Trans. Amer. math. Soc. **60**, 355--451 (1946).

BLEICHER, M. N., and G. B. PRESTON: Abstract linear dependence relations. Publ. Math. Debrecen **8**, 55--63 (1961).

BOUGAYEV, N. V.: Theory of numerical derivatives. Moscow, 1870--1873, pp. 1--222.

BRUIJN, N. G. DE: Generalization of Polya's fundamental theorem in enumerative combinatorial analysis. Indagationes math. **21**, 59--69 (1959).

CHUNG, K.-L., and L. T. C. HSU: A combinatorial formula with its application to the theory of probability of arbitrary events. Ann. math. Statistics **16**, 91--95 (1945).

DEDEKIND, R.: Gesammelte Mathematische Werke, volls. I--II--III. Hamburg: Deutsche Math. Verein. (1930).

DELSARTE, S.: Fonctions de Möbius sur les groupes abéliens finis. Ann. of Math., II. Ser. **49**, 600--609 (1948).

DILWORTH, R. P.: Proof of a conjecture on finite modular lattices. Ann. of Math., II. Ser. **60**, 359--364 (1954).

DIRAC, G. A.: On the four-color conjecture. Proc. London math. Society, III. Ser. **13**, 193 to 218 (1963).

DOWKER, C. H.: Homology groups of relations. Ann. of Math., II. Ser. **56**, 84--95 (1952).

DUBREIL-JACOTIN, M.-L., L. LESIEUR et R. CROISOT: Leçons sur la théorie des treilles des structures algebriques ordonnées et des treilles géometriques. Paris: Gauthier-Villars 1953.

EILENBERG, S., and N. STEENROD: Foundations of algebraic topology. Princeton: University Press 1952.

FARY, I.: On straight-line representation of planar graphs. Acta Sci. math. Szeged 11. 229—233 (1948).

FELLER, W.: An introduction to probability theory and its applications, second edition. New York: Wiley 1960.

FRANKLIN, P.: The four-color problem. Amer. J. Math. 44, 225—236 (1922).

FRÉCHET, M.: Les probabilités associées à un système d'évenements compatibles et dépendants. Actualitées scientifiques et industrielles, nos. 859 et 942. Paris: Hermann 1940 et 1943.

FRONTERA MARQUÉS, B.: Una función numérica en los retículos finitos que se anula para los retículos reducibles. Actas de la 2a, Reunión de matemáticos españoles. Zaragoza 103 - 111 1962.

FRUCHT, R., and G.-C. ROTA: La función de Möbius para el retículo di particiones de un conjunto finito. To appear in Scientia (Chile).

GOLDBERG, K., M. S. GREEN and R. E. NETTLETON: Dense subgraphs and connectivity. Canadian J. Math. 11 (1959).

GOLOMB, S. W.: A mathematical theory of discrete classification. Fourth Symposium in Information Theory, London, 1961.

GREEN, M. S., and R. E. NETTLETON: Möbius function on the lattice of dense subgraphs. J. Res. nat. Bur. Standards 64B, 41—47 (1962).

— — Expression in terms of modular distribution functions for the entropy density in an infinite system. J. Chemical Physisc 29, 1365—1370 (1958).

HADWIGER, H.: Eulers Charakteristik und kombinatorische Geometrie. J. reine angew. Math. 194, 101—110 (1955).

HALL, PHILIP: A contribution to the theory of groups of prime power order. Proc. London math. Soc., II. Ser. 36, 39—95 (1932).

— The Eulerian functions of a group. Quart. J. Math. Oxford Ser. 134—151, 1936.

HARARY, F.: Unsolved problems in the enumeration of graphs. Publ. math. Inst. Hungar. Acad. Sci. 5, 63—95 (1960).

HARDY, G. H.: Ramanujan. Cambridge: University Press 1940.

—, and E. M. WRIGHT: An introduction to the theory of numbers. Oxford: University Press 1954.

HARTMANIS, J.: Lattice theory of generalized partitions. Canadian J. Math. 11, 97—106 (1959).

HILLE, E.: The inversion problems of Möbius. Duke math. J. 3, 549—568 (1937).

HSU, L. T. C.: Abstract theory of inversion of iterated summation. Duke math. J. 14, 465 to 473 (1947).

— On Romanov's device of orthogonalization. Sci. Rep. Nat. Tsing Hua Univ. 5, 1—12 (1948).

— Note on an abstract inversion principle. Proc. Edinburgh math. Soc. (2) 9, 71—73 (1954).

JACKSON, F. H.: Series connected with the enumeration of partitions. Proc. London math. Soc., II. Ser. 1, 63—88 (1904).

— The q-form of Taylor's theorem. Messenger of Mathematics 38, 57—61 (1909).

JÓNSSON, B.: Lattice-theoretic approach to projective and affine geometry. Symposium on the Axiomatic Method. Amsterdam, North-Holland Publishing Company, 1959, 188—205.

—, and A. TARSKI: Direct decomposition of finite algebraic systems. Notre Dame Mathematical lectures, no. 5. Indiana: Notre Dame 1947.

KAC, M., and J. C. WARD: A combinatorial solution of the two-dimensional Ising model. Phys. Review 88, 1332—1337 (1952).

KAPLANSKI, I., and J. RIORDAN: The problème des ménages. Scripta math. 12, 113—124 (1946).

KLEE, V.: The Euler characteristic in combinatorial geometry. Amer. math. Monthly 70, 119—127 (1963).

LAZARSON, T.: The representation problem for independence functions. J. London math. Soc. **33**, 21–25 (1958).

MACLANE, S.: A lattice formulation of transcendence degrees and p-bases. Duke math. J. **4**, 455–468 (1938).

MACMILLAN, B.: Absolutely monotone functions. Ann. of Math., II. Ser. **60**, 467–501 (1954).

MÖBIUS, A. F.: Über eine besondere Art von Umkehrung der Reihen. J. reine angew. Math. **9**, 105–123 (1832).

ORE, O.: Theory of graphs. Providence: American Mathematical Society 1962.

POLYA, G.: Kombinatorische Anzahlbestimmungen für Gruppen, Graphen und chemische Verbindungen. Acta math. **68**, 145–253 (1937).

RADO, R.: Note on independence functions. Proc. London math. Soc., III. Ser. **7**, 300–320 (1957).

READ, R. C.: The enumeration of locally restricted graphs, I. J. London math. Soc. **34**, 417 to 436 (1959).

REDFIELD, J. H.: The theory of group-reduced distributions. Amer. J. Math. **49**, 433–455 (1927).

REVUZ, ANDRÉ: Fonctions croissantes et mesures sur les espaces topologiques ordonnés. Ann. Inst. Fourier **6** 187–268 (1955).

RIORDAN, J.: An introduction to combinatorial analysis. New York: Wiley 1958.

ROMANOV, N. P.: On a special orthonormal system and its connection with the theory of primes. Math. Sbornik, N. S. **16**, 353–364 (1945).

ROTA, G.-C.: Combinatorial theory and Möbius functions. To appear in Amer. math. Monthly.

— The number of partitions of a set. To appear in Amer. math. Monthly.

RYSER, H. J.: Combinatorial Mathematics. Buffalo: Mathematical Association of America 1963.

SCHÜTZENBERGER, M. P.: Contribution aux applications statistiques de la théorie de l'information. Publ. Inst. Stat. Univ. Paris, **3**, 5–117 (1954).

TARSKI, A.: Ordinal algebras. Amsterdam: North-Holland Publishing Company 1956.

TOUCHARD, J.: Sur un problème de permutations. C. r. Acad. Sci., Paris, **198**, 631–633 (1934).

TUTTE, W. T.: A contribution to the theory of chromatic polynomials. Canadian J. Math. **6**, 80–91 (1953).

— A class of Abelian group. Canadian J. Math. **8**, 13–28 (1956).

— A homotopy theorem for matroids, I. and II. Trans. Amer. math. Soc. **88**, 144–140 (1958).

— Matroids and graphs. Trans. Amer. math. Soc. **90**, 527–552 (1959).

WARD, M.: The algebra of lattice functions. Duke math. J. **5**, 357–371 (1939).

WEISNER, L.: Abstract theory of inversion of finite series. Trans. Amer. math. Soc. **38**, 474–484 (1935).

— Some properties of prime-power groups. Trans. Amer. math. Soc. **38**, 485–492 (1935).

WHITNEY, H.: A logical expansion in mathematics. Bull. Amer. math. Soc. **38**, 572–579 (1932).

— Characteristic functions and the algebra of logic. Ann. of Math., II. Ser. **34**, 405–414 (1933).

— The abstract properties of linear dependence. Amer. J. Math. **57**, 507–533 (1935).

WIELANDT, H.: Beziehungen zwischen den Fixpunktzahlen von Automorphismengruppen einer endlichen Gruppe. Math. Z. **73**, 146–158 (1960).

WINTNER, A.: Eratosthenian Averages. Baltimore (privately printed) 1943.

Department of Mathematics
Massachusetts Institute of Technology
Cambridge 39, Massachusetts

(Received September 2, 1963)

Reprinted from *American Journal of Mathematics*,
volume 57, pages 800 through 804,
by permission of the Johns Hopkins University Press.

The Homology Groups of A Lattice

JON FOLKMAN

Communicated by G.-C. ROTA

Summary. In [2], Rota defined homology groups for certain subsets, called *cross-cuts*, of a lattice. He showed that the value of the Euler characteristic associated with this homology theory depends only on the lattice, not on the choice of the cross-cut. It was conjectured that the homology groups themselves depend only on the lattice. In this paper we prove Rota's conjecture. Using this result, we determine the structure of the homology groups of an important class of lattices called geometric lattices.

1. Introduction. By a lattice with 0 and 1 we mean a lattice L with elements $0, 1 \in L$ such that $0 \leqq x \leqq 1$ for all $x \in L$. If L is a lattice with 0 and 1, a *cross-cut* of L is a set $C \subset L$ such that

(1.1) $0, 1 \notin C$,
(1.2) If $x, y \in C$ then $x \nless y$ and $y \nless x$,
(1.3) any finite chain $x_1 < x_2 < \cdots < x_n$ in L can be extended to a chain which contains an element of C.

In [2], Rota defined homology groups $H_n(C)$ for a cross-cut C of a finite lattice with 0 and 1. He showed that the Euler characteristic, $\chi(C) = \text{rank } H_0(C) - \text{rank } H_1(C) + \text{rank } H_2(C) - \cdots$, obtained from this homology theory is an invariant of the lattice. In fact, $\chi(C) = 1 + \mu(0, 1)$, where μ is the Mobius function of the lattice. Rota conjectured that the homology groups themselves are invariants of the lattice, *i.e.*, they do not depend on the choice of the cross-cut C. In this paper we extend the definition of homology groups to cross-cuts of a possibly infinite lattice and prove Rota's conjecture in this more general case.

2. Coverings of abstract complexes. In this section we establish an isomorphism between the homology groups of a complex K and the homology groups of the nerve of a suitable covering of K. This isomorphism is used to prove our main result. Our notation is that of [1].

* This research is sponsored by the United States Air Force under Project RAND—Contract No. AF 49(638) -700.

Let K be a (nonempty) abstract simplicial complex and let $\{L_\alpha\}_{\alpha \epsilon A}$ be a family of subcomplexes of K which cover K. We define a simplicial map f from K', the barycentric subdivision of K, into $N = N(\{L_\alpha\}_{\alpha \epsilon A})$, the nerve of the covering $\{L_\alpha\}_{\alpha \epsilon A}$, as follows: For each simplex σ of K let $f(\sigma) = \alpha \epsilon A$, where σ is a simplex of L_α. The function f maps the simplexes of K, which are the vertices of K', to A, which is the set of vertices of N. To verify that f defines a simplicial map $f: K' \to N$, let $(\sigma_0, \cdots, \sigma_n)$ be a simplex of K'. Let $f(\sigma_i) = \alpha_i$, $i = 0, \cdots, n$. Since $(\sigma_0, \cdots, \sigma_n)$ is a simplex of K', $\sigma_0 \subset \sigma_1 \subset \cdots \subset \sigma_n$. Therefore $\sigma_0 \subset \sigma_i$, which is a simplex of L_{α_i}, $i = 0, \cdots, n$; so σ_0 is a simplex of $L_{\alpha_0} \cap \cdots \cap L_{\alpha_n}$. Hence $(\alpha_0, \cdots, \alpha_n)$ is a simplex of N and f is simplicial. The function f is not uniquely defined, but it is trivial to show that any two such maps are contiguous and hence induce the same homomorphisms of homology.

Theorem 2.1. *Suppose the covering $\{L_\alpha\}_{\alpha \epsilon A}$ has the property that for any $\alpha_1, \cdots, \alpha_n \epsilon A$, $n = 1, 2, \cdots$, $L_{\alpha_1} \cap \cdots \cap L_{\alpha_n}$ has the homology of a point or is empty. Then the map $f: K' \to N$ induces isomorphisms of homology.*

Proof. If A contains only one element, α, then $K = L_\alpha$ has the homology of point and N is a point; so the result is clear. Assume the theorem when A contains n elements, $n \geqq 1$, and let $A = \{\alpha_1, \cdots, \alpha_{n+1}\}$. Let $K_1 = L_{\alpha_1} \cup \cdots \cup L_{\alpha_n}$, $N_1 = N(\{L_{\alpha_1}, \cdots, L_{\alpha_n}\})$. Let N_2 be the closed star of α_{n+1} in the complex N. We may assume that $f: K' \to N$ is chosen so that if σ is a simplex of K_1, $f(\sigma) \epsilon \{\alpha_1, \cdots, \alpha_n\}$. Then $f(K_1') \subset N_1$ and $(f \mid K_1')_* : H_i(K_1') \to H_i(N_1)$ is an isomorphism by the inductive assumption. Also

$$f(L'_{\alpha_{n+1}}) \subset N_2 \quad \text{and} \quad (f \mid L'_{\alpha_{n+1}})_* : H_i(L'_{\alpha_{n+1}}) \to H_i(N_2)$$

is an isomorphism because $L'_{\alpha_{n+1}}$ and N_2 both have the homology of a point.

Let $M_{\alpha_i} = L_{\alpha_i} \cap L_{\alpha_{n+1}}$, $i = 1, \cdots, n$. Then $\{M_{\alpha_1}, \cdots, M_{\alpha_n}\}$ is a covering of $K_1 \cap L_{\alpha_{n+1}}$ which satisfies the hypothesis of the theorem. Now

$$f(K_1' \cap L'_{\alpha_{n+1}}) \subset N_1 \cap N_2 = N(\{M_{\alpha_1}, \cdots, M_{\alpha_n}\})$$

and $K_1' \cap L'_{\alpha_{n+1}} = (K_1 \cap L_{\alpha_{n+1}})'$; so by the inductive assumption

$$(f \mid K_1' \cap L'_{\alpha_{n+1}})_* : H_i(K_1' \cap L'_{\alpha_{n+1}}) \to H_i(N_1 \cap N_2)$$

is an isomorphism.

Since $K' = K_1' \cup L'_{\alpha_{n+1}}$ and $N = N_1 \cup N_2$, we have the following commutative diagram, where the rows are portions of Mayer-Vietoris' exact sequences.

$$\begin{array}{ccccccc}
H_i(K_1' \cap L'_{\alpha_{n+1}}) & \to & H_i(K_1') \oplus H_i(L'_{\alpha_{n+1}}) & \to & H_i(K) & \to \\
\downarrow (f \mid K_1' \cap L'_{\alpha_{n+1}})_* & & \downarrow (f \mid K_1')_* \oplus (f \mid L'_{\alpha_{n+1}})_* & & \downarrow f_* & \\
H_i(N_1 \cap N_2) & \to & H_i(N_1) \oplus H_i(N_2) & \to & H_i(N) & \to
\end{array}$$

$$\begin{array}{ll} \to H_{i-1}(K_1' \cap L_{\alpha+1}') \to & H_{i-1}(K_1') \oplus H_{i-1}(L_{\alpha+1}') \\ \quad\downarrow (f \mid K_1' \cap L_{\alpha+1}')_* & \downarrow (f \mid K_1')_* \oplus (f \mid L_{\alpha+1}')_* \\ \to H_{i-1}(N_1 \cap N_2) \quad \to & H_{i-1}(N_1) \oplus H_{i-1}(N_2). \end{array}$$

All vertical maps except f_* are known to be isomorphisms; so by the five-lemma we conclude that $f_*: H_i(K') \to H_i(N)$ is an isomorphism.

Note that we have tacitly assumed that $L_{\alpha+1}$ and $K_1 \cap L_{\alpha+1}$ are nonempty. If $L_{\alpha+1} = \varnothing$ then $K = K_1$, $N = N_1$, and the above statement reduces to the inductive assumption. If $K_1 \cap L_{\alpha+1} = \varnothing$ then $N_1 \cap N_2 = \varnothing$; so the map $(f \mid K_1' \cap L_{\alpha+1}')_*$ is a map between trivial groups and the above argument still holds.

We have established the theorem by induction when the index set A is finite. Now let A be arbitrary. For each finite, nonempty set $S \subset A$, let $K_S = \bigcup_{\alpha \in S} L_\alpha$. Let $N_S = N(\{L_\alpha\}_{\alpha \in S})$. Let $i_S : K_S' \subset K'$ and $j_S : N_S \subset N$ denote the inclusion maps. Let $f_S : K_S' \to N_S$ be a simplicial map defined in the same manner as $f : K' \to N$. By what we have already shown, $f_{S*} : H_i(K_S') \to H_i(N_S)$ is an isomorphism. It is trivial to show that $j_S f_S$ and $f i_S : K_S' \to N$ are contiguous; so $j_{S*} f_{S*} = f_* i_{S*}$. If $S \subset T$, let $i_{S,T} : K_S' \subset K_T'$ and $j_{S,T} : N_S \subset N_T$ denote the inclusions. The following statements follow immediately from the fact that any chain on an abstract simplicial complex involves only finitely many simplexes.

(a) For any $x \in H_i(K')$ there is a finite set $S \subset A$ and a $y \in H_i(K_S')$ with $x = i_{S*}(y)$.

(b) If $S \subset A$, S finite, and $x \in H_i(K_S')$ with $i_{S*}(x) = 0$, then there is a finite $T \subset A$ with $S \subset T$ and $i_{S,T*}(x) = 0$.

(a) and (b) also hold with K' replaced by N. Now let $x \in H_i(K')$ and suppose $f_*(x) = 0$. By (a), $x = i_{S*}(y)$, $y \in H_i(K_S')$. $0 = f_* i_{S*}(y) = j_{S*} f_{S*}(y)$; so for some finite $T \supset S$, $j_{S,T*} f_{S*}(y) = 0$ by (b). $f_{T*} i_{S,T*}(y) = j_{S,T*} f_{S*}(y) = 0$; so $i_{S,T*}(y) = 0$. Hence $x = i_{S*}(y) = i_{T*} i_{S,T*}(y) = 0$; so f_* is $1 - 1$.

Now let $x \in H_i(N)$. By (a), $x = j_{S*}(y)$, $y \in H_i(N_S)$, where S is finite. Since f_{S*} is an isomorphism, $y = f_{S*}(z)$, $z \in H_i(K_S')$. Hence $x = j_{S*} f_{S*}(z) = f_* i_{S*}(z)$; so f_* is onto. This completes the proof of 2.1.

3. Invariance of the homology groups of a lattice. Let L be a lattice with 0 and 1 and let C be a cross-cut of L. We say a finite subset $\{x_1, \cdots, x_n\} \subset C$ *spans* in case $x_1 \wedge \cdots \wedge x_n = 0$ and $x_1 \vee \cdots \vee x_n = 1$. Here $x \wedge y$ denotes the largest element $\leqq x$ and y, and $x \vee y$ denotes the smallest element $\geqq x$ and y.

Let $K(C)$ be the abstract simplicial complex whose vertices are elements of C and whose simplexes are all finite subsets of C which do not span. Define $H_i(C) = H_i(K(C))$. Let $K(L)$ be the abstract simplicial complex whose vertices are the elements of $L - \{0, 1\}$ and whose simplexes are all finite simply ordered subsets of $L - \{0, 1\}$. Define $H_i(L) = H_i(K(L))$.

Theorem 3.1. $H_i(C) \cong H_i(L)$.

Proof. Let $K = K(L)$. For each $x \in C$ let L_x be the subcomplex of K consisting of all simplexes $\{y_0, \cdots, y_n\}$ such that the set $\{y_0, \cdots, y_n, x\}$ is simply ordered. By property 1.3 of a cross-cut, the family $\{L_x\}_{x \in C}$ is a covering of K. Let $x_1, \cdots, x_n \in C$. Let $x = x_1 \wedge \cdots \wedge x_n$ and $y = x_1 \vee \cdots \vee x_n$. If

$$\{y_0, \cdots, y_p\}, \qquad y_0 < y_1 < \cdots < y_p,$$

is a simplex of $L_{x_1} \cap \cdots \cap L_{x_n}$, then for each $j = 1, \cdots, n$ let $i(j) = \inf(\{p+1\} \cup \{i \mid x_j < y_i\})$. If $i(j) < i(k)$ then $x_j < y_{i(j)} \leqq x_k$, contradicting property (1.2) of a cross-cut. Hence, $i(j) = i_0$ is independent of j. If $i < i_0$ then $y_i \leqq x$. If $i \geqq i_0$ then $y \leqq y_i$. If we let $\Omega_{x,y}$ be the subcomplex of K consisting of all simplexes whose vertices lie in the set $\{z \mid 0 < z \leqq x \text{ or } y \leqq z < 1\}$, then the above shows that $L_{x_1} \cap \cdots \cap L_{x_n} \subset \Omega_{x,y}$. But clearly

$$\Omega_{x,y} \subset L_{x_1} \cap \cdots \cap L_{x_n};$$

so $L_{x_1} \cap \cdots \cap L_{x_n} = \Omega_{x,y}$. Now $\Omega_{0,1} = \emptyset$. If $0 < x$ then $x \leqq x_1 < 1$ by property 1.1 of a cross-cut; so $\{x\}$ is a simplex of $\Omega_{x,y}$. Similarly, if $y < 1$ then $\{y\}$ is a simplex of $\Omega_{x,y}$. Hence $L_{x_1} \cap \cdots \cap L_{x_n} = \Omega_{x,y} = \emptyset$ if and only if $x = 0$ and $y = 1$, which is true if and only if $x_1, \cdots, x_n$ spans. Therefore

$$N(\{L_x\}_{x \in C}) = K(C).$$

If $x \neq 0$ then $\Omega_{x,y}$ = closed star of x in $\Omega_{x,y}$. If $y \neq 1$ then $\Omega_{x,y}$ = closed star of y in $\Omega_{x,y}$. Hence $L_{x_1} \cap \cdots \cap L_{x_n} = \Omega_{x,y}$ is empty or has the homology of a point; so the covering $\{L_x\}_{x \in C}$ satisfies the hypothesis of 2.1. Therefore $H_i(K') \cong H_i(K(C))$. But $H_i(K') \cong H_i(K)$; so $H_i(L) = H_i(K(L)) \cong H_i(K') \cong H_i(K(C)) = H_i(C)$.

Corollary 3.2. If C and C′ are two cross-cuts of L, then $H_i(C) \cong H_i(C')$.

Proof. $H_i(C) \cong H_i(L) \cong H_i(C')$.

4. Geometric lattices. Let L be a finite lattice with 0 and 1. We say L satisfies the chain condition if, for each $x \in L$, every maximal chain, $0 = y_0 < y_1 < \cdots < y_r = x$, from 0 to x, has the same length, $r = r(x)$. We call $r(x)$ the *rank of* x. Note that the length of a chain is one less than the number of elements it contains; so $r(0) = 0$. We call $r(1)$ the *rank of* L.

By a *geometric lattice* we mean a finite lattice L, with 0 and 1, such that

(4.1) L satisfies the chain condition;

(4.2) if $x \in L$ and $x > 0$ then $x = x_1 \vee \cdots \vee x_p$, where $r(x_1) = \cdots = r(x_p) = 1$;

(4.3) if $x, y \in L$ then $r(x \vee y) + r(x \wedge y) \leqq r(x) + r(y)$. (For equivalent definitions and terminology, see Rota, [2].)

The homology groups of a geometric lattice have a particularly simple structure, as the following shows.

Theorem 4.1. Let L be a geometric lattice of rank n and let μ denote the Mobius function of L. Let F(k) denote the free Abelian group of rank k.

If $n = 0, 1$, *then* $H_i(L) = 0$ *for all* i.

$$\textit{If}\quad n = 2,\quad \textit{then}\quad H_i(L) = \begin{cases} F(\mu(0, 1) + 1), & \text{if } i = 0, \\ 0, & \text{if } i \neq 0. \end{cases}$$

$$\textit{If}\quad n > 2,\quad \textit{then}\quad H_i(L) = \begin{cases} F(|\mu(0, 1)|), & \text{if } i = n - 2, \\ F(1), & \text{if } i = 0, \\ 0, & \text{if } i \neq 0,\ n - 2. \end{cases}$$

Proof. If $n = 0, 1$, then L contains no elements except 0 and 1. Hence $K(L) = \emptyset$ and the conclusion follows. If $n = 2$, then any chain in L contains at most 3 elements; so it contains at most one element from $L - \{0, 1\}$. Therefore, $L(K)$ is a zero-dimensional complex; so $H_0(L)$ is a free Abelian group and $H_i(L) = 0$ for $i \neq 0$. By the theorem of Rota quoted in the introduction, $\mu(0, 1) + 1 = \operatorname{rank} H_0(L) - \operatorname{rank} H_1(L) + \cdots = \operatorname{rank} H_0(L)$. Hence, $H_0(L) = F(\mu(0, 1) + 1)$.

Now suppose $n > 2$. Let C be the set of all elements of L of rank 1. Then C is a cross-cut of L. Let $p \leq n - 2$ and let $x_0, \cdots, x_p \in C$. By (4.3),

$$r(x_0 \vee \cdots \vee x_p) \leq r(x_0 \vee \cdots \vee x_{p-1}) + r(x_p) \leq r(x_0 \vee \cdots \vee x_{p-2}) + r(x_{p-1}) + r(x_p) \leq \cdots \leq r(x_0) + r(x_1) + \cdots + r(x_p) = p + 1 \leq n - 1 < n = r(1).$$

Hence, $x_0 \vee \cdots \vee x_p \neq 1$; so $x_0, \cdots, x_p$ do not span. Therefore, $(x_0, \cdots, x_p)$ is a simplex of $K(C)$. Let $S(C)$ be the complex whose vertices are the elements of C and whose simplexes are all subsets of C. Then $K^{n-2}(C) = S^{n-2}(C)$, where $K^{n-2}(C)$ and $S^{n-2}(C)$ are the $n - 2$ dimensional skeletons of $K(C)$ and $S(C)$. Therefore, if $i < n - 2$, we have

$$\begin{aligned} H_i(K(C)) &= H_i(K^{n-2}(C)) = H_i(S^{n-2}(C)) \\ &= H_i(S(C)) = \begin{cases} F(1), & \text{if } i = 0, \\ 0, & \text{if } i \neq 0. \end{cases} \end{aligned}$$

Hence,

$$H_i(L) = H_i(K(C)) = \begin{cases} F(1), & \text{if } i = 0, \\ 0, & \text{if } 0 < i < n - 2. \end{cases}$$

Since L has rank n, every chain in L contains at most $n + 1$ elements; so every chain in $L - \{0, 1\}$ contains at most $n - 1$ elements. Therefore, $K(L)$ has dimension at most $n - 2$. Hence $H_{n-2}(L) = H_{n-2}(K(L))$ is a free Abelian group, and $H_i(L) = H_i(K(L)) = 0$ for $i > n - 2$. Now,

$$\begin{aligned} 1 + \mu(0, 1) &= \operatorname{rank}\ H_0(L) - \operatorname{rank}\ H_1(L) + \cdots \\ &= \operatorname{rank}\ H_0(L) + (-1)^{n-2} \operatorname{rank}\ H_{n-2}(L) \\ &= 1 + (-1)^{n-2} \operatorname{rank}\ H_{n-2}(L); \end{aligned}$$

so rank $H_{n-2}(L) = |\mu(0, 1)|$. Therefore, $H_{n-2}(L) = F(|\mu(0, 1)|)$, and the theorem is proved.

References

[1] Eilenberg, S., & N. E. Steenrod, *Foundations of Algebraic Topology*, Princeton University Press, Princeton, New Jersey, 1952.

[2] Rota, Gian-Carlo, On the Foundations of Combinatorial Theory, I. Theory of Möbius Functions, *Zeitschrift für Wahrsheinlichkeitstheorie und Verwandte Gebiete*, 2(1964) 340–368.

The Rand Corporation
Santa Monica, Cal.

Modular Elements of Geometric Lattices

RICHARD P. STANLEY

1. Modular Elements

Let L be a finite geometric lattice of rank n with rank function r. (For definitions, see e.g., [3, Chapter 2], [4], or [1, Chapter 4].) An element $x \in L$ is called a *modular element* if it forms a modular pair with every $y \in L$, i.e., if $a \leqslant y$ then $a \vee (x \wedge y) = (a \vee x) \wedge y$. Recall that in an upper semimodular lattice (and thus in a geometric lattice) the relation of being a modular pair is symmetric; in fact (x, y) is a modular pair if and only if $r(x)+r(y)=r(x \vee y)+r(x \wedge y)$ [1, p. 83]. Every point (atom) of a geometric lattice is a modular element. If every element of L is modular, then L is a modular lattice. The main object of this paper is to show that a modular element of L induces a factorization of the characteristic polynomial of L. This is done in Section 2. First we discuss some other aspects of modular elements.

The following theorem provides a characterization of modular elements.

THEOREM 1. *An element $x \in L$ is modular if and only if no two complements of x are comparable.*

Proof. If x is modular and x' is a complement of x, then $r(x')=n-r(x)$. Hence all the complements of x have the same rank and are incomparable.

Conversely, assume x is not modular. Then there are elements $y<z$ such that $x \wedge y = x \wedge z$ and $x \vee y = x \vee z$. Let $p_1, p_2, \ldots, p_n$ be a basis for L such that $p_1, p_2, \ldots, p_s$ is a basis for $x \wedge y$; $p_1, p_2, \ldots, p_t$ is a basis for y; $p_1, p_2, \ldots, p_u$ is a basis for z; and $p_1, p_2, \ldots, p_v$ is a basis for $x \vee y$. Thus $0 \leqslant s < t < u < v \leqslant n$, $r(x \wedge y)=s$, $r(y)=t$, $r(z)=u$, $r(x \vee y)=v$. Let

$$x' = p_{s+1} \vee p_{s+2} \vee \cdots \vee p_t \vee p_{v+1} \vee p_{v+2} \vee \cdots \vee p_n$$
$$x'' = x' \vee p_{t+1} \vee p_{t+2} \vee \cdots \vee p_u .$$

It is easily seen that x' and x'' are both complements of x with $x' < x''$. □

In particular, an element x with a unique complement x' is modular. For a stronger result, recall that an element x is said to be in the *center* of an ordered set P with 0 and 1 if $P=[0, x] \times [0, x']$ for some x' [1, p. 67]. For an element x to be in the center of a geometric lattice L, each of the following conditions is necessary and sufficient:

(1) x is *distributive* (because x is complemented, cf. [1, p. 69]), i.e., for any y, $z \in L$, the sublattice generated by x, y, z is distributive.

(2) x has a *unique complement*. This result appears to be new; Curtis Greene has

Presented by G. Birkhoff. Received December 5, 1970. Accepted for publication in final form March 22, 1971.

in fact proved the more general result (unpublished) that the intersection (meet) of all the complements of any element of a geometric lattice L is in the center of L.

(3) x is a *separator* of L [3, Chapter 12], i.e., for any point p and any copoint q not containing p, either $p \leqslant x$ or $x \leqslant q$.

(4) x is a *standard element* of L [1, p. 69] (this follows from (2) since a complemented standard element has a unique complement). The concept of standard elements is due to G. Grätzer.

2. The Characteristic Polynomial

The characteristic polynomial $p_L(\lambda)$ [4] of a geometric lattice L is defined by

$$p_L(\lambda) = \sum_{y \in L} \mu(0, y)\, \lambda^{n-r(y)},$$

where μ denotes the Möbius function of L (see [4]).

This polynomial was first considered by G. D. Birkhoff, while its connection with Möbius functions was noted by Garrett Birkhoff. If $x \in L$, the characteristic polynomial of the segment $[0, x]$ is denoted $p_x(\lambda)$.

The main result of this paper is the following factorization theorem:

THEOREM 2. *If x is a modular element of a finite geometric lattice L of rank n, then*

$$p_L(\lambda) = p_x(\lambda) \left[\sum_{b:\, x \wedge b = 0} \mu(0, b)\, \lambda^{n-r(x)-r(b)} \right]$$

The expression in brackets may be thought of as the characteristic polynomial of the order ideal $C(x) = \{b \mid x \wedge b = 0\}$. $C(x)$ will have a 1 if and only if x has a unique complement x' in L. In this case $C(x) = [0, x']$ and $L = [0, x] \times [0, x']$. Thus when x has a unique complement Theorem 2 is trivial, since $p_{L_1 \times L_2} = p_{L_1} p_{L_2}$.

To prove Theorem 2, we first prove two lemmas. The first lemma is a special case of some results of Schwan on modular pairs (see [1, Section IV.2]), but for the sake of completeness we include a proof. It is to be assumed throughout that x is a modular element of the finite geometric lattice L, and that L has rank n.

LEMMA 1. *For any $a \in L$, the map*

$$\sigma_a : [a, a \vee x] \to [a \wedge x, x]$$

defined by $\sigma_a(y) = x \wedge y$ is an isomorphism with inverse $\tau_a(y) = a \vee y$.

Proof. Clearly σ_a and τ_a are order-preserving. By modularity of x, it is immediate that if $y \in [a \wedge x, x]$, then $\sigma_a \tau_a(y) = x \wedge (a \vee y) = (x \wedge a) \vee y = y$. Also if $y \in [a, a \vee x]$, then $\tau_a \sigma_a(y) = a \vee (x \wedge y) = (a \vee x) \wedge y = y$, and the proof follows. □

LEMMA 2. For any $y \in L$, $x \wedge y$ is a modular element of $[0, y]$.

Proof. Let $a \in [0, y]$ and let $b \leqslant a$. We need to show $b \vee ((x \wedge y) \wedge a) = (b \vee (x \wedge y)) \wedge a$. Using the modularity of x, we have

$$\begin{aligned}(b \vee (x \wedge y)) \wedge a &= ((b \vee x) \wedge y) \wedge a = (b \vee x) \wedge a = b \vee (x \wedge a) \\ &= b \vee (x \wedge (y \wedge a)) = b \vee ((x \wedge y) \wedge a). \quad \square\end{aligned}$$

Proof of Theorem 2. By Crapo's Complementation Theorem [2], if $a \in [0, y]$ then

$$\mu(0, y) = \sum_{a', a''} \mu(0, a') \zeta(a', a'') \mu(a'', y),$$

where a' and a'' are complements of a in $[0, y]$, and ζ is the zeta function of $[0, y]$. Choosing $a = x \wedge y$, then by Lemma 2 all the complements of a have the same rank and hence are incomparable. Thus

$$\mu(0, y) = \sum \mu(0, b) \mu(b, y), \tag{1}$$

where the sum is over all complements b of $x \wedge y$ in $[0, y]$, i.e., over all $b \in L$ satisfying $0 \leqslant b \leqslant y$, $b \wedge (x \wedge y) = 0$, $b \vee (x \wedge y) = y$. Now $b \wedge (x \wedge y) = b \wedge x$, and by the modularity of x, $b \vee (x \wedge y) = (b \vee x) \wedge y$. Thus the sum (1) is over all $b \in L$ satisfying $b \wedge x = 0$ and $y \in [b, b \vee x]$. Hence

$$\begin{aligned} p_L(\lambda) &= \sum_{y \in L} \mu(0, y) \lambda^{n-r(y)} \\ &= \sum_{y \in L} \sum_{\substack{b \wedge x = 0 \\ y \in [b, b \vee x]}} \mu(0, b) \mu(b, y) \lambda^{n-r(y)} \\ &= \sum_{b \wedge x = 0} \sum_{y \in [b, b \vee x]} \mu(0, b) \mu(b, y) \lambda^{n-r(y)} \end{aligned}$$

Now by Lemma 1, as y ranges over $[b, b \vee x]$, $z = x \wedge y$ ranges over the isomorphic interval $[b \wedge x, x] = [0, x]$, and $\mu(b, y) = \mu(0, z)$. Moreover $r(y) = r(b) + r(z)$. Therefore

$$\begin{aligned} P_L(\lambda) &= \sum_{b \wedge x = 0} \sum_{z \in [0, x]} \mu(0, a) \mu(0, z) \lambda^{n-r(b)-r(z)} \\ &= \left[\sum_{z \in [0, x]} \mu(0, z) \lambda^{r(x)-r(z)}\right] \left[\sum_{b \wedge x = 0} \mu(0, b) \lambda^{n-r(b)-r(x)}\right] \\ &= p_x(\lambda) \left[\sum_{b \wedge x = 0} \mu(0, b) \lambda^{n-r(b)-r(x)}\right]. \quad \square \end{aligned}$$

3. Examples

As a special case of Theorem 2, suppose x is modular copoint, and that exactly α points (atoms) a of L do not lie below x. Then these points a, together with 0, are the only elements b of L satisfying $b \wedge x = 0$. Moreover, $\mu(0, a) = -1$, $\mu(0, 0) = 1$, so

$$p_L(\lambda) = p_x(\lambda)\left[\sum_a (-1) + \lambda\right] = p_x(\lambda)(\lambda - \alpha).$$

Thus if L contains a maximal chain $0 = x_0 < x_1 < \cdots < x_n = 1$ such that each x_{i-1} is modular in $[0, x_i]$ and such that exactly α_i atoms of $[0, x_i]$ do not lie below x_{i-1}, then $p_L(\lambda) = (\lambda - \alpha_1)(\lambda - \alpha_2)\ldots(\lambda - \alpha_n)$. It is easy to show that the condition that each x_{i-1} is modular in $[0, x_i]$ is equivalent to the condition that each x_i is modular in L. A class of geometric lattices with such a 'modular maximal chain' is the following: Let $\alpha_1, \alpha_2, \ldots, \alpha_n$ be any collection of positive integers with $\alpha_n = 1$. Let $p_1, \ldots, p_n$ be n independent points, and on the line $p_i p_n$ $(1 \leqslant i \leqslant n-1)$ insert an additional $\alpha_i - 1$ points. The geometric lattice L of flats of this geometry contains a modular maximal chain, and $p_L(\lambda) = (\lambda - \alpha_1)(\lambda - \alpha_2)\ldots(\lambda - \alpha_n)$.

At this point it is natural to ask for a characterization of the modular elements of various geometric lattices. We state such a characterization when L is the lattice of contractions of a finite graph. Recall that a contraction of a graph G may be regarded as a partition π of the vertices of G, such that the subgraph H induced by each block B of π is connected [3, Chapter 6].

THEOREM 3. *Let L be the lattice of contractions of a doubly connected finite graph G. Then $\pi \in L$ is a modular element of L if and only if the following conditions hold:*

(i) *At most one block B of π contains more than one vertex of G.*

(ii) *Let H be the subgraph induced by the block B of* (i). *Let K be any connected component of the subgraph induced by $G - B$, and let H_1 be the graph induced by the set of vertices in H which are connected to some vertex in K. Then H_1 is a clique (complete subgraph) of G.* □

The proof is of a routine nature and will be omitted.

If G is not doubly connected, then the lattice of contractions of G is a direct product of the lattices of contractions of the maximal doubly connected subgraphs of G, so Theorem 3 easily extends to arbitrary finite graphs G.

REFERENCES

[1] Garrett Birkhoff, *Lattice Theory*, Third Edition (American Mathematical Society, 1967).

[2] Henry Crapo, *The Möbius Function of a Lattice*, J. Combinatorial Theory *1* (1966), 126–131.

[3] Henry Crapo, and Gian-Carlo Rota, *On the Foundations of Combinatorial Theory: Combinatorial Geometries*, M.I.T. Press.

[4] Gian-Carlo Rota, *On the Foundations of Combinatorial Theory, I: Theory of Möbius Functions*, Z. Wahrscheinlichkeitstheorie *2* (1964), 340–368.

M.I.T., Cambridge, Mass. U.S.A.

A Counterexample to the Generalization of Sperner's Theorem*

R. P. DILWORTH AND CURTIS GREENE

California Institute of Technology, Pasadena, California 91109

Communicated by Gian-Carlo Rota

Received February 21, 1969

It has been conjectured that the analog of Sperner's theorem on non-comparable subsets of a set holds for arbitrary geometric lattices, namely, that the maximal number of non-comparable elements in a finite geometric lattice is max $w(k)$, where $w(k)$ is the number of elements of rank k. It is shown in this note that the conjecture is not true in general. A class of geometric lattices, each of which is a bond lattice of a finite graph, is constructed in which the conjecture fails to hold.

INTRODUCTION

A theorem of Sperner [1] states that $\max_k \binom{n}{k}$ is the maximal number of non-comparable subsets in a set of n elements. The analogous theorem for finite projective geometries asserts that the maximal number of non-comparable subspaces is $\max_k w(k)$, where $w(k)$ is the number of subspaces of dimension k. This theorem has been proved by Harper [2]. Since the lattices of subsets of a set and the lattice of subspaces of a projective space are examples of geometric lattices, it has been conjectured by Rota that the analog of Sperner's theorem holds in any geometric lattice, namely, that the maximal number of noncomparable elements in a finite geometric lattice L is $\max_k w(k)$, where $w(k)$ is the number of elements of rank k in L. It follows from a result of Baker [3] that the conjecture holds for all geometric lattices in which the number of covering elements is constant for elements of a given rank and dually.

It will be shown in this note that the conjecture is not true in general. In fact, we shall construct a class of geometric lattices each of which is a bond lattice of a finite graph, in which the conjecture fails to hold. The

* This research was partially supported by NSF Grant GP 8423

lattice of smallest order which is so constructed and which contradicts the conjecture contains 60,073 elements.

This class of geometric lattices also exhibits another unusual property. For each n, there is a bond lattice of dimension n in the class such that, by removing a single point, another geometric lattice is obtained which is also the bond lattice of a graph. In particular, for each n there exist two bond lattices of dimension n which have isomorphic structure above rank 1 but which are not isomorphic. This contrasts with the theorem of McLaughlin [4, Theorem 1.3] that the partially ordered set of points and dual points in a geometric lattice completely characterizes the lattice.

THE COUNTEREXAMPLES

Let G be a finite graph with no trivial cycles and no double edges. A subset E of the set of edges of G is *closed* if it contains every edge whose vertices are joined by a path formed from the edges of E. It is well known that the closed subsets form a geometric lattice (i.e., a semimodular point lattice) $L(G)$ in which the points correspond to the signleton sets formed from the edges of G.

Consider the graph G_n shown in Figure 1. Let $p_0 = \{(a, b)\}$ be the point

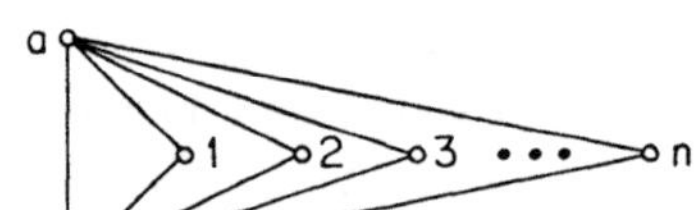

FIGURE 1

of $L(G_n)$ corresponding to the edge (a, b). If A is a closed subset of $E(G_n)$ containing the edge (a, b), then, for each i, A either contains both of the edges (a, i) and (b, i) or neither. Thus A is determined by the subset

$$A^* = \{i \mid (a, i), (b, i) \in A\}.$$

It follows that the quotient lattice $1/p_0$ of $L(G_n)$ is isomorphic to B_n, the Boolean algebra of all subsets of $\{1, 2,..., n\}$. Thus the rank of A in $L(G_n)$ is one more than the number of elements in A^*. Next let B be a closed subset not containing (a, b). Then, for each i, B can contain at most one of the two edges (a, i) and (b, i). Thus B is determined by the subset

$$B^* = \{i \mid \text{either } (a, i) \in B \text{ or } (b, i) \in B\}$$

and a choice function which for each $i \in B^*$ selects one of the two edges (a, i), (b, i). The rank of B in $L(G_n)$ is clearly equal to the number of elements in B^*.

It follows from the observations of the previous paragraph that the number $w(k)$ of elements of rank k in $L(G_n)$ is given by

$$w(k) = 2^k \binom{n}{k} + \binom{n}{k-1}.$$

It is easily verified that the sequence $w(k)$ is unimodal and that the value of k for which $w(k)$ reaches its maximum is approximately $2n/3$. Since $1/p_0$ is isomorphic to B_n, the rank having the most elements containing p_0 is approximately $n/2$.

Now let L be an arbitrary finite geometric lattice. Furthermore let $w(k)$ denote the number of elements of rank k in L and, if p is a point of L, let $w_p(k)$ denote the number of elements of rank k in L which lie above p. Finally let l be such that $w(l) = \max_k w(k)$.

In order to produce a counterexample to the analog of Sperner's theorem it suffices to find a geometric lattice L and a point p such that $w_p(l-1) > w_p(l)$. For let T_l denote the set of elements of rank l which do not lie above p and let S_{l-1} be the elements of rank $l-1$ which lie above p. Then $T_l \cup S_{l-1}$ is clearly a non-comparable subset of L and

$$\begin{aligned} | T_l \cup S_{l-1} | &= | T_l | + | S_{l-1} | \\ &= w(l) - w_p(l) + w_p(l-1) > w(l) = \max_k w(k). \end{aligned}$$

Now, for $L(G_n)$, l is approximately $2/3n$. Since the rank at which the maximum number of elements containing p_0 occurs is approximately $n/2$, it follows that, for large n, $w_{p_0}(l-1) > w_{p_0}(l)$ and hence the geometric lattices $L(G_n)$ furnish counterexamples for all large n. In particular, for $n = 10$, $l = 7$, we have

$$w_{p_0}(6) = 252 > 210 = w_{p_0}(7)$$

and $L(G_{10})$ is a counterexample. The order of $L(G_{10})$ is 60,073.

Sublattices of $L(G_n)$

Let G_n' denote the graph G_n with the edge (a, b) deleted (Figure 2). If A' is a closed subset of G_n', then A' may be viewed as a subset of G_n. The closure of A' in G_n provides a natural mapping of $L(G_n')$ into $L(G_n)$.

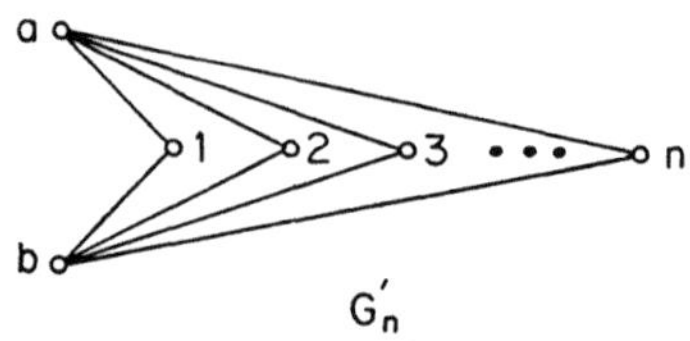

FIGURE 2

Since a closed subset of G_n containing any two of (a, b), (a, i), (b, i) must also contain the third, it is easily seen that the mapping of $L(G_n')$ into $L(G_n)$ is injective and the image set is $L(G_n) - \{(a, b)\}$. Thus $L(G_n')$ is isomorphic to the lattice $L(G_n)$ with the point $\{(a, b)\}$ deleted and hence $L(G_n)$ and $L(G_n')$ are not isomorphic. On the other hand $L_2(G_n) \simeq L_2(G_n')$, where $L_2(G_n)$ and $L_2(G_n')$ denote the partially ordered sets of elements of rank $\geqslant 2$ in $L(G_n)$ and $L(G_n')$ respectively.

It is interesting to note that the graphs G_n are the only graphs whose bond lattices have this property.

REFERENCES

1. E. SPERNER, Ein Satz über Untermengen einer endlichen Menge, *Math. Z.* **27** (1928), 544–548.
2. L. H. HARPER, The Morphology of Geometric Lattices (to appear).
3. K. A. BAKER, A Generalization of Sperner's Lemma (to appear).
4. J. E. MCLAUGHLIN, Structure Theorems for Relatively Complemented Lattices, *Pacific J. Math.* **3** (1953), 197–208.

WHITNEY NUMBER INEQUALITIES FOR GEOMETRIC LATTICES

THOMAS A. DOWLING[1] AND RICHARD M. WILSON[2]

ABSTRACT. Let L be a finite geometric lattice of rank r, and for $i = 0, 1, \cdots, r$, let W_i denote the number of elements of L with rank i. For $1 \le k \le r-2$, we have $W_1 + W_2 + \cdots + W_k \le W_{r-k} + \cdots + W_{r-2} + W_{r-1}$ with equality if and only if the lattice L is modular. We give two further results concerning matchings of lattice elements of rank $\le k$ into those of rank $\ge r-k$, and observe that a middle term can be interpolated in the above inequality.

1. **Introduction.** For a finite geometric lattice L of rank r, we denote the number of lattice elements of rank i by $W_i = W_i(L)$. $W_0, W_1, \cdots, W_r$ are the *Whitney numbers* of the second kind. Of course, $W_0 = W_r = 1$.

There are several interesting conjectures concerning the Whitney numbers of geometric lattices. Foremost among these is the unimodality conjecture of G.-C. Rota which asserts that $W_j \ge \min\{W_i, W_k\}$ whenever $i \le j \le k$. This is known to be true for partition lattices [9], [11]. Another conjecture asserts that $W_k \le W_{r-k}$ whenever $k \le r/2$ in a geometric lattice of rank r. This would imply that $W_0 \le W_1 \le W_2 \le \cdots \le W_{[r/2]}$ (by applying the above conjecture to truncations of the lattice). All of these conjectures are valid for geometric lattices with at most eight points [2] and "perfect matroid designs" [13].

In [1], Kelly and Basterfield proved that $W_1 \le W_{r-1}$ for geometric lattices of rank r. Another proof is given by C. Greene [7] who added the result that $W_1 = W_{r-1}$ holds if and only if the lattice is modular, and also gave several results concerning the matching of the points into the copoints. A third proof, by entirely different techniques, is given by Greene in [8].

In this note we prove that $W_1 + W_2 + \cdots + W_k \le W_{r-k} + \cdots + W_{r-2} + W_{r-1}$

Presented to the Society, January 25, 1973 under the title *A Whitney number inequality for geometric lattices*; received by the editors June 26, 1973 and, in revised form, December 13, 1973.

AMS (MOS) subject classifications (1970). Primary 05B35; Secondary 05A20,

1 This research supported in part by NSF Grant GP 38951 (O.S.U.R.F. Project No. 3684-A1).

2 This research supported in part by NSF Grant GP 28943 (O.S.U.R.F. Project No. 3228-A1).

for geometric lattices of rank r with equality holding for some k, $1 \leq k \leq r-2$, if and only if the lattice is modular. This provides some justification for the empirical observation that geometric lattices are "top heavy" and is not unexpected in view of Crapo and Rota's assertion [4] that "Roughly speaking... everything which 'happens' in a geometric lattice also 'happens' somewhere at the top of the lattice."

Our proof appeals to elementary linear algebra and Möbius inversion. The inequality is proved by showing that a certain set of $W_{r-k} + \cdots + W_r$ vectors spans a $(W_0 + \cdots + W_k)$-dimensional subspace of a vector space $V(L)$ introduced in §3. These techniques also allow easy derivations of the generalizations of Greene's results concerning matchings of points to copoints (Theorems 2 and 3 of §5). In Theorem 4 of §5, we give a further result to indicate possible variations of the methods. A corollary of this result is used as a lemma in a further paper [6] where the authors give lower bounds for W_k in terms of W_1 and the rank r of a geometric lattice.

2. **Preliminaries.** Definitions and results required in the sequel are summarized in this section. A detailed treatment of geometric lattices may be found in [3] or [4].

A *geometric lattice* is a lattice L of finite height in which $x \gtrdot y$ (x covers y) if and only if $x = y \vee p$ for some point $p \not\leq y$. Here a *point* is a lattice element covering the minimum element 0 of L. The maximum element of a lattice L is denoted by 1, and a *copoint* is a lattice element covered by 1.

The *rank function* ρ of a geometric lattice L is uniquely determined by

$$\rho(0) = 0 \quad \text{and} \quad \rho(x) = \rho(y) + 1 \quad \text{whenever } x \gtrdot y,$$

and satisfies the *semimodular law*

$$\rho(x) + \rho(y) \geq \rho(x \vee y) + \rho(x \wedge y). \tag{1}$$

A geometric lattice is *modular* when $\rho(x) + \rho(y) = \rho(x \vee y) + \rho(x \wedge y)$ for all $x, y \in L$. The *rank* of L is the integer $r = \rho(1)$.

For $a \leq b$ in a geometric lattice, the *interval* $[a, b] = \{x \in L : a \leq x \leq b\}$ is also a geometric lattice. When $a \leq x \leq b$ in a geometric lattice, there always exist *modular complements* of x in the interval $[a, b]$, i.e., lattice elements y such that $x \wedge y = a$, $x \vee y = b$, and $\rho(x) + \rho(y) = \rho(a) + \rho(b)$.

The *Möbius function* [12] of a finite lattice L is the integer-valued function $\mu = \mu_L$ on $L \times L$ with the properties $\mu(x, y) = 0$ unless $x \leq y$,

$\mu(x, x) = 1$, and $\Sigma_{x\leq z\leq y}\mu(x, z) = \Sigma_{x\leq z\leq y}\mu(z, y) = 0$ whenever $x < y$. From these properties follows the principle of *Möbius inversion*: Given functions f, g from L into an additive abelian group, $f(x) = \Sigma_{y\geq x} g(y)$ for all x implies $g(x) = \Sigma_{y\geq x}\mu(x, y)f(y)$, and $f(x) = \Sigma_{y\leq x} g(y)$ implies $g(x) = \Sigma_{y\leq x}\mu(y, x)f(y)$.

We shall require the fact [12, Theorem 3] that for a geometric lattice, $\mu(x, y) \neq 0$ for $x \leq y$.

In [5, Lemma 2.2], Dilworth showed that in a modular geometric lattice, $W_k = W_{r-k}$ for all k.

3. **A vector space.** Given a finite lattice L, we introduce the free vector space $V(L)$ (over the rationals Q) generated by the lattice elements. Formally, we may take $V(L)$ to be the set of all mappings from L into Q, with the usual addition and scalar multiplication. For each lattice element x, let $I_x \in V(L)$ denote the characteristic function of the singleton subset $\{x\}$, i.e.,

$$I_x(y) = \begin{cases} 1 & \text{if } y = x, \\ 0 & \text{otherwise.} \end{cases}$$

Clearly, the vectors $\{I_x : x \in L\}$ form a basis for $V(L)$.

For each $x \in L$, let J_x and K_x denote, respectively, the characteristic functions of the subsets $\{y \in L : y \vee x = 1\}$ and $\{y \in L : y \leq x\}$. That is,

$$J_x = \sum_{y:y\vee x=1} I_y \quad \text{and} \quad K_x = \sum_{y:y\leq x} I_y.$$

The linear relations between the three sets of vectors $\{I_x\}$, $\{J_x\}$, $\{K_x\}$ are given in the following lemma in terms of the Möbius function μ of the lattice L.

Lemma 1. *Let L be a finite lattice. Then for each $x \in L$, the following equations hold in $V(L)$:*

(i) $I_x = \Sigma_{y:y\leq x}\mu(y, x)K_y$.

(ii) $J_x = \Sigma_{y:y\geq x}\mu(y, 1)K_y$.

(iii) $\mu(x, 1)K_x = \Sigma_{y:y\geq x}\mu(x, y)J_y$.

(iv) *If $\mu(a, 1) \neq 0$ for all $a \in L$, then*

$$I_x = \sum_y \lambda(x, y)J_y \quad \text{where } \lambda(x, y) = \sum_{a:a\leq x\wedge y} \frac{\mu(a, x)\mu(a, y)}{\mu(a, 1)}.$$

Proofs. For (i), apply Möbius inversion to the definition of K_x. For (ii), observe that

$$\sum_{y:y\geq x} \mu(y,1)K_y = \sum_{y:y\geq x} \mu(y,1) \sum_{z:z\leq y} I_z = \sum_z \Bigg(\sum_{y:y\geq x\vee z} \mu(y,1)\Bigg) I_z = J_x$$

since

$$\sum_{y:1\geq y\geq x\vee z} \mu(y,1) = \begin{cases} 1 & \text{if } x \vee z = 1, \\ 0 & \text{otherwise.} \end{cases}$$

Applying Möbius inversion to (ii), we arrive at (iii).

Assuming $\mu(a,1)$ is never zero, (i) and (iii) give

$$I_x = \sum_{a:a\leq x} \mu(a,x)K_a = \sum_{a:a\leq x} \frac{\mu(a,x)}{\mu(a,1)} \sum_{y:y\geq a} \mu(a,y)J_y = \sum_y \lambda(x,y)J_y.$$

Remark. With $x = 0$ in Lemma 1(iii) and $0 \neq a \in L$,

$$\mu(0,1)K_0(a) = \sum_y \mu(0,y)J_y(a), \quad \text{or } 0 = \sum_{y:y\vee a=1} \mu(0,y).$$

This is one form of *Weisner's theorem* [12, p. 351].

4. The main theorem.

Theorem 1. *For any finite geometric lattice L, the Whitney numbers satisfy*

$$W_1 + W_2 + \cdots + W_k \leq W_{r-k} + \cdots + W_{r-2} + W_{r-1}, \tag{2}$$

where r is the rank of L and $1 \leq k \leq r-1$. If equality holds for some k, $1 \leq k \leq r-2$, then the lattice L is modular.

Proof. Let U_k be the subspace of $V(L)$ spanned by the vectors $\{I_x : \rho(x) \leq k\}$, and let $\pi: V(L) \to U_k$ be the projection associating to each mapping $L \to Q$ of $V(L)$ its restriction to the subset $\{x \in L : \rho(x) \leq k\}$. That is, π is the linear mapping $V(L) \to U_k$ defined by

$$\pi(I_x) = \begin{cases} I_x & \text{if } \rho(x) \leq k, \\ 0 & \text{otherwise.} \end{cases}$$

Note that if $\rho(y) < r - k$, then

$$\pi(J_y) = \sum_{x:x\vee y=1} \pi(I_x) = 0,$$

since $x \vee y = 1$ implies $\rho(x) \geq \rho(1) + \rho(x \wedge y) - \rho(y) > k$ by the semimodular law (1).

For each $x \in L$ with $\rho(x) \le k$, we have from Lemma 1(iv),

$$I_x = \pi(I_x) = \sum_y \lambda(x, y)\pi(J_y) = \sum_{y:\rho(y)\ge r-k} \lambda(x, y)\pi(J_y).$$

It follows that the $W_{r-k} + \cdots + W_{r-1} + W_r$ vectors $\{\pi(J_y):\rho(y) \ge r-k\}$ span the subspace U_k which has dimension $W_0 + W_1 + \cdots + W_k$. Since $W_0 = W_r = 1$, the stated inequality (2) is established.

We now assume that equality holds in (2) for some k, $1 \le k \le r-2$. Then, evidently, the vectors $\{\pi(J_y):\rho(y) \ge r-k\}$ form a basis for the subspace U_k. For $x \in L$ with $\rho(x) = k+1$, we have from Lemma 1(iv),

$$0 = \pi(I_x) = \sum_{y:\rho(y)\ge r-k} \lambda(x, y)\pi(J_y),$$

and hence $\lambda(x, y) = 0$ whenever $\rho(y) \ge r-k$.

Now $x \wedge y = 0$ would imply that $\lambda(x, y) = \mu(0, x)\mu(0, y)/\mu(0, 1) \ne 0$. Thus our assumption of equality implies in particular that $x \wedge y > 0$ whenever $\rho(x) = k+1$ and $\rho(y) = r-k$. The proof of Theorem 1 is thus completed by the following lemma.

Lemma 2. *Let L be a geometric lattice of rank r and $1 \le k \le r-2$. If $x \wedge y > 0$ whenever $\rho(x) = k+1$ and $\rho(y) = r-k$, then the lattice L is modular.*

Proof. Consider first the case $k = 1$. If L is not modular, there exist elements $x, y \in L$ with $\rho(x) + \rho(y) > \rho(x \wedge y) + \rho(x \vee y)$. The image of a maximal chain C in $[x \wedge y, y]$ under the order homomorphism $z \mapsto x \vee z$ is then a maximal chain in $[x, x \vee y]$ with repeated elements, so there must exist three consecutive members $z_1 \lessdot z_2 \lessdot z_3$ of C with $x \vee z_1 \lessdot x \vee z_2 = x \vee z_3$. We may therefore assume without loss of generality (replace x by $x \vee z_1$, y by z_3) that $\rho(x \vee y) - \rho(x) = 1$, $\rho(y) - \rho(x \wedge y) = 2$. Let c be a modular complement of $x \vee y$ in the interval $[x, 1]$ and let l be a modular complement of $x \wedge y$ in the interval $[0, y]$. Then $\rho(c) = r-1$, $\rho(l) = 2$ and

$$l \wedge c = l \wedge y \wedge c = l \wedge y \wedge (x \vee y) \wedge c = l \wedge y \wedge x = 0,$$

contrary to hypothesis.

Consider now the general case. The hypothesis remains unchanged if we replace k by $r-k-1$, so we shall assume $k \le r/2$.

Suppose for contradiction, that L is not modular. Then there exists a copoint c and a line l such that $l \wedge c = 0$. Choose $z \ge l$ with $\rho(z) = k+1$ and put $a = z \wedge c$. We have $\rho(a) \le k$. There are two cases.

Case 1. $\rho(a) \leq k-1$. Let y be a modular complement of a in the interval $[0, c]$. Then $\rho(y) \geq r-k$, and $z \wedge y = z \wedge c \wedge y = a \wedge y = 0$, a contradiction.

Case 2. $\rho(a) = k$. Choose $b \geq z$ with $\rho(b) = 2k$ and let x be a modular complement of z in the interval $[l, b]$. Let y be a modular complement of b in $[a, 1]$. Then $\rho(x) = k+1$, $\rho(y) = r-k$, and

$$x \wedge y = x \wedge b \wedge y = x \wedge a = x \wedge z \wedge c = l \wedge c = 0,$$

contrary to hypothesis.

5. Further results.

Theorem 2. *Let L be a finite lattice such that $\mu(a, 1) \neq 0$ for all $a \in L$. Then there exists a permutation $f:L \to L$ such that $x \vee f(x) = 1$ for all $x \in L$.*

Proof. By Lemma 1(iv), $\{J_x : x \in L\}$ is a basis for $V(L)$. Hence the matrix whose rows and columns are indexed by L, the entry in row x and column y being

$$J_x(y) = \begin{cases} 1 & \text{if } x \vee y = 1, \\ 0 & \text{otherwise}, \end{cases}$$

is nonsingular. Some term in the determinant expansion does not vanish, i.e. for some permutation f of L, $J_x(f(x)) = 1$ for all $x \in L$.

(Note that, necessarily, $f(0) = 1$ and $f(1) = 0$.)

Remark. The inequality of Theorem 1 is an immediate consequence of Theorem 2 since for a geometric lattice L, f maps elements of rank $\leq k$ injectively into elements of rank $\geq r-k$.

By applying Theorem 2 to the dual lattice L^*, we have

Theorem 2*. *Let L be a finite lattice such that $\mu(0, a) \neq 0$ for all $a \in L$. Then there exists a permutation $f^*:L \to L$ such that $f^*(0) = 1$, $f^*(1) = 0$, and $x \wedge f^*(x) = 0$ for all $x \in L$.*

This answers affirmatively a conjecture of D. G. Kelly [10] that the assertion of Theorem 2* holds for geometric lattices.

An example of a nongeometric lattice for which Theorems 2 and 2* apply is the lattice of faces of a convex polytope, where $\mu(E, F)$ is $(-1)^{\dim(F)-\dim(E)}$ for faces $E \leq F$.

Theorem 3. *Let* L *be a finite geometric lattice of rank* r *and* $0 \leq k \leq r$. *Then there exists an injection* $g: \{x \in L: \rho(x) \leq k\} \rightarrow \{y \in L: \rho(y) \geq r-k\}$ *such that* $x \leq g(x)$ *for all* x *in the domain of* g.

Proof. With the notation as in the proof of Theorem 1, we have seen that $\{\pi(J_y): \rho(y) \geq r-k\}$ spans the subspace U_k. By Lemma 1 (ii),

$$\pi(J_y) = \sum_{z: z \geq y} \mu(z, 1)\pi(K_z),$$

so, evidently, $\{\pi(K_z): \rho(z) \geq r-k\}$ also spans U_k. Hence the matrix whose rows are indexed by $\{x \in L: \rho(x) \leq k\}$ and columns by $\{y \in L: \rho(y) \geq r-k\}$, the entry in row x and column y being

$$K_y(x) = \begin{cases} 1 & \text{if } x \leq y, \\ 0 & \text{otherwise}, \end{cases}$$

has rank equal to the number of rows. The existence of the injection g now follows, since some maximal square submatrix is nonsingular.

Remark. From Lemma 1 and with $T = \{y \in L: \rho(y) \geq r-k\}$,

$$\mu(0, 1)I_0 = \sum_{y \in T} \mu(0, y)\pi(J_y) = \sum_z \left(\sum_{y \leq z, y \in T} \mu(0, y) \right) \mu(z, 1)\pi(K_z).$$

The coefficient of $\mu(0, 1)\pi(K_1)$ in this sum is $\sum_{y \in T} \mu(0, y) = \mu_{L'}(0, 1)$, where L' is the kth truncation of L, i.e. the geometric lattice obtained by identifying all elements of L with rank $\geq r-k$. In particular, the coefficient of $\pi(K_1)$ is nonzero. This remark can be used to see that in Theorem 3, the injection g can be chosen so that $g(0) = 1$ (whenever $k < r$).

For a geometric lattice L, let $\mathrm{Top}_k(L)$ and $\mathrm{Bot}_k(L)$ denote, respectively, the sum of the top and bottom $k+1$ Whitney numbers of L, i.e.

$$\mathrm{Bot}_k(L) = W_0 + W_1 + \cdots + W_k,$$

$$\mathrm{Top}_k(L) = W_r + W_{r-1} + \cdots + W_{r-k}, \quad \text{where } r = \mathrm{rank}(L).$$

Theorem 4. *Let* L *be a finite geometric lattice of rank* r *and fix* $a \in L$ *with* $\rho(a) \leq r-k$. *Then*

$$\mathrm{Top}_k(L) \geq \mathrm{Top}_k(L) - \mathrm{Top}_k([a, 1]) + \mathrm{Bot}_k([a, 1]) \geq \mathrm{Bot}_k(L).$$

Proof. Applying Theorem 1 to the geometric lattice $[a, 1]$, we have

$\mathrm{Top}_k([a, 1]) \geq \mathrm{Bot}_k([a, 1])$ and the first inequality follows.

For $z \in L$, define $A_z = \sum_{x:x \vee a = z} I_x \in V(L)$. Then whenever $y \geq a$ in L,

$$K_y = \sum_{z:a \leq z \leq y} A_z. \tag{3}$$

To prove the second inequality, it will suffice to show that the set of vectors

$$S = \{\pi(K_y): \rho(y) \geq r - k,\ y \not\geq a\} \cup \{\pi(A_z): \rho(z) \leq \rho(a) + k,\ z \geq a\}$$

spans the subspace U_k of $V(L)$. From the proof of Theorem 3, $\{\pi(K_y): \rho(y) \geq r - k\}$ spans U_k, so it remains only to show that $\pi(K_y)$ is in the span of S for $y \geq a$. But this follows from equation (3) above and the observation that the semimodular law (1) implies $\pi(A_z) = 0$ whenever $\rho(z) > \rho(a) + k$.

With $k = 1$ and a taken to be a point, we have the

Corollary. *In a finite geometric lattice of rank ≥ 2, the number of lines (rank 2 lattice elements) on a given point plus the number of copoints not on that point cannot be less than the total number of points.*

REFERENCES

1. J. G. Basterfield and L. M. Kelly, *A characterization of sets of n points which determine n hyperplanes*, Proc. Cambridge Philos. Soc. 64 (1968), 585–588. MR 38 #2040.

2. J. E. Blackburn, H. H. Crapo and D. A. Higgs, *A catalogue of combinatorial geometries*, University of Waterloo, Waterloo, Ontario, 1969.

3. G. Birkhoff, *Lattice theory*, 3rd ed., Amer. Math. Soc. Colloq. Publ., vol. 25, Amer. Math. Soc., Providence, R. I., 1967. MR 37 #2638.

4. H. H. Crapo and G.-C. Rota, *On the foundations of combinatorial theory: Combinatorial geometries*, M. I. T. Press, Cambridge, Mass., 1970. (preliminary edition). MR 45 #74.

5. R. P. Dilworth, *Proof of a conjecture on finite modular lattices*, Ann. of Math. (2) 60 (1954), 359–364. MR 16, 106.

6. T. A. Dowling and R. M. Wilson, *The slimmest geometric lattices*, Trans. Amer. Math. Soc. 196 (1974), 203–215.

7. C. Greene, *A rank inequality for finite geometric lattices*, J. Combinatorial Theory 9 (1970), 357–364. MR 42 #1727.

8. ———, *Inequalities for geometric lattices*, Proc. Conf. on Möbius Algebras (H. Crapo and G. Roulet, editors), University of Waterloo, Waterloo, Ont., 1971.

9. L. H. Harper, *Stirling behaviour is asymptotically normal*, Ann. Math. Statist. 38 (1967), 410–414. MR 35 #2312.

10. D. G. Kelly, *Disjoining permutations in finite boolean algebras*, Utilitas Mathematica 3 (1973), 65–74.

11. E. Lieb, *Concavity properties and a generating function for Stirling numbers*, J. Combinatorial Theory 5 (1968), 203–206. MR 37 #6195.

12. G.-C. Rota, *On the foundations of combinatorial theory.* I: *Theory of Möbius functions*, Z. Wahrscheinlichkeitstheorie und Verw. Gebiete 2 (1964), 340–368. MR 30 #4688.

13. P. Young, U. S. R. Murty and J. Edmonds, *Equicardinal matroids and matroid designs*, Proc. Second Chapel Hill Conference on Combinatorial Mathematics and its Applications, University of North Carolina, Chapel Hill, N. C., 1970.

DEPARTMENT OF MATHEMATICS, OHIO STATE UNIVERSITY, COLUMBUS, OHIO 43210

Chapter IV
The Tutte decomposition

1. Contractions and deletions

"Simple ideas are often the most powerful." This adage is best exemplified in matroid theory by the method of Tutte (-Grothendieck) decomposition. This method has its origin in the following recursion formula (due to Foster[1]; see the concluding note in Whitney [32]) for the chromatic polynomial $P(L; \lambda)$ of a graph L. Let L be a graph and A an edge in L linking two distinct vertices. Let L'_A be the graph obtained from L by deleting the edge A and L''_A the graph obtained from L by contracting A. Then

$$P(L; \lambda) = P(L'_A; \lambda) - P(L''_A; \lambda).$$

In addition to the chromatic polynomial, the number of spanning forests and the number of nowhere-zero flows also satisfy similar recursions. These examples suggest that a general theory of functions $W(L)$ defined on graphs taking values in an abelian group and satisfying the following two conditions $(*)$:

$W(L_1) = W(L_2)$	whenever $L_1 \cong L_2$, and
$W(L) = W(L'_A) + W(L''_A)$	for any edge A of L linking two distinct vertices,

would be rewarding. This theory was first constructed in the paper "A ring in graph theory" by Tutte, the first paper reprinted in this chapter. Although this paper is concerned exclusively with graphs, many of its results extend to matroids.

In §1, W-functions and V-functions are defined. A *W-function* is a function defined on graphs taking values in an abelian group and satisfying the conditions $(*)$. A *V-function* is a function V on graphs taking values in a commutative ring satisfying $(*)$ and the additional condition

---

[1] Some papers in graph theory by R.M. Foster are [32, 52, 62].

$$V(L_1 L_2) = V(L_1)\, V(L_2)$$

where $L_1 L_2$ is the product (i.e. disjoint union) of the two graphs L_1 and L_2. The three classical examples of V-functions are described and a summary of the paper is given at the end of this section. Topological rather than graph theoretic terminology is sometimes used in this paper; however, very little algebraic topology is used. In particular, by defining $p_0(L)$ to be the number of connected components and $p_1(L)$ to be the nullity of L, the appeal to homology in Equation (1) can be avoided.

In §2, a ring R satisfying a "universal property" for V-functions is constructed. The ring R is closely related to the Grothendieck ring in algebraic K-theory. (See, for example, Bass [68], Chap. 7, §1.) It may be instructive to describe in some detail this relation. To do so, we construct the simplest example, the Grothendieck ring $K_0(C)$ of a field C. Our notation is chosen to conform with Tutte's notation. If X is a vector space (over C), the *isomorphism class* $\boldsymbol{X}$ of X is the class of all vector spaces isomorphic to X. Let $\mathcal{F}(C)$ be the set of all isomorphism classes of finite dimensional vector spaces and let B be the free abelian group generated by $\mathcal{F}(C)$. An exact sequence of vector spaces is a sequence of linear transformations $X \to Y \to Z$ such that $X \to Y$ is injective and $Y \to Z$ is surjective. To each exact sequence, we associate the element

$$\boldsymbol{X} - \boldsymbol{Y} + \boldsymbol{Z}$$

in B. Let B' be the subgroup in B generated by all such elements. The *Grothendieck group* $K_0(C)$ of the field C is the quotient group B/B'. If X is a vector space, we (still) denote by $\boldsymbol{X}$ its isomorphism class considered as an element of $K_0(C)$. The Grothendieck group can be made into a ring with the multiplication given by

$$\boldsymbol{X}\boldsymbol{Y} = \text{the isomorphism class of } X \otimes Y,$$

where $X \otimes Y$ is the tensor product of X and Y. To see that this multiplication is well-defined on $K_0(C)$, observe that if $X \to Y \to Z$ is an exact sequence, then so is $X \otimes U \to Y \otimes U \to Z \otimes U$. (This is not necessarily true for modules over a ring.) Thus, the subgroup B' is in fact an ideal in the ring B equipped with the above multiplication. The quotient ring B/B' is the *Grothendieck ring,* also denoted by $K_0(C)$, of the field C. A function V defined from $\mathcal{F}(C)$ to a commutative ring H is said to be *additive* if

$V(X_1) = V(X_2)$ whenever $X_1 \cong X_2$,

$V(X) - V(Y) + V(Z) = 0$ for every exact sequence $X \to Y \to Z$,

and

$V(X \otimes Y) = V(X)\, V(Y)$.

By construction, the Grothendieck ring has the following *universal* property:

> V is an additive function if and only if there exists a ring homomorphism h: $K_0(C) \to H$ such that $V(X) = h(\boldsymbol{X})$.

In fact, the Grothendieck ring is isomorphic to the integers $\mathbb{Z}$. To prove this, observe that if $Y \cong X \oplus Z$, $X \to Y$ is the canonical injection, and $Y \to Z$ the canonical projection, then $X \to Y \to Z$ is an exact sequence. Hence, if $Y \cong X \oplus Z$, then $\boldsymbol{Y} = \boldsymbol{X} + \boldsymbol{Z}$ in $K_0(C)$. But every C-vector space of finite dimension n is isomorphic to the n-fold direct sum $C \oplus \ldots \oplus C$. We conclude that $K_0(C)$ is generated by the isomorphism class $\boldsymbol{C}$ consisting of all the vector spaces of dimension one: thus, $K_0(C) \cong \mathbb{Z}$. Finally, observe that since every homomorphism h of $\mathbb{Z}$ to a ring H is determined by its value $h(1)$ at 1 in $\mathbb{Z}$, the universal property of $K_0(C)$ implies that every additive function defined from $\mathcal{F}(C)$ to a ring H is of the form $\dim(X)g$ for some element g in H. (Here, $\dim(X)$ is the dimension of X.)

Returning to the Tutte decomposition, observe that there is a natural injection of the edges of L'_A into the edges of L and a natural surjection of the edges of L not including A onto the edges of L''_A. Thus, the graphs L'_A, L, L''_A can be thought of as forming an exact sequence $L'_A \to L \to L''_A$ (of a rather special kind). Similarly, V-functions can be thought of as additive functions for graphs and the ring R defined in §2 which is universal for V-functions (Theorem I) as the Grothendieck ring for graphs. We should remark that Tutte's construction of R predates Grothendieck's construction of $K_0(C)$ by several years. Because of this, we use the briefer term, "Tutte decomposition", to refer to this theory rather than the more usual "Tutte-Grothendieck decomposition." §2 concludes with the result that every isomorphism class of graphs can be written in R as a polynomial with integer coefficients and no constant term in the elementary graphs $\boldsymbol{y}_r$. (The elementary graph $\boldsymbol{y}_r$ is the isomorphism class of the graph consisting of a single vertex and r loops.)

In §3, the V-function $Z(L)$ is defined. It can be thought of as a "nullity generating function" for the subgraphs of L. The function $Z(L)$ is used in §4 to show that the ring R is isomorphic to the ring R_0 of all polynomials in the indeterminates z_0, z_1, z_2, ... with integer coefficients and no constant term

(Theorem V). In the next section, the V-functions which are topologically invariant (i.e. invariant under subdivisions of edge) are characterized in terms of an ideal in the ring R.

Although the V-function $Z(L)$ and much of §§3-5 can be interpreted in terms of matroids, work on the Tutte decomposition for matroids has been concentrated on the dichromatic or Tutte polynomial. (A quick introduction to the Tutte decomposition for matroids can be found in §4 of [IV. 2].) The *dichromatic polynomial* $Q(L; t, z)$ is the polynomial in t and z obtained by setting $z_i = t z^i$ in the V-function $Z(L)$. Its basic properties are developed in §6. The first result (Theorem IX) describes the behavior of $Q(L; t, z)$ under duality. Next, two results (Theorem X and XI) yielding interpretations of evaluations of the dichromatic polynomial in terms of the number of *α-colorings* (i.e. colorings of the vertices) and *β-colorings* (i.e. nowhere-zero flows) in a graph are proved. These results indicate that the dichromatic polynomial of a planar graph is a mixture of the chromatic polynomial of the graph and the chromatic polynomial of its dual.

The last section, §7, is concerned with F-functions. These are functions defined on *cubical networks,* which are essentially graphs in which every vertex has degree 2 or 3. (Note that "nodes" and "arcs" are not synonyms for "vertices" and "edges" in this section.) The basic idea behind F-functions can perhaps be best seen by looking at the specific example π_0. For simplicity, we shall work only with cubical networks in which all the vertices have degree 3, i.e. regular graphs of degree 3. For such cubical networks, a *subnetwork* is a subgraph in which every vertex in the full network has an incident edge in the subgraph. The function $\pi_0(N)$ is defined to be the number of subnetworks T in N containing no vertex of odd degree. This condition implies that every vertex in T must have degree 2 and thus. $\pi_0(N)$ is the number of 2-factors (or "solutions to Petersen's problem") in N. Now, given a cubical network N, the cubical networks $\overline{N}$, N'_X and N'_Z are defined by making the changes shown in Fig. 1 in §7 of the paper. The function $\pi_0(N)$ satisfies the property:

$$\pi_0(N) + \pi_0(N'_X) = \pi_0(\overline{N}) + \pi_0(N'_Z).$$

(To see this, observe that there is a direct one-to-one correspondence between the 2-factors of N and N'_X, and the 2-factors of $\overline{N}$ and N'_Z.) This property, apart from signs, is the defining property of an F-function. §7 is devoted to the general theory of F-functions and includes two results (Theorems XII and XIII) on the connection between F-functions and topologically invariant W-functions.

Much work has been done on the Tutte decomposition since this pioneering paper. The following is a list of papers on theoretical aspects of the Tutte decomposition. (For some applications, see the §§2 and 3 of this commentary.)

Survey: Brylawski [82].
General theory of the Tutte decomposition for matroids: Bender et al. [84], Brylawski [72, 72′].
Tutte decomposition and series-parallel networks: Brylawski [71].
Dichromatic and related polynomials for graphs: Arrowsmith and Jaeger [82], Beissinger [82], Berman [77, 78], Farrell [79], Joyce [84], Martin [78], Rosenstiehl and Read [78], Smith [69], Tutte [54, 67, 75, 80].
Dichromatic polynomial for matroids: Crapo [69], Heron [72], Las Vergnas [80], Oxley and Welsh [79], Smith [78].
Beta invariant: Crapo [67], Oxley [82].
Reconstruction: Brylawski [81], Tutte [79].
Codichromatic graphs: Tutte [74].
Other decompositions: Kung [78], Stanley [73].

References

Arrowsmith, D.K. and Jaeger, F.: On the enumeration of chain in regular chain groups, *J. Combin. Theory Ser. B 32*(1982), 75-89.

Bass, H.: *Algebraic K-theory,* Benjamin, New York, 1968.

Beissinger, J.S.: On external activity and inversions in trees, *J. Combin. Theory Ser. B 33*(1982), 87-92.

Bender, E., Viennot, G. and Williamson, S. G.: Global analysis of the delete-contract recursion for graphs and matroids, *Linear and Multilinear Algebra 15*(1984), 133-160.

Berman, G.: The dichromate and orientations of a graph, *Canad. J. Math. 29*(1977), 947-956.

Berman, G.: Decomposition of graph functions, *J. Combin. Theory Ser. B 25*(1978), 151-165.

Brylawski, T.: A combinatorial model for series-parallel networks, *Trans. Amer. Math. Soc. 154*(1971), 1-22.

Brylawski, T.: The Tutte-Grothendieck ring, *Algebra Universalis 2*(1972), 375-388.

Brylawski, T.: A decomposition for combinatorial geometries, *Trans. Amer. Math. Soc. 171*(1972), 235-282.

BRYLAWSKI, T.: Hyperplane reconstruction of the Tutte polynomial of a geometric lattice, *Discrete Math. 35*(1981), 25-38.

BRYLAWSKI, T.: The Tutte polynomial, Part I: General theory, *Matroid Theory and its Applications (Proc. C.I.M.E., Varenna, 1980),* Liguori, Naples, 1982.

CRAPO, H.H.: A higher invariant for matroids, *J. Combin. Theory 2*(1967), 406-417.

CRAPO, H.H.: The Tutte polynomial, *Aequationes Math. 3*(1969), 211-229.

FARRELL, E.J.: On a general class of graph polynomials, *J. Combin. Theory Ser. B 26*(1979), 111-122.

FOSTER, R.M.: Geometrical circuits of electrical networks, *Trans. Amer. Inst. Elect. Engrs. 52*(1932), 309-317.

FOSTER, R.M.: Topologic and algebraic considerations in network synthesis, *Proc. Sympos. Modern Network Synthesis, New York, 1952,* pp. 8-18, Polytechnic Inst. of Brooklyn, New York, 1952.

FOSTER, R.M.: Academic and theoretical aspects of circuit theory, *Proc. IRE 50*(1962), 866-871.

HERON, A.P.: Matroid polynomials, *Combinatorics* (Proc. Conf. Combinatorial Math., Math. Inst., Oxford, 1972), pp. 164-202, Inst. Math. Appl., Southend-on-Sea, 1972.

JOYCE, D.: Generalized chromatic polynomials, *Discrete Math. 50*(1984), 51-62

KUNG, J.P.S.: The Rédei function of a relation, *J. Combin. Theory Ser. A 29*(1980), 287-296.

LAS VERGNAS, M.: On the Tutte polynomial of a morphism of matroids, *Combinatorics 79, Part I, Ann. Discrete Math. 8*(1980), 7-20.

MARTIN, P.: Remarkable valuation of the dichromatic polynomial of planar multigraphs, *J. Combin. Theory Ser. B 24*(1978), 318-324.

OXLEY, J.G.: On Crapo's beta invariant for matroids, *Stud. Appl. Math. 66*(1982), 267-277.

OXLEY, J.G. and WELSH, D.J.A.: The Tutte polynomial and percolation, *Graph Theory and Related Topics (Proc. Conf., Univ. Waterloo, Waterloo, Ont., 1977),* pp. 329-339, Academic Press, New York and London, 1979.

ROSENSTIEHL, P. and READ, R.C.: On the principal edge tripartition of a graph, *Ann. Discrete Math. 3*(1978), 195-226.

SMITH, C.A.B.: Map colourings and linear mappings, *Combinatorial Mathematics and its Applications (Proc. Conf., Oxford, 1969),* pp. 259-283, Academic Press, London. 1969.

SMITH, C.A.B.: On Tutte's dichromatic polynomial, *Advances in Graph Theory* (B. Bollobás, ed.) *Ann. Discrete Math. 3*(1978), 247-257.

STANLEY, R.P.: A Brylawski decomposition for finite ordered sets, *Discrete Math.* *4*(1973), 77-82.
TUTTE, W.T.: A contribution to the theory of chromatic polynomials, *Canad. J. Math.* *6*(1954), 80-91
TUTTE, W.T.: On dichromatic polynomials, *J. Combin. Theory* *2*(1967), 301-302.
TUTTE, W.T.: Codichromatic graphs, *J. Combin. Theory Ser. B* *16*(1974), 168-175.
TUTTE, W.T.: The dichromatic polynomial, *Proc. Fifth British Combinatorial Conf. (Univ. Aberdeen, Aberdeen, 1975)*, pp. 605-635, Congressus Numerantium no. 15, Utilitas Math., Winnipeg, Man., 1976.
TUTTE, W.T.: All the king's horses. A guide to reconstruction, *Graph Theory and Related Topics (Proc. Conf., Univ. Waterloo, Waterloo, Ont., 1977)*, pp. 15-33, Academic Press, 1979.
TUTTE, W.T.: 1-factors and polynomials, *Europ. J. Combin.* *1*(1980). 77-87.
WHITNEY, H.: The coloring of graphs, *Annals of Math. (2)* *33*(1933), 688-718.

2. The critical problem

The second paper reprinted in this chapter is "Weight enumeration and the geometry of linear codes" by Greene. In this paper, Greene discusses the intimate connection between matroid theory and the theory of linear error-correcting codes. In particular, using the Tutte decomposition and the critical problem, it is shown that the weight enumerator of a linear code is an evaluation of the Tutte polynomial of the matroid formed by the column vectors of the generator matrix of the code. From this result, an elegant proof is given of the MacWilliams identity relating the weight enumerators of a linear code and its dual. This very accessible paper is almost self-contained. The only exception is that some knowledge of the critical problem is assumed. We shall give a brief exposition of the critical problem.

Let S be a set of non-zero vectors in a vector space $V_n(q)$ of dimension n over the finite field $GF(q)$ with q elements. An ordered m-tuple $(H_1, \dots, H_m)$ of hyperplanes (i.e. subspaces of codimension one) is said to *distinguish* the set S if

$$H_1 \cap \dots \cap H_m \cap S = \phi.$$

The *critical problem* for the set S is to find the minimum number c for which there exists a c-tuple of hyperplanes distinguishing S. The number c is called the *critical exponent* of S. It is often useful to rephrase the critical problem in terms of linear functionals. Let $(L_1, \ldots, L_m)$ be an m-tuple of linear functionals (i.e. linear functions from $V_n(q)$ to the field $GF(q)$). The m-tuple $(L_1, \ldots, L_m)$ is said to *distinguish* S if for every vector p in S, there exists a linear functional L_i such that $L_i(p) \neq 0$. The critical problem for S is to find the minimum number c for which there exists a c-tuple of linear functionals distinguishing S. To see the equivalence of these two formulations, let $\ker L_i$ be the hyperplane on which L_i is zero and observe that $(L_1, \ldots, L_m)$ distinguishes S if and only if the m-tuple of hyperplanes $(\ker L_1, \ldots, \ker L_m)$ distinguishes S.

Theorem 1 (Crapo-Rota): Let S be a set of non-zero vectors in $V_n(q)$, G the matroid on the set S given by linear dependence, r the rank function of G, and $L(G)$ the lattice of closed sets of G. Then, the number of m-tuples of linear functionals distinguishing S equals

$$q^{m(n-r(S))}\chi_{L(G)}(q^m) = \sum_{a \in L(G)} \mu(0, a)\, q^{m(n-r(a))},$$

where $\chi_{L(G)}(\lambda)$ is the characteristic polynomial and μ the Möbius function of the geometric lattice $L(G)$.

Proof: Let a be a closed set in the matroid G and a^* the subspace in $V_n(q)$ spanned by the vectors in a. The number of linear functionals whose kernel contains a equals the number of vectors in the quotient space $V_n(q)/a^*$, which in turn equals $q^{n-r(a)}$. Hence, the number of m-tuples $(L_1, \ldots, L_m)$ such that

$$a \subseteq \ker L_1 \cap \ldots \cap \ker L_n \cap S$$

equals $q^{m(n-r(a))}$. From this, we obtain

$$q^{m(n-r(a))} = \sum_{b \in L(G),\, b \geqslant a} |\{(L_1, \ldots, L_m) : \ker L_1 \cap \ldots \cap \ker L_m \cap S = b\}|.$$

By Möbius inversion (see [III. 1]),

$$|\{(L_1, \ldots, L_m) : \ker L_1 \cap \ldots \cap \ker L_m \cap S = \phi\}| = \sum_{a \in L(G)} \mu(0, a)\, q^{m(n-r(a))}.$$

This proves the theorem. ■

(The same technique yields a proof of Proposition 3.1 in Greene's paper.)

Using this theorem, another formulation of the critical problem for S is: find the minimum number c for which

$$\chi_{L(G)}(q^c) \neq 0 .$$

Since the lattice $L(G)$ of closed sets depends only on the matroid structure G on the set S, the characteristic polynomial, and hence the critical exponent, also depend only on the matroid G.

The critical problem may be regarded as a (somewhat weaker) geometric analogue of the coloring problem for graphs. More precisely, let Γ be a graph with vertex set $N(\Gamma)$. A (*proper*) *coloring* of Γ using k colors is a function from $N(\Gamma)$ to a set of size k such that $f(u) \neq f(v)$ whenever $\{u, v\}$ is an edge in Γ. To see the connection with the critical problem, let V be the vector space of formal linear combinations

$$\sum_{v \in N(\Gamma)} a_v v$$

of vertices with coefficients a_v in the finite field $GF(2)$ with two elements. Let $E(\Gamma)$ be the subset of vectors in V of the form $u + v$, where $\{u, v\}$ is an edge in the graph Γ.

Theorem 2 (Crapo-Rota): The number of proper colorings of Γ using 2^m colors equals the number of m-tuples of linear functionals distinguishing $E(\Gamma)$.

Proof: We may assume that the 2^m colors are from the set $[GF(2)]^m$ of m-tuples with coordinates in $GF(2)$. Let $L = (L_1, \ldots, L_m)$ be a m-tuple of linear functionals distinguishing $E(\Gamma)$. From L, we construct the coloring $L^\dagger$ from $N(\Gamma)$ to $[GF(2)]^m$ by defining

$$L^\dagger(v) = (L_1(v), \ldots, L_m(v)).$$

To see that $L^\dagger$ is proper, observe that if $\{u, v\}$ is an edge, then (as $(L_1, \ldots, L_m)$ distinguishes $E(\Gamma)$) there exists a linear functional L_i such that $L_i(u+v) \neq 0$, or, $L_i(u) \neq L_i(v)$. Hence, $L^\dagger(u) \neq L^\dagger(v)$. Since this construction is reversible, the theorem is proved. ∎

Another instance of a critical problem is the *fundamental problem of linear coding theory*:

Given n and t, find the maximum number k for which there exists an (n, k)-code over $GF(q)$ (i.e. a subspace in $V_n(q)$ of dimension k) with no non-zero vector of Hamming weight less than or equal to t.

(The *Hamming weight* of a vector x is the number of non-zero coordinates in x.) This problem is equivalent to the critical problem for the punctured t-ball $S_{n,t}$ defined by

$$S_{n,t} = \{x \in V_n(q):\ x \neq 0 \text{ and } x \text{ has Hamming weight at most } t\}.$$

See Dowling [71] for more details. (For other uses of matroid theory in coding theory, see Blake and Mullin [76].)

We end this section with a list of papers on the critical problem. (Papers on graph colorings and papers on blocking sets and other related concepts in projective geometry are not included unless directly relevant.)

Origin: Crapo and Rota [70], Chap. 16.
Related development: Tutte [66, 69, 69′].
Extensions and generalizations: (Intersection theory) Brylawski [79, 81], Brylawski et al. [80]; (Rédei functions) Kung [80], Kung, Murty and Rota [80].
Bounds for the critical exponent: Asano et al. [84], Lindström [78], Oxley [78, 83].
Packing and covering: Mullin and Stanton [79], Oxley [78, 80], Seymour [80].
Special cases: (Flows in graphs) Jaeger [79]; see also Tutte [54], Seymour [81]. (Binary matroids without $M(K_5)$, F_7 or F_7^* as minors) Walton and Welsh [80]. (Transversal matroids) Whittle [84].
Constructive approach: Jaeger [81].
Applications: (Coding theory) Dowling [71]. (Artin's conjecture on primitive roots) Matthews [77].
Uniform critical exponents and growth rates: Kung [86].
Survey: Welsh [79, 80].

References

Asano, T., Nishizeki, T., Oxley, J.G. and Saito, N.: A note on the critical problem for matroids, *Europ. J. Combin.* 5(1984), 93-97.

Blake, I.F. and Mullin, R.C.: *An introduction to algebraic and combinatorial coding theory,* Academic Press, New York, 1976.

Brylawski, T.: Intersection theory for embeddings of matroids into uniform geometries, *Stud. Appl. Math. 61*(1979), 211-244.

Brylawski, T.: Intersection theory for graphs, *J. Combin. Theory Ser. B 30* (1981), 233-246.

Brylawski, T., Lo Re, P.M., Mazzocca, F. and Olanda, D.: Alcune applicazioni della teorie dell' intersezione alle geometrie di Galois, *Ricerche di Matematica 29*(1980), 65-84.

Crapo, H.H. and Rota, G.-C.: *On the foundations of combinatorial theory: Combinatorial geometries* (Prelim. ed.), M.I.T. Press, Cambridge, Mass., 1970.

Dowling, T.A.: Codes, packings and the critical problem, *Atti del Convegno di Geometria Combinatoria e sue applicazioni (Univ. Perugia, Perugia, 1970),* pp. 209-224, Ist. Mat. Univ. Perugia, 1971.

Jaeger, F.: Flows and generalized coloring theorems in graphs, *J. Combin. Theory Ser. B 26*(1979), 205-216.

Jaeger, F.: A constructive approach to the critical problem for matroids, *Europ. J. Combin. 2*(1981), 137-144.

Kung, J.P.S.: The Rédei function of a relation, *J. Combin. Theory Ser. A 29* (1980), 287-296.

Kung, J.P.S.: Growth rates and critical exponents of classes of binary geometries, *Trans. Amer. Math. Soc. 293*(1986), 837-859.

Kung, J.P.S., Murty, M.R. and Rota, G.-C.: On the Rédei zeta function, *J. Number Theory 12*(1980), 421-536.

Lindström, B.: On the chromatic number of regular matroids, *J. Combin. Theory Ser. B 24*(1978), 367-369.

Matthews, K.R.: An example from power residues of the critical problem of Crapo and Rota, *J. Number Theory 9*(1977), 203-208.

Mullin, R.C. and Stanton, R.G.: A covering problem in binary spaces of finite dimension, *Graph Theory and Related Topics (Proc. Conf. Univ. Waterloo, Waterloo, Ont., 1977),* pp. 315-327, Academic Press, New York, 1979.

Oxley, J.G.: Colouring, packing and the critical problem, *Quart. J. Math. Oxford Ser. (2) 29*(1978), 11-22.

Oxley, J.G.: Cocircuits coverings and packings for binary matroids, *Math. Proc. Cambridge Philos. Soc. 83*(1978), 347-351.

Oxley, J.G.: On a covering problem of Mullin and Stanton for binary matroids, *Aequationes Math. 19*(1980), 104-112.

Oxley, J.G.: On a matroid identity, *Discrete Math. 44*(1983), 55-60.

Seymour, P.D.: Packing and covering with matroid circuits, *J. Combin. Theory Ser. B 28*(1980), 237-242.

Seymour, P.D.: Nowhere-zero 6-flows, *J. Combin. Theory Ser. B 30*(1981), 130-135.

Tutte, W.T.: A contribution to the theory of chromatic polynomials, *Canad. J. Math. 6*(1954), 80-91.

Tutte, W.T.: On the algebraic theory of graph colorings, *J. Combin. Theory 1*(1966), 15-50.

Tutte, W.T.: A geometrical version of the four color problem, *Combinatorial Mathematics and its Applications (Proc. Conf., Univ. North Carolina, Chapel Hill, NC., 1967),* pp. 553-560, Univ. North Carolina Press, Chapel Hill, N.C., 1969.

Tutte, W.T.: Projective geometry and the 4-color problem, *Recent Progress in Combinatorics (Proc. Third Waterloo Conf. on Combinatorics, 1968),* pp. 199-207, Academic Press, New York, 1969.

Walton, P.N. and Welsh, D.J.A.: On the chromatic number of binary matroids, *Mathematika 27*(1980), 1-9.

Welsh, D.J.A.: Colouring problems and matroids, *Surveys in Combinatorics (Proc. Seventh British Combinatorial Conf., Cambridge, 1979),* pp. 229-257, London Math. Soc. Lecture Note Ser. No. 38, Cambridge Univ. Press, Cambridge, 1979.

Welsh, D.J.A.: Colourings, flows and projective geometry, *Nieuw Arch. Wisk. (3) 28*(1980), 159-176.

Whittle, G.P.: On the critical exponents of transversal matroids, *J. Combin. Theory Ser. B 37*(1984), 94-95.

3. Cutting up space

The simplest non-trivial example of the problem of counting the faces of an arrangement of hyperplanes is the following classical problem in enumeration. Let $\ell_1, \ell_2, \ldots, \ell_n$ be n lines "in general position" in the real plane $\mathbb{R}^2$. Let c_n be the number of connected components in the complement $\mathbb{R}^2 \setminus (\ell_1 \cup \ldots \cup \ell_n)$. Find a formula for c_n. One way to solve this problem is to find a recursion for c_n. Remove the line ℓ_n and consider the complement $\mathbb{R}^2 \setminus (\ell_1 \cup \ldots \cup \ell_{n-1})$. As the line ℓ_n is in general position relative to the other lines, removing it reduces the number of connected components by n. Hence, the numbers c_n satisfy the recursion

$$c_n = c_{n-1} + n$$

with the initial condition $c_0 = 1$. Solving this, we obtain

$$c_n = 1 + \binom{n+1}{2} = 1 + n + \binom{n}{2}.$$

This situation contains all the essential features of the general theory of Euclidean arrangements of hyperplanes. To see this, we construct the *cut-intersection semilattice* of the lines $\ell_1, \ldots, \ell_n$. This is the semilattice whose elements are those intersections

$$\bigcap_{j \in S} \ell_j, \quad S \subseteq \{1, 2, \ldots, n\}$$

which are non-empty, ordered by reverse inclusion. Since the lines $\ell_1, \ldots, \ell_n$ are in general position, their cut-intersection semilattice is a semilattice of rank 2 with the elements of rank 2 the intersections $\ell_i \cap \ell_j$, $i \neq j$, the elements of rank 1 the lines ℓ_i, and the minimum $\mathbb{R}^2$, the intersection of the empty collection of lines. This is a semilattice in which every interval is a geometric lattice. Its characteristic polynomial is

$$\chi(\lambda) = \lambda^2 - n\lambda + \binom{n}{2}.$$

Our formula for c_n can be rewritten as

$$c_n = |\chi(-1)|.$$

This is no accident, for it is true in general that every interval of the cut-intersection semilattice L of a collection of hyperplanes $H_1, \ldots, H_n$ in $\mathbb{R}^d$ is a geometric lattice. Moreover, the number of connected components $c(H_1, \ldots, H_n)$ in the complement $\mathbb{R}^d \setminus (H_1 \cup \ldots \cup H_n)$ is

$$|\chi_L(-1)|,$$

where $\chi_L(\lambda)$ is the characteristic polynomial of the cut-intersection semilattice L. Since the characteristic polynomial satisfies the Tutte decomposition, a natural way to prove this is to proceed by induction using the recursion

$$c(H_1, \ldots, H_n) = c(H_1, \ldots, H_{n-1}) + c(H_1', \ldots, H_{n-1}'),$$

where H_i' is the intersection of H_i with H_n and $H_1', \ldots, H_{n-1}'$ are considered to be hyperplanes in the $(d-1)$-dimensional space H_n. For the earlier example of lines in $\mathbb{R}^2$, this is simply the recursion $c_n = c_{n-1} + n$.

The connection among characteristic polynomials, geometric lattices, and arrangements of hyperplanes first appeared in the paper "Facing up to arrangements: Face-count formulas for partitions of space by hyperplanes" by Zaslavsky. Because of its length, only an extract is reprinted here. The topics include the basic facts about arrangements of hyperplanes and their cut-intersection semilattices (§§0, 1), formulas for the number of k-dimensional faces of Euclidean and projective arrangements (§§ 2A, 2B), and their proofs using Euler's formula (§3AB) and the Tutte decomposition (§§4A, 4B). The principal omissions are the formula for the number of bounded faces (§2C) and its proofs (§§3C, 4C), the interpretation of the Crapo beta invariant as the number of bounded regions (§7), and a theory of bounded faces (§§8, 9). In addition, we have omitted several applications and special cases of the counting formulas (§§5, 6).

The extract is self-contained with one exception. In Lemma 4A4 (§4A), the result (due to Brylawski) that the magnitude of the Möbius function $\mu(0, 1)$ satisfies the Tutte decomposition is used without proof. For the reader's convenience, we indicate here how the proof goes. By definition, the Möbius function $\mu_G(0, 1)$ of a matroid G is the value $\mu(0, 1)$ of the Möbius function of the lattice of flats of G if G has no loops and 0 otherwise. Using the Corollary on p. 348 and Proposition 1, §5, in [III. 1], we have for a matroid G with no loops,

$$\mu_G(0, 1) = \sum_A (-1)^{|A|},$$

the sum being over all spanning sets A in G. Hence, by Theorem 4, §7, in [III. 1],

$$|\mu_G(0, 1)| = (-1)^{\operatorname{rank}(G)} \sum_A (-1)^{|A|} = T_G(1, 0),$$

where $T_G(x, y)$ is the Tutte polynomial defined in §2 of [IV. 2]. Thus, by Theorem 4.2 in [IV. 2], $|\mu_G(0, 1)|$ satisfies the Tutte decomposition.

A list of related papers is given in the references.

References

BRYLAWSKI, T.: A combinatorial perspective on the Radon convexity theorem, *Geometriae Dedicata 5*(1976), 459-466.

CARTIER, P.: Les arrangements d'hyperplanes: Un chapitre de géométrie combinatoire, *Séminaire Bourbaki, 1980/81,* pp. 1-22 (No. 561), Lecture Notes in Math., Vol. 901, Springer-Verlag, Berlin and New York, 1981.

CORDOVIL, R.: Sur l'évaluation $t(M;\ 2,\ 0)$ du polynôme de Tutte d'un matroïde et une conjecture de B. Grünbaum relative aux arrangement de droites du plan, *Europ. J. Combin. 1*(1980), 317-322.

CORDOVIL, R., LAS VERGNAS, M. and MANDEL, A.: Euler's relations, Möbius functions and matroid identities, *Geometriae Dedicata 12*(1982), 147-162.

EDELMAN, P.: A partial order on the regions of R^n dissected by hyperplanes, *Trans. Amer. Math. Soc., 283*(1984), 617-632.

GOOD, I.J. and TIDEMAN, T.N.: Stirling numbers and a geometric structure from voting theory, *J. Combin. Theory Ser. A 23*(1977), 34-45.

GREENE, C. and ZASLAVSKY, T.: On the interpretation of Whitney numbers through arrangements of hyperplanes, zonotopes, non-Radon partitions, and orientations of graphs, *Trans. Amer. Math. Soc. 280*(1983), 97-126.

JAMBU M. and TERAO, H.: Free arrangements of hyperplanes and supersolvable lattices, *Advances in Math. 52*(1984), 248-258.

ORLIK, P. and SOLOMON, L.: Combinatorics and topology of complements of hyperplanes, *Invent. Math. 56*(1980), 167-189.

TERAO, H.: Arrangement of hyperplanes and their freeness, I and II, *J. Faculty Sci., Univ. Tokyo, Sci. IA, 27*(1980), 293-312 and 313-320.

TERAO, H.: Generalized exponents of a free arrangements of hyperplanes and Shepard-Todd-Brieskorn formula, *Invent. Math. 63*(1981), 159-179.

ZASLAVSKY, T.: Maximal dissections of a simplex, *J. Combin. Theory Ser. A 20*(1976), 244-257.

ZASLAVSKY, T.: A combinatorial analysis of topological dissections, *Advances in Math. 25*(1977), 267-285.

ZASLAVSKY, T.: Arrangements of hyperplanes; matroids and graphs, *Proc. Tenth. S.E. Conf. on Combinatorics, Graph Theory and Computing (Boca Raton, 1979),* Vol. II, pp. 895-911, Utilitas Math., Winnipeg, Man., 1979.

ZASLAVSKY, T.: The geometry of root systems and signed graphs, *Amer. Math. Monthly 88*(1981), 88-105.

ZASLAVSKY, T.: The slimmest arrangements of hyperplanes: I. Geometric lattices and projective arrangements, *Geometriae Dedicata 14*(1983), 243-259.

ZASLAVSKY, T.: The slimmest arrangements of hyperplanes: II. Basepointed geometric lattices and Euclidean arrangements, *Mathematika 28*(1981), 169-190

Reprinted from *American Journal of Mathematics*,
volume 57, pages 800 through 804,
by permission of the Johns Hopkins University Press.

Weight Enumeration and the Geometry of Linear Codes

By Curtis Greene

The problem of computing the weight distribution of a linear code is proved to be a special case of the problem of finding the Tutte polynomial (or generalized dichromatic polynomial) of a combinatorial geometry. Many other connections between the two fields are discussed.

---

1. Introduction

The close connection between combinatorial geometries and linear codes has long been apparent to workers in both fields. Somewhat less well known is the fact that, when one's attention is suitably restricted—by looking only at linear representations of geometries—the two fields exactly coincide. Many problems in coding theory can naturally be regarded as problems in combinatorial geometry, and vice versa. In this note, we consider one such problem—computing the weight distribution of a code—as a problem in combinatorial geometry.

Our main result is that the weight enumerator of a code is a special case (or evaluation) of a more general object called the *Tutte polynomial* of a geometry. This polynomial, closely related to the dichromatic polynomial of a graph, has been extensively studied by Tutte [9, 10], Crapo [3,4], Brylawski [2], and others. Although the Tutte polynomial is, in general, just as hard to compute as the weight enumerator of a code, it is hoped that something may be gained by linking the two theories.

As an example of the parallel relationship between codes and geometries, we show how to derive the MacWilliams formula [8], which relates the weight enumerator of a code to that of its dual code, from a corresponding duality of the Tutte polynomial. The latter duality arises in a very natural way, since the Tutte polynomial can be thought of as a "mixture" of the chromatic polynomial of a graph and the chromatic polynomial of its dual (when the geometry comes from a planar graph).

This leads to an obvious connection between coloring problems and certain problems in coding theory. In fact, the fundamental problem of combinatorial geometry—called the *critical problem* by Rota—can be translated easily into a question about codes. However, it seems to be of little interest from a practical point of view, since it involves *maximizing* the *maximum* weight of a code vector over suitable extension fields. This of course has little bearing on the fundamental problem of linear coding theory, which is to maximize the minimum weight of a code vector. Perhaps this explains why the critical problem has never been the subject of direct investigation by coding theorists (except in another sense not related to the present discussion—see Dowling [6]).

2. Notation and terminology

For basic information on combinatorial geometries, we refer the reader to [5], and for more on codes, to [1] or [7].

Throughout this paper, let F_q denote a finite field of q elements. To each $k\times n$ matrix M of rank k over F_q we associate the following two objects:

(1) the row space of M, denoted by $U(M)$, and
(2) the column geometry of M, denoted by $G(C_M)$.[1]

The subspace $U(M)$ is an (n,k) linear code over F_q—that is (by definition), a subspace of $F_q^{(n)}$ of dimension k. $G(C_M)$ is a geometry of rank k with n points.

If U is any (n,k) linear code over F_q, then $U=U(M)$ for any matrix M whose row vectors form a basis of U. The matrix M is called a *generating matrix* for U. It is an easy exercise to show that $G(C_M)$ depends only on U and not on M. We define two representations of a geometry $G(S)$ in $F_q^{(k)}$ to be equivalent if they are related by a nonsingular linear transformation of $F_q^{(k)}$. The above considerations establish a one-to-one correspondence between (n,k) linear codes over F_q and equivalence classes of representations of geometries of rank k with n points.

If $U\subseteq F_q^{(n)}$ is any (n,k) code, then its *dual code* $U^\perp$ is the $(n,n-k)$ code consisting of all vectors $v\in F_q^{(n)}$ such that $u\cdot v=0$ for all $u\in U$. If $G(S)$ is any geometry of rank k on a set S of n points, then its *dual geometry* $G^*(S)$ is the geometry on S whose bases are complements of bases of $G(S)$. The rank of $G^*(S)$ is clearly $n-k$. One can easily check that, if U is generated by a matrix M and $U^\perp$ is generated by a matrix $M^\perp$, then $G^*(C_M)=G(C_{M^\perp})$. It is occasionally useful to note that $G(C_{M^\perp})$ depends only on $G(C_M)$ and not on M.

If G is a combinatorial geometry, we denote by $L(G)$ the lattice of flats—or closed subsets—of G. The *characteristic polynomial* $\chi_G(\lambda)$ of G is defined by

$$\chi_G(\lambda)=\sum_{a\in L(G)}\mu(0,a)\lambda^{r(G)-r(a)}. \tag{2.1}$$

[1]That is, $G(C_M)$ is the combinatorial geometry whose underlying set is the set of columns (or column positions) of M, with dependence taken to be ordinary linear dependence. Actually, these structures are *pregeometries* in the sense of [5] (or *matroids* in the sense of [13]), but we will avoid making a distinction in this paper. On the other hand, we distinguish sharply between the abstract *geometry* determined by a set of vectors and the *representation* provided by the vectors themselves. A geometry consists solely of a set with an abstract relation of dependence among its elements.

[Here, r denotes the rank function of G, and μ is the Mobius function of $L(G)$.] If no element of G depends on the empty set, then

$$\chi_G(\lambda) = \sum_{A \subseteq G} (-1)^{|A|} \lambda^{r(G) - r(A)}, \tag{2.2}$$

where A ranges over all subsets (closed or not) of G. The *Tutte polynomial* $T_G(x,y)$ is defined as follows:

$$T_G(x,y) = \sum_{A \subseteq G} (x-1)^{r(G)-r(A)} (y-1)^{|A|-r(A)}. \tag{2.3}$$

The Tutte polynomial has many remarkable properties, and has been extensively studied by several authors [2, 4, 9]. Among the immediate consequences of (2.3) are the following:

$$T_G(1-\lambda, 0) = (-1)^{r(G)} \chi_G(\lambda), \tag{2.4}$$

$$T_G(0, 1-\lambda) = (-1)^{r(G^*)} \chi_{G^*}(\lambda), \tag{2.5}$$

$$T_G(y,x) = T_{G^*}(x,y). \tag{2.6}$$

The latter two relations come from the fact that, if $G = G(S)$, with rank function r, and G^* has rank function r^*, then, for any subset $A \subseteq S$,

$$r^*(A) = |A| + r(S-A) - r(G).$$

We will return to the Tutte polynomial in Sec. 4.

If $U \subseteq F^{(n)}$ is an (n,k) linear code over F, and $u \in U$, then the (Hamming) *weight* of u is defined as the number of nonzero components of u, and is denoted by $w(u)$. The *weight enumerator* $A_U(z)$ of U is the polynomial

$$A_U(z) = \sum_{u \in U} z^{w(u)} = \sum_{i=0}^{n} A_i z^i,$$

where A_i is the number of vectors $u \in U$ with $w(u) = i$.

3. Computation of $A(z)$

In Sec. 1, we alluded to the fact that $A_U(z)$ depends only on the column geometry $G(C_M)$, for any choice of a generating matrix M of U. The actual computation has an especially simple form, which we derive in this section.

The main idea is the following, which is well known in connection with the "critical problem" for combinatorial geometries (see [5]).

PROPOSITION 3.1 (Crapo-Rota). *Let S be a set of points in a vector space V over F_q. Let $G = G(S)$ be the geometry associated with S, and let $L(G)$ denote the lattice of flats of G. If $A \subseteq S$, then the number of linear functionals on V which* *vanish on A but not on $S-A$ is precisely*

$$\begin{cases} \chi_{1/A}(q) & \text{if } A \text{ is a flat of } G \\ 0 & \text{otherwise} \end{cases}$$

(Here $1/A$ denotes the interval sublattice of $L(G)$ consisting of all flats which contain A.)

We omit the proof, which consists of a straightforward inclusion-exclusion argument involving either (2.1) or (2.2).

The foregoing proposition can be translated into a statement about codes as follows: if U is an (n,k) code with generating matrix M, we regard the columns of M as points in $F_q^{(k)}$, and associate linear functionals with k-dimensional row vectors. Since U consists of all n-tuples of the form αM, where α is a k-tuple, we have

PROPOSITION 3.2. *Let U be an (n,k) code over F_q, with generating matrix M. If A is any set of columns of M, then the number of vectors $u \in U$ which have zeros in precisely the components corresponding to A is*

$$\begin{cases} \chi_{1/A}(q) & \text{if } A \text{ is a flat of } G(C_M), \\ 0 & \text{otherwise.} \end{cases}$$

Proposition 3.2 leads immediately to a general formula for the weight enumerator.

PROPOSITION 3.3. *Let U be an (n,k) linear code over F_q, and let $G=G(C_M)$ be the geometry associated with any generating matrix M of U. Then*

$$A(z)=\sum_{A \in L(G)} z^{n-|A|}\chi_{1/A}(q) \tag{3.3}$$

Although (3.3) permits an immediate evaluation of $A(z)$ once the polynomials $\chi_{1/A}(z)$ are known, the computation of χ is not an easy matter in general. Nevertheless, there are many classes of geometries for which χ is well known, and some of them correspond to familiar codes. We give two examples:

Example 3.4. Dual Hamming codes. If U is a $\big((q^k-1)/q-1,k\big)$ code with a generating matrix M whose columns consist of one nonzero representative from each of the $(q^k-1)/(q-1)$ one-dimensional subspaces of $F_q^{(k)}$, then $G(C_M)$ is the full projective geometry $\mathrm{PG}_{k-1}(F_q)$. For any flat A of dimension $l<k$, it is well known that

$$\chi_{1/A}(\lambda)=(\lambda-1)(\lambda-q)\cdots(\lambda-q^{k-l-1})$$

and hence

$$\begin{aligned} A_U(z) &= 1+(q-1)\frac{q^k-1}{q-1}z^{(q^k-1)/(q-1)-(q^{k-1}-1)/(q-1)} \\ &= 1+(q^k-1)z^{q^{k-1}}. \end{aligned}$$

Example 3.5. Optimal codes. If U is an (n,k) code, let d denote the minimum weight of a nonzero codeword $u \in U$. By (3.3), $d = n - \alpha$, where α is the cardinality of the largest (proper) flat of $L(G)$. However, since G has rank k, there must exist proper flats of size at least $k-1$, so that $\alpha \geqslant k-1$ and $d \leqslant n-k+1$. A code U for which $d = n-k+1$ is called an optimal code. The corresponding geometry is a truncated Boolean algebra $B_n^{(k)}$—that is, a geometry on a set of n points whose independent sets are those sets of size $\leqslant k$. It is easy to see that

$$\chi_{B_n^{(k)}}(\lambda) = \sum_{i=0}^{k-1} (-1)^i \binom{n}{i} \lambda^{k-i} + (-1)^k \binom{n-1}{k-1}.$$

Since every interval $1/A$ is also a truncated Boolean algebra (in fact $B_{n-a}^{(k-a)}$ if $|A| = a$), the weight enumerator of U can be computed immediately:

$$A_U(z) = 1 + \sum_{a=0}^{k-1} z^{n-a} \binom{n}{a} \left\{ \sum_{i=0}^{k-a-1} (-1)^i \binom{n-a}{i} q^{k-a-i} + (-1)^{k-a} \binom{n-a-1}{k-a-1} \right\}.$$

Next, we consider what happens to the weight distribution of a code if the column geometry is fixed but the field is allowed to vary. For example, one can consider a given code over various extensions of the original field. (It is easy to see that, in this case, the underlying geometry remains the same.) The following is an immediate consequence of (3.3):

PROPOSITION 3.6. *Let G be a geometry of rank k with n points. If G has a linear representation over F_q, let $A_i(G,q)$ denote the ith coefficient of $A_U(z)$ for any (n,k) code $U \subseteq F_q^{(n)}$ whose column geometry is G.* [*By* (3.3), *$A_U(z)$ depends only on G and q.*] *Then, as a function of q, $A_i(G,q)$ is a polynomial of degree m, where*

$$m = \max\{n - r(A) \mid A \in L(G) \text{ and } |A| = n - i\}.$$

Of course, there may be many values of q for which no F_q-representation exists (and hence no code $U \subseteq F_q^{(n)}$ exists). In fact, it is a difficult and unsolved problem to determine exactly which geometries G have representations over a given field F_q.

Let U be an (n,k) code over F_q, and let $F_q\alpha$ be any finite extension of F_q. Denote by $U_{(\alpha)}$ the (n,k) code over $F_{q\alpha}$ generated by U as a subset of $F_{q\alpha}^{(n)}$. From (3.3), we have

PROPOSITION 3.7. *Let U be an (n,k) code over F_q; then, for sufficiently large values of α, $U_{(\alpha)}$ contains a vector of weight n.*

To determine exactly how large α must be is, in general, an extremely difficult problem. In fact, it is entirely equivalent to the "critical problem" for combinatorial geometries [5]:

3.8. *Critical problem for geometries.* Let V be a vector space over F_q, and let G be a geometry represented by a set of vectors $S \subseteq V$. Determine

$$\gamma(S) = \max\{\dim W \mid W \text{ is a subspace of } V \text{ and } S \cap W = \emptyset\}.$$

3.9. *Critical problem of codes*. Let $U \subseteq F_q^{(n)}$ be an (n,k) code. Determine

$$c(U) = \min\{\alpha \mid U_{(\alpha)} \text{ contains a vector of weight } n\}.$$

To make precise our assertion that these two problems are the same, we state the following:

PROPOSITION 3.10. *Let M be a known matrix of rank k, with entries in F_q. If S is the set of columns of M, and U is the row space of M, then*

$$c(U) = k - \gamma(S) = c.$$

[The common value c is called the *critical exponent* of the geometry G(S).]

Proof: We refer to [5], where it is shown that $k - \gamma(S) = \min\{\alpha \mid \chi_{G(S)}(q^\alpha) \neq 0\}$. Clearly, by (3.3), this is the same as $c(U)$.

Remark: It is not difficult to see what is going on directly. In fact, as is shown in [5], $k - \gamma(S)$ is the smallest value of c for which there exists a sequence $f_1, \ldots, f_c$ of linear functionals on V with the property that, for every $s \in S$, there exists at least one f_i with $f_i(s) \neq 0$. In passing from Proposition 3.1 to 3.2, we associated linear functionals on $F_q^{(k)}$ with k-tuples over F_q, and hence with linear combinations of codewords. It is an easy matter to extend this correspondence and associate sequences of α linear functionals on $F_q^{(k)}$ with k-tuples over $F_q\alpha$ (since $F_{q\alpha}$ has dimension α over F_q). We leave it as an exercise for the reader to construct a direct proof of (3.10) in this way.

The critical problems for codes can be restated in such a way that no reference to extension fields is made:

3.11. *If $U \subseteq F_q^{(n)}$ is an (n,k) code, determine the minimum number c such that U contains c codewords whose supports (nonvanishing coordinates) cover all the coordinate positions.*

It is easy to see that $c = c(U)$ as defined above.

Dowling [6] has shown that another version of the critical problem (for a particular class of geometries) is equivalent to the usual "fundamental problem of coding theory"—namely, the problem of maximizing the minimum weight of an (n,k) code. Dowling's problem and the one presented here are not equivalent.

One of the most important applications of the critical problem for geometries is to the problem of *graph-coloring*. In a trivial way, every coloring problem can be translated into a coding problem by considering the row space of the incidence matrix of the graph [over various extension fields of GF(2)]. For example, if Γ is a graph with n edges, then Γ has a four-coloring if and only if the row space [over GF(4)] of the incidence matrix of Γ has a vector of weight n. (This idea forms the basis of Tutte's geometrical version of the four-color problem [11]. In its present form it was first stated by Veblen [12].) This observation should give some idea of the difficulty of (3.9) and (3.11) in general.

4. The Tutte-Grothendieck decomposition of a matrix

In this section we derive another (equivalent) formula for the weight enumerator, using an entirely different approach. The idea is due originally to Tutte [9], but has been thoroughly developed and generalized recently by Brylawski [2]. For simplicity, we will outline the theory as it applies to geometries represented by matrices, although Brylawski's results hold much more generally. Ultimately, we show how the weight enumerator of a code U can be computed by evaluating the Tutte polynomial of the geometry $G(C_M)$, where M is a generating matrix of U.

DEFINITION 4.1. Let M be a $k \times n$ matrix over F_q, and let α be any nonzero column of M. A *reduction of M with respect to α* is a matrix M^* which is row-equivalent to M and has the form

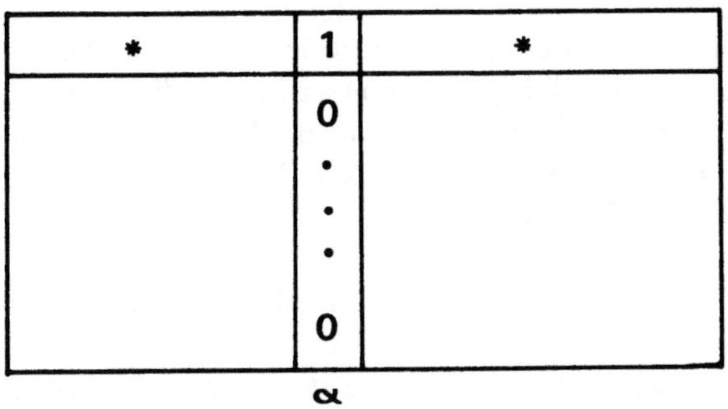

Figure 1.

We assume further that the removal of column α does not decrease the rank of the matrix. Of course, such a reduction may not always exist. For any reduction M^*, we denote by $M^*-\alpha$ the matrix obtained from M^* by removing α, and denote by M^*/α the result of removing both α and the top row of M^*.

If α happens to be the first column, the picture looks like this:

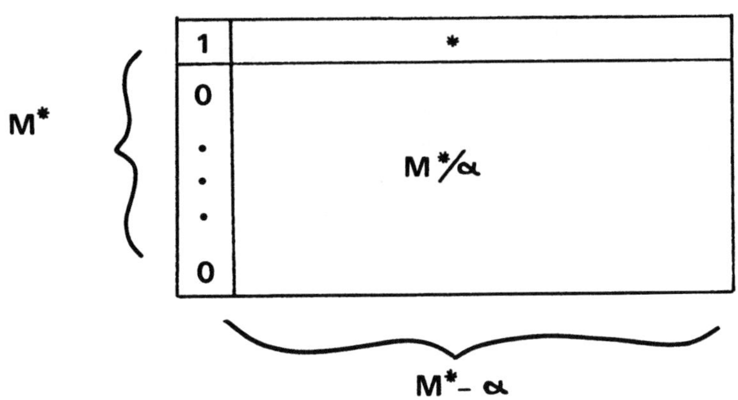

Figure 2.

It is an elementary exercise to check that the geometries $G(C_{M^*-\alpha})$ and $G(C_{M^*/\alpha})$ do not depend on the particular reduction M^*. In fact, the *codes* generated by the rows of $M^*-\alpha$ and M^*/α depend only on M and α and not on the choice of M^*.

DEFINITION 4.2. Let $\mathfrak{M}_q$ denote the class of all matrices over F_q. A

Tutte-Grothendieck invariant on $\mathfrak{M}_q$ (see [2]) is a function f defined for all matrices $M \in \mathfrak{M}_q$, taking values in a commutative ring, and satisfying

(1) $f(M)$ depends only on $G(C_M)$.
(2) For any reduction M^* of M by a column α,

$$f(M)=f(M^*-\alpha)+f(M^*/\alpha).$$

(3) For any two matrices M_1 and M_2,

$$f(M_1 \oplus M_2)=f(M_1)f(M_2).$$

The importance of Tutte-Grothendieck invariants derives from the following theorem, which gives a striking characterization of all invariants. It is essentially a translation of a result of Brylawski into the language of matrix theory.

THEOREM 4.2 (Brylawski [2]). *Let f be a Tutte-Grothendieck invariant, and let M be any matrix in* $\mathfrak{M}_q$. *Then*

$$f(M)=T_{G(C_M)}(f(M_1),f(M_0)),$$

where M_1 *and* M_0 *are* 1×1 *matrices of rank* 1 *and rank* 0, *respectively.*

We omit the proof of Theorem 4.2, but suggest to the skeptical reader the following sketch: first check that $T_{G(C_M)}(f(M_1),f(M_0))$ is an invariant which agrees with f on the matrices M_1 and M_0. Induction then shows trivially that the two invariants must agree on all matrices.

Our next objective is to extract the weight enumerator of a code from the Tutte polynomial of a generating matrix. Unfortunately, the weight enumerator itself is not an invariant, but it comes close, since the relation

$$A_{U(M_1\oplus M_2)}(z)=A_{U(M_1)}(z)A_{U(M_2)}(z)$$

holds trivially for any two matrices M_1 and M_2. However, condition (2) of definition 4.2 fails to hold. Nevertheless, we have the following

PROPOSITION 4.3. *Let M be any matrix over F, and let* α *be any nonzero column of M. If M is a reduction of M, then*

$$A_{U(M)}=(1-z)A_{U(M^*/\alpha)}+zA_{U(M^*-\alpha)}.$$

Proof: The code words in $U(M)$ fall into two classes: those which have a zero in position α, and those which do not. The former have a weight distribution identical to $U(M^*/\alpha)$, and the latter correspond to vectors in $U(M^*-\alpha)$ but not in $U(M^*/\alpha)$, with a single nonzero coordinate added. (Here we use the fact that $M^*-\alpha$ has the same rank as M^*.) Hence the relation

$$A_{U(M)}=A_{U(M^*/\alpha)}+z\left\{A_{U(M^*-\alpha)}-A_{U(M^*/\alpha)}\right\}$$

holds, and the proposition follows.

Our main result is the following:

THEOREM 4.4. *The function f, defined for any $k \times n$ matrix M over F by*

$$f(M)=\frac{1}{(1-z)^k z^{n-k}}A_{U(M)}(z),$$

is a Tutte-Grothendieck invariant. In fact,

$$f(M)=T_{G(C_M)}\left(\frac{1+(q-1)z}{1-z},\frac{1}{z}\right)$$

Proof: The fact that f is a T-G invariant follows immediately from Proposition 4.3. To verify the second assertion, it is only necessary, by Theorem 4.2, to check that $f(M_1)=[1+(q-1)z]/(1-z)$ and $f(M_0)=1/z$, both of which are trivial.

COROLLARY 4.5. *If U is an (n,k) code over F, and M is any generating matrix for U, then*

$$A_U(z)=(1-z)^k z^{n-k} T_{G(C_M)}\left(\frac{1+(q-1)z}{1-z},\frac{1}{z}\right).$$

We leave it to the reader to check that this result agrees with (3.3) [by substituting the appropriate values of x and y into (2.3)].

As an application of Corollary 4.5, we prove the following:

THEOREM 4.6 (MacWilliams [7]). *Let U be an (n,k) code over F, and let $U^{\perp}$ be its dual code. Then*

$$A_{U^{\perp}}(z)=q^{-k}[1+(q-1)z]^n A_U\left(\frac{1-z}{1+(q-1)z}\right).$$

Proof: Replacing z by $(1-z)/[1+(q-1)z]$ has the effect of interchanging $[1+(q-1)z]/(1-z)$ and $1/z$, and also $1-z$ and $qz/[1+(q-1)z]$. Hence, by (2.6) and (4.5),

$$A_U\left(\frac{1-z}{1+(q-1)z}\right)=\frac{q^k z^k}{[1+(q-1)z]^k}\frac{(1-z)^{n-k}}{[1+(q-1)z]^{n-k}}T_{G(C_M)}\left(\frac{1}{z},\frac{1+(q-1)z}{1-z}\right)$$

$$=\frac{q^k}{[1+(q-1)z]^n}(1-z)^{n-k}z^k T_{G(C_{M^{\perp}})}\left(\frac{1+(q-1)z}{1-z},\frac{1}{z}\right)$$

$$=\frac{q^k}{[1+(q-1)z]^n}A_{U^{\perp}}(z).$$

This completes the proof.

References

1. E. R. Berlekamp, *Algebraic Coding Theory*, McGraw-Hill, New York, 1968.
2. T. Brylawski, A decomposition for combinatorial geometries, to be published.
3. H. Crapo, Möbius inversion in lattices, *Arch. Math.* **19**, 38 (1968).
4. H. Crapo, The Tutte polynomial, *Acquationes Math.* **3**, 211 (1969).
5. H. Crapo and G.-C. Rota, *Combinatorial Geometries*, M.I.T. Press, Cambridge, 1971.
6. T. Dowling, Codes, packing, and the critical problem, *Atti del Convegno di Geometria Combinatoria e sua Applicazioni*, Perugia, 1971, p. 209.
7. J. H. Van Lint, *Coding Theory*, Springer, Heidelberg, 1971.
8. J. MacWilliams, A theorem on the distribution of weights in a systematic code, *Bell Syst. Tech. J.* **42**, 654 (1962).
9. W. T. Tutte, A ring in graph theory, *Proc. Camb. Phil. Soc.* **43**, 26 (1947).
10. W. T. Tutte, On dichromatic polynomials, *J. Comb. Theory* **2**, 301 (1967).
11. W. T. Tutte, On the algebraic theory of graph coloring, *J. Comb. Theory* **1**, 15 (1966).
12. O. Veblen, An application of modular equations in analysis situs, *Ann. Math.* **14**, 86 (1912).
13. H. Whitney, On the abstract properties of linear dependence, *Am. J. Math.* **57**, 509 (1935).

MASSACHUSETTS INSTITUTE OF TECHNOLOGY

(Received May 27, 1975)

Reprinted from *American Journal of Mathematics*,
volume 57, pages 800 through 804,
by permission of the Johns Hopkins University Press.

Thomas Zaslavsky

Facing up to Arrangements: Face-Count Formulas for Partitions of Space by Hyperplanes (Extract)

0. Introduction to arrangements

Take a finite set of hyperplanes (subspaces with dimension $d-1$) of a Euclidean or projective d-space. When these hyperplanes are removed, the remainder of the space falls into components, each one a d-dimensional open polyhedron (not necessarily bounded). The original set of hyperplanes, called *cuts,* together with the set of all the k-dimensional open *faces* of all these polyhedra for $-1 \leqslant k \leqslant d$, is known as an *arrangement of hyperplanes* of the space. Since the set of all faces forms a partition of the space – indeed a cell-complex decomposition – an arrangement is often called a *partition of space by hyperplanes.*[1]

The center of interest in the theory of arrangements is the combinatorial properties of the faces, particularly questions of enumeration. One can ask, for instance, how many *regions* (d-faces) there are, or how many have a given combinatorial type. Perhaps the best-known problem of this kind is to count the number of regions of an arrangement of lines. The solution is classical (it is no longer possible to name a discoverer!): it counts the regions in terms of either the number of vertices having each degree k, or the number of lines containing k vertices. In a Euclidean space parallelisms must be allowed for as well. There has been no similar solution in higher dimensions, giving the number of regions in terms, say, of the numbers of lines containing k vertices, planes containing j lines or k vertices, etc. Indeed up to now the only general solutions in higher dimensions, so far as I know, have been the subtractive formulas of Roberts [F] for E^3; and Winder's enumeration [PH] of the regions of an arrangement of hyperplanes of R^d, centered on the origin, in terms of the "degeneracy" of sets of cuts.

However a restricted problem has been solved repeatedly. An arrangement is called *simple* if its cuts are in general position. The number of regions,

[1] For background on arrangements, see especially Grünbaum's [CP] or [AH]. Arrangements of lines and planar generalizations are discussed in Grünbaum [AS]. His definition differs trivially from mine in that he requires there be no point belonging to every cut.

or k-faces, or bounded faces, has been computed by many authors[2] for Euclidean and projective spaces of various dimensions. The dimension of the space and the number of cuts do not determine the combinatorial type of an arrangement. Yet in every case, Euclidean, projective, or bounded, the result has been a formula that depends only on those two numbers. It is this fact that the present paper will explain and generalize to all arrangements.

To describe the generalization I will have to compare two kinds of equivalence of arrangements. Arrangements are *combinatorially equivalent,* or *isomorphic,* if there is a one-to-one correspondence of their faces preserving incidence. This is an isomorphism of the cell-complex decomposition of the space due to the arrangement. A weaker notion is *geometric equivalence,* which means that the linear relations (and parallelisms, in the Euclidean case) of the cuts are the same. A formal definition is most neatly stated in terms of lattice theory. Let A be an arrangement – pedantically, the cuts of an arrangement, but we will be casual about the distinction – in the space E^d or P^d. The *cut-intersection semilattice* of A is the set of subspaces

$$L(A) = \{ \cap T\colon\ T \subseteq A \text{ and } \cap T \neq \phi\} \quad \text{if Euclidean,}$$

$$L(A) = \{ \cap T\colon\ T \subseteq A\} \quad \text{if projective,}$$

ordered by *reverse inclusion*:

$$s \leqslant t \Leftrightarrow s \supseteq t \qquad \text{for } s, t \in L(A).$$

Geometric equivalence of the arrangements A and B means precisely that $L(A) \simeq L(B)$. One can show that any combinatorial equivalence actually extends to an isomorphism of cut-intersection semilattices, thus $A \simeq B$ implies $L(A) \simeq L(B)$. To see that the reverse implication is false, observe that there is more than one isomorphism type of simple arrangement of 5 or more lines in the Euclidean plane, or 6 or more in the projective plane, or of as few as 4 lines arranged with parallelisms in E^2 (cf. Grünbaum [CP, pp. 394ff.]).

Here, then, is the general law for counting faces: for each k, *the number of k-faces of an arrangement depends only on its geometric equivalence class,*

[2] Grünbaum [AH, p. 46] gives a long list. Steiner's solution [T] for the Euclidean plane and three-space (even allowing parallelisms) and Schläfli's [VK] for all Euclidean spaces are two of the earliest. Buck's paper [P] is short and complete, giving the numbers of k-faces in E^d and P^d and of bounded k-faces in E^d for all k, d, and numbers of cuts.

not its combinatorial isomorphism type. (The exact formulas for the Euclidean and projective cases – the main theorems of this extract – are presented in Section 2.) That this law specializes to just the known fact about simple arrangements is apparent, as soon as one notes that the cut-intersection lattice of a simple arrangement of *n* cuts in *d*-space is of a type (a truncated Boolean algebra) that is entirely specified by the parameters *d* and *n*.

1. First facts about arrangements

1A. The lattice and rank of an arrangement.

Let us begin by looking more closely at the semilattice of cut intersections of an arrangement A. The join and meet operations in $L(A)$ can be stated explicitly:

$$s \vee t = s \cap t \qquad \text{for } s, t \in LA,$$

defined when A is projective, or A is Euclidean and $s \cap t \neq \phi$;

$$s \wedge t = \bigcap \{h \in A\colon\ h \supseteq s \cup t\} \qquad \text{for } s, t \in LA.$$

The *rank* of $t \in LA$ is $\lambda(t) = \lambda_{LA}(t) =$ the length of the longest chain in LA below t. The *rank* of A is

$$r(A) = r(L(A)) = \max_{t \in LA} \lambda(t).$$

The "smallest" member of LA is the whole space E^d or P^d, denoted $0 = 0_{LA}$ (rank $= 0$). There may be no single "largest" member of LA; but if there is, it is $\cap A$, denoted $1 = 1_{LA}$. Then $r(A) = \lambda(1_{LA})$.

A Euclidean arrangement with 1 – that is, which has a point common to all its cuts – we will call *central.* (The subspace $\cap E$, common to all the cut intersections, may be called the *center* of E.) Sometimes it is handy to have a top element among the cut intersections; then one uses the full *cut-intersection lattice*

$$L_1(A) = \{\cap T\colon\ T \subseteq A\}$$

$$= \begin{cases} LA & \text{if } A \text{ is central or projective,} \\ LA \cup \{1_{L_1A}\} & \text{if } A \text{ is non-central.} \end{cases}$$

In L_1A, both meet and join are always defined: if there is no join in LA, there is the join $s \vee t = 1_{L_1A}$ in L_1A. The rank of L_1A equals $r(L_1A) = \lambda(1_{L_1A})$, which is $rA + 1$ if A is a non-central Euclidean arrangement, rA otherwise.

The members of LA, being subspaces, obey the modular law

$$\dim t + \dim u = \dim [t, u] + \dim t \cap u,$$

provided A is projective or $t \cap u \neq \phi$. From this one can prove that

$$\lambda(t) = d - \dim t \qquad \text{for } t \in LA \tag{1.1}$$

and that LA satisfies the *semimodular law*

$$\lambda t + \lambda u \geqslant \lambda(t \wedge u) + \lambda(t \vee u)$$

for $t, u \in LA$ such that $t \vee u$ is defined. Consequently every interval of LA is a geometric lattice. (For geometric lattices, see Birkhoff [LT3] or Crapo-Rota [CG].) This implies that $\lambda(t)$ is the length of *every* maximal chain in LA below t.

1 B. The lattice and the geometry of an arrangement.

As I have mentioned geometric lattices, it is appropriate to discuss an alternative way of looking at $L(A)$. A geometric lattice may be regarded as the lattice of closed subsets of its set of atoms. This is the approach of combinatorial incidence geometry (expounded in Crapo-Rota's treatise [CG]).

For an arrangement A, let

$$A(t) = \{h \in A\colon\ h \supseteq t\} \qquad \text{for any subspace } t.$$

Then the operation

$$T \mapsto A(\cap T) \qquad \text{for } T \subseteq A$$

defines a closure on A. The set of cuts of A with this structure is called the *geometry* of the arrangement A and is denoted $G(A)$. Its lattice of closed sets (ordered by inclusion) is denoted $LG(A)$. $LG(A)$ is in one-to-one, order-preserving correspondence with $L_1(A)$, for we have inverse mappings

$$\begin{aligned} t &\mapsto A(t) \quad && \text{for } t \in L_1(A),\\ T &\mapsto \cap T \quad && \text{for } T \in LG(A), \end{aligned} \tag{1.2}$$

such that $r(A(t)) = \lambda t$, $r(T) = \lambda(\cap T)$ if $\cap T \neq \phi$.

For the kind of geometry we will be doing in this paper, the subspace approach with $L(A)$ is usually more convenient. But there are times when it is useful to have the other. Perhaps the most interesting is for the understanding of how $L(A)$, or $L_1(A)$, expresses the linear dependencies and parallelisms of the cuts. A cut h is dependent on $h_1, \ldots, h_n$ if and only if h is in the closure of $\{h_1, \ldots, h_n\}$. For a Euclidean arrangement E, parallelism is related to the non-existence in $L(E)$ of the join of two closed subsets of cuts. Thus one might say that the import of our main theorems is that the number of faces of each dimension depends only on the linear dependences and parallelisms of the cuts; and the same for the bounded regions of a Euclidean arrangement.

1 C. The Möbius function and two latticial polynomials.

Our theorems about arrangements enumerate their faces in terms of three lattice-theoretic functions. The first is the *Möbius function*, which is defined recursively on any finite partially ordered set (e.g., $L(A)$ or $L_1(A)$) by:

$$\begin{aligned} &\mu(s, s) = 1, \\ &\mu(s, t) = -\sum_{s \leq u < t} \mu(s, u) \qquad \text{if } s < t, \qquad (1.3) \\ &\mu(s, t) = 0 \qquad \text{if } s \not\leq t. \end{aligned}$$

The sign of $\mu(s, t)$ for $s \leq t$ in a geometric lattice is $(-1)^{\lambda t - \lambda s}$ (Rota [FCT, §7, Theorem 4]), whence the same is true in any cut-intersection semilattice.

With the Möbius function and the rank, one can define the *characteristic polynomial*

$$\chi_{LA}(y) = \sum_{t \in LA} \mu_{LA}(0, t)\, y^{rA - \lambda t}.$$

Likewise any quotient semilattice

$$L(A)/s = \{t \in LA\colon\ t \geq s\}$$

also has a rank function, which is

$$\lambda_{LA/s}(t) = \lambda_{LA}(t) - \lambda_{LA}(s),$$

so a characteristic polynomial. Furthermore $L_1(A)$ has its characteristic polynomial which works out to

$$\chi_{L_1A}(y) = \begin{cases} \chi_{LA}(y) & \text{if } A \text{ is central or projective} \\ y\chi_{LA}(y) + \mu_{L_1A}(0, 1) & \text{if } A \text{ is non-central.} \end{cases}$$

These remarks apply equally to the third function, the *Möbius polynomial* of $L(A)$:

$$\begin{aligned} M_{LA}(x, y) &= \sum_{s,t \in LA} \mu_{LA}(s, t) x^{\lambda s} y^{rA - \lambda t} \\ &= \sum_{s \in LA} x^{\lambda s} \chi_{LA/s}(y), \end{aligned} \tag{1.4}$$

and $M_{L_1A}(x, y)$ similarly defined. Notice that the characteristic polynomial satisfies $\chi(y) = M(0, y)$.

Incidentally, the Möbius polynomial is a special case of a polynomial discussed by Crapo [MI, §5]. It is the coboundary polynomial of the identity map of $L(A)$.

1 D. Direct sum, induced arrangement, projectivization.

Before getting to the main theorems, I want to describe three important operations on arrangements. First of all suppose an arrangement A has partition into $B \cup C$ such that

$$r(B \cup C) = r(B) + r(C). \tag{1.5}$$

Then A is called the *direct sum* of B and C, written $B \oplus C$. This definition is valid for both Euclidean and projective arrangements. Some of its properties are proved in Lemma 4A3. Notice that since $r(B \cup C) \leq rB + rC$ for any arrangements B and C, the definition can be shortened to

$$\begin{aligned} &A = B \cup C, \\ &r(A) \geq r(B) + r(C). \end{aligned} \tag{1.5a}$$

Figure 1 illustrates direct sum.

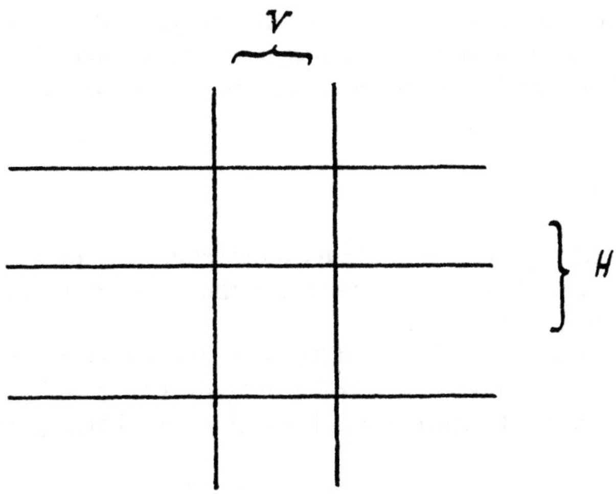

Fig. 1. An arrangement of lines. It is the direct sum $H \oplus V$ of the parallel horizontal lines H and the parallel vertical lines V.

A multiple direct sum, $A = B_1 \oplus B_2 \oplus \cdots \oplus B_n$, is defined analogously to (1.5) or (1.5a).

Secondly, let s be any subspace of the whole space E^d or P^d. The arrangement A *induces* on s an arrangement

$$A_s = \{h \cap s\colon \ h \in A \text{ and } \dim h \cap s = \dim s - 1\}.$$

When $s \in LA$ it is easy to check that

$$\begin{aligned} L(A_s) &= L(A)/s, \\ r(A_s) &= r(A) - \lambda(s), \\ \chi_{A_s}(y) &= \chi_{LA/s}(y). \end{aligned} \tag{1.6}$$

Third is the construction of the projective equivalent of a Euclidean arrangement. Let t be any subspace of E^d. The *projective completion* of t is the projective subspace spanned by t, written $t' = [t]_P$. For any set S of Euclidean subspaces, call

$$S' = \{t'\colon \ t \in S\} \tag{1.7}$$

the projective completion of S. Any E can be embedded in projective space, by embedding E^d in P^d through adjoining ∞, the ideal hyperplane, and taking the projective completion E'. But as an arrangement, E' does not really represent E, for when we pass into projective space some of the unbounded regions of E may no longer be separate. Thus we define the *projectivization* of E to be

$$E_P = E' + \infty .$$

E_P has the same regions as E, and almost the same faces, only needing adjustment for the ideal faces of E_P, which are the faces of the arrangement induced on ∞, that is $(E_P)_\infty = E'_\infty$.

This construction clarifies the difference between central and non-central Euclidean arrangements: E is central if and only if $1_{E'} \nleq \infty$. There is a simple connection with the structure of E_P. Look at a central arrangement I:

$$rI' = d - \dim \cap I' = d - (1 + \dim \cap I_P) = r(I_P) - r(\{\infty\}).$$

Therefore I_P is the direct sum of I' and $\{\infty\}$. It is easy to see that on the other hand if $E_P = E' \oplus \{\infty\}$, then $\cap E \neq \phi$.

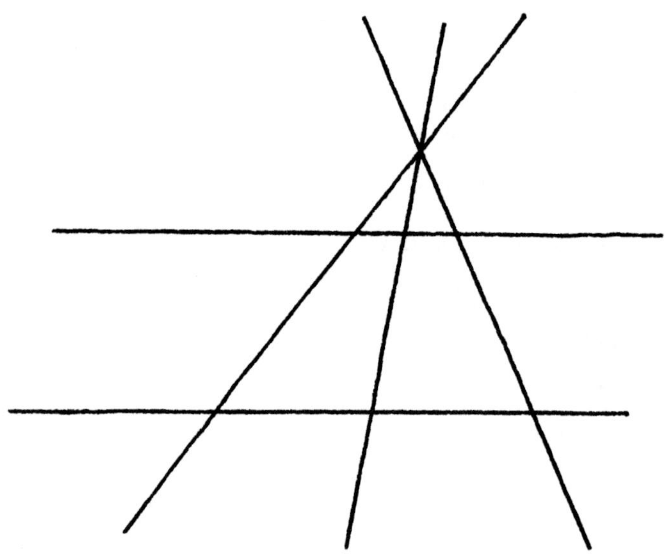

Figure 2. A Euclidean arrangement of 5 lines. It has 14 regions (4 of them bounded), 20 edges (10 bounded), and 7 vertices. By adding the line at infinity, we get a projective arrangement of 6 lines with 14 regions, 24 edges, and 11 vertices.

2. The main theorems

2A. The Euclidean case.

It is time to state the first main theorem. Let E be an arrangement of hyperplanes in E^d. Let

$$c(E) = \text{the number of regions of } E,$$

$$f_k(E) = \text{the number of } k\text{-dimensional faces of } E,$$

for $0 \leqslant k \leqslant d$. The *f-polynomial* of E, the generating polynomial of the face-counting numbers f_k, is

$$f_E(x) = f_d(E) + f_{d-1}(E)x + \cdots + f_0(E)x^d$$

$$= \sum_{k=0}^{d} f_k(E)x^{d-k}.$$

The reason for making f_{d-r} the coefficient of x^r will be apparent.

Theorem A. Let E be a Euclidean arrangement of hyperplanes. The number of its regions is

$$c(E) = \sum_{t \in LE} |\mu_{LE}(0, t)| = (-1)^{rE} \chi_{LE}(-1).$$

Its f-polynomial is

$$f_E(x) = (-1)^{rE} M_{LE}(-x, -1).$$

We will see two proofs of the formula for $c(E)$ in Sections 3AB and 4A. For now let us find an alternative way to express Theorem A; then I will show how to derive the f_k from the formula for c.

Corollary 2.1. If E is a non-central Euclidean arrangement, then

$$c(E) = |\chi_{L_1 E}(-1) - \chi_{L_1 E}(0)|$$

and

$$f_E(x) = (-1)^{rL_1E}[M_{L_1E}(-x, -1) - M_{L_1E}(-x, 0)].$$

That $\mu_{L_1E}(s, 1) = -\sum\{\mu_{LE}(s, t): t \in LE, t \geq s\}$ for $s \in LE$ when E is non-central by (1.3), implies

$$\begin{aligned} M_{L_1E}(x, y) &= y M_{LE}(x, y) - M_{LE}(x, 1) + x^{rE+1}, \\ M_{LE}(x, y) &= \frac{1}{y}[M_{L_1E}(x, y) - M_{L_1E}(x, 0)]. \end{aligned} \tag{2.1}$$

The corollary is a consequence of the first of these. ■

To derive Theorem A for $f_E(x)$, consider that each k-face of E lies in a unique k-dimensional member of LE, so by Theorem A for $c(E)$ and (1.6),

$$\begin{aligned} f_k(E) &= \sum\{c(E_t): t \in LE, \dim t = k\} \\ &= (-1)^{rE-r}\sum\{\chi_{LE/t}(-1): t \in LE, \lambda t = r\} \end{aligned}$$

for $0 \leq r = d - k \leq rE$. Inserting this formula in the f generating polynomial and comparing with (1.4), one obtains the result. ■

Thus we see that the face numbers of a Euclidean arrangement are determined by its cut-intersection semilattice, which is in turn determined by its cut-intersection lattice and its centrality. It is not sufficient to know the lattice alone. If a non-central arrangement N and a central arrangement I have the same lattice $L_1(N) \simeq L(I)$, then Theorem A gives the difference of their region numbers as

$$c(I) - c(N) = |\mu_{LI}(0, 1)|,$$

which never is zero, by Rota's [FCT, §7, Theorem 4] and the geometricity of $L(I)$.

2 B. The projective case.

The f generating polynomial of an arrangement in projective space P^d is defined to be

$$f_A(x) = \sum_{r=0}^{rA} f_{d-r}(A)x^r = \sum_{k=-1}^{d} f_k(A)x^{d-k}.$$

Theorem B. The face numbers of a projective arrangement A are given by

$$f_A(x) = \tfrac{1}{2}\,[x^{rA} + (-1)^{rA} M_{LA}(-x, -1)].$$

If $A \neq \phi$, the number of regions is

$$c(A) = \tfrac{1}{2} \sum_{t \in LA} |\mu_{LA}(0, t)| = \tfrac{1}{2}\,(-1)^{rA} \chi_{LA}(-1).$$

Thus the numbers of regions and faces of a projective arrangement depend only on the lattice of cut intersections.

Proofs of Theorem B appear in Sections 3AB and 4B, where $c(A)$ is evaluated. The expression for $f_A(x)$ can be established on the lines of the proof in Section 2A for Euclidean arrangements. ■

3. Quick proofs (Eulerian method)

There is a close reciprocal connection between the main Theorems A and B and the famous Euler relation, that for any cell complex decomposition of a topological space X (as for example an arrangement of hyperplanes of Euclidean or projective space),

$$f_0 - f_1 + f_2 - \cdots \pm f_d = \varkappa(X). \tag{3.1}$$

The number $\varkappa(X)$, the *combinatorial Euler number* of X, is independent of the decomposition; and when X is compact it is equal to the *Euler characteristic* $\chi(X)$. This theorem is proved in algebraic topology. There are also geometric proofs of special cases, like the one for convex polytopes in Grünbaum [CP].

Consider for Euclidean arrangements. If the Euler formula (3.1) for these arrangements, and the value of $\varkappa(E^d)$, have been established – either through appeal to algebraic topology or by a direct proof – then an application of Möbius inversion quickly proves Theorem A. This proof, which we will carry out in Section 3AB, generalizes the method used by Buck [P] for simple arrangements. On the other hand if Theorem A is proved by combinatorial geometry from first principles, as in Section 4A, then (3.1) for arrangements in E^d is a trivial corollary. Thus our Theorem A in effect is a new non-topological proof of Euler's relation – for arrangements of hyperplanes only.

The reason that passage from the one theorem to the other is easy is that the Tutte-Grothendieck method we use in Section 4 to prove Theorem A is similar to the usual combinatorial-geometric sort of proof of the Euler relation. What we do in part is to clarify (if that is the word) the various geometric and lattice-theoretic aspects of the method. The same remarks hold for Theorem B.

3AB. Proof of the whole-space cases.

Let X be Euclidean or projective d-space and let A denote an arrangement of hyperplanes in X. The Euler numbers are

$$\varkappa(P^d) = \tfrac{1}{2}\,(1 + (-1)^d) = \begin{cases} 0 & \text{if } d \text{ is odd,} \\ 1 & \text{if } d \text{ is even,} \end{cases}$$

$$\varkappa(E^d) = (-1)^d.$$

In order to do a Möbius inversion, we have to express the left-hand side of (3.1) as a sum over $L(A)$. Every k-face is a region of exactly one k-dimensional member of $L(A)$. Thus

$$f_k(A) = \sum\{c(A_s):\ s \in LA \text{ and } \dim s = k\}$$

and (3.1) becomes

$$\sum_{s \in LA} (-1)^{\dim s} c(A_s) = \varkappa(X). \tag{3.2}$$

One more requirement must be satisfied: (3.2) must be a function of a lattice element. But every $t \in LA$ is a Euclidean (resp., projective) space, like X; so that substituting t for X, and correspondingly A_t for A, and remembering (1.6) $L(A_t) = LA/t$, we obtain

$$\sum_{s \in LA:\ s \geqslant t} (-1)^{\dim s} c(A_s) = \varkappa(t), \qquad \text{for } t \in LA. \tag{3.3}$$

Now we are ready for Möbius inversion. The inverse of (3.3) is

$$\sum_{t \in LA:\ t \geqslant s} \varkappa(t)\mu(s, t) = (-1)^{\dim s} c(A_s), \qquad \text{for } s \in LA. \tag{3.4}$$

Setting $s = X = E^d$, (3.4) takes the form

$$\sum_{t \in LA} (-1)^{\dim t} \mu_{LA}(0, t) = (-1)^d c(A),$$

which is equivalent to the first formula of Theorem A. Setting $s = X = P^d$, we obtain

$$\tfrac{1}{2}\Big[\sum_{t \in LA} \mu_{LA}(0, t) + \sum_{t \in LA} (-1)^{\dim t} \mu_{LA}(0, t)\Big] = (-1)^d c(A).$$

Provided that A is not void, LA has a top element $1 > 0$, hence the first sum is 0. The remaining part is the formula of Theorem B. ■

4. The long proofs (Tutte-Grothendieck method)

The combinatorial proof of the main theorems is based on the concept, introduced originally by Brylawski [DCG] for combinatorial incidence geometries, of a *Tutte-Grothendieck invariant.* For arrangements this is any function defined for all Euclidean arrangements, or for all projective arrangements, with the following properties:

$$A \simeq B \Rightarrow f(A) = f(B),$$

$$f(A \oplus B) = f(A)f(B),$$

$$f(A+h) = f(A) + f(A_h) \qquad \text{when } h \text{ is not a summand of } A+h.$$

The basic result – which really is obvious – is that any two Tutte-Grothendieck invariants of arrangements that agree on empty and single-cut arrangements agree for all arrangements.

Brylawski showed that the Möbius function and the characteristic polynomial of a geometry, with correct choices of sign, are Tutte-Grothendieck invariants. (It was the latter observation that led me to discover the region-counting formulas.) As we will see, the characteristic polynomial (up to sign) is a Tutte-Grothendieck invariant of either kind of arrangements; and $c(E)$ is of Euclidean arrangements, while $2c(A)$ is of projective ones. Comparing c with the characteristic polynomial will enable us to establish Theorems A and B.

4A. Proof of the Euclidean case.

First we must find the effect on $c(E)$ of modifying E.

Lemma 4A1. If $h \notin E$, then $c(E+h) = c(E) + c(E_h)$.

For the proof, consider the regions of E.
If P is a region not met by h, it is a region of $E+h$.

If P is met by h, it is cut into three parts: two open subsets of E^d, which are regions of $E+h$, and the open subset $P \cap h$ of h, which is a region of E_h. (This is because P is a convex open set.) Contrariwise if Q is a region of E_h, then $Q = P \cap h$ for some region P of E. For if Q met no region of E, then $Q \subseteq$ some $h' \in E$, thus $h = [Q] = h' \in E$, contradicting the assumption $h \notin E$. While if Q meets P, then $Q \subseteq P$, whence $Q = P \cap h$.

Thus each region of E corresponds to either one region of $E+h$, or to two regions of $E+h$ and one region of E_h; and this correspondence exhausts the regions of $E+h$ and of E_h. That is, the lemma follows. ■

Lemma 4A2. $c(E \oplus F) = c(E)c(F)$.

For the proof, introduce coordinates into E^d. Any affine subspace s translates to a linear subspace s_0 through the origin. So let $E_0 = \{h_0 \colon h \in E\}$. If t is a top element of $L\,E$, then[3] it is parallel to, or contained in, every cut of E, so $t_0 \subseteq \cap E_0$. Actually $t_0 = \cap E_0$, since $t = \cap T \Rightarrow t_0 = \cap T_0 \supseteq \cap E_0$. Let $u_0 = \cap F_0$. Then

$$t_0 \cap u_0 = \bigcap (E \oplus F)_0.$$

From (1.5), (1.1), and the modular law for linear subspaces,

$$\begin{aligned} \dim E^d + \dim t_0 \cap u_0 &= \dim \mathrm{t}_0 + \dim u_0 \\ &= \dim(t_0 + u_0) + \dim t_0 \cap u_0. \end{aligned}$$

Hence t_0 and u_0 together linearly span E^d. Let a_0 and b_0 be linear subspaces complementary to t_0 and u_0. Choose coordinates

$$\begin{aligned} &x_1, \ldots, x_{d-rE-rF} && \text{for } t_0 \cap u_0, \\ &y_1, \ldots, y_{rE} && \text{for } a_0, \\ &z_1, \ldots, z_{rF} && \text{for } b_0. \end{aligned}$$

The equations of the cuts of $E \oplus F$ now have the simple forms

$$\begin{aligned} &\alpha_g + \sum_{j=1}^{rE} \alpha_{gj} y_j = 0 && \text{for } g \in E, \\ &\beta_h + \sum_{k=1}^{rF} \beta_{hk} z_k = 0 && \text{for } h \in F. \end{aligned}$$

[3] (Editor's note) For more details, see Lemma 2D1 on p. 27 of the complete paper.

Each region P of E is described by the inequalities

$$\varepsilon_P(g)\,[\alpha_g + \sum_j \alpha_{gj} y_j] > 0, \qquad \text{all } g \in E, \tag{4.1a}$$

for a sign function ε_P: $E \to \{\pm 1\}$ which leaves (4.1a) with the solution set P; and each region Q of F by

$$\varepsilon_Q(h)\,[\beta_h + \sum_k \beta_{hk} z_k] > 0, \qquad \text{all } h \in F, \tag{4.1b}$$

for an ε_Q: $F \to \{\pm 1\}$. From (4.1) it is clear that any region of E and any region of F meet in a region of $E \oplus F$. Conversely each region of $E \oplus F$ arises in this way. That proves the lemma. ■

Lemmas 4A1 and 4A2 show that $c(E)$ is a Tutte-Grothendieck invariant of Euclidean arrangements. Next we have to see that $(-1)^{rE}\chi_{LE}(y)$ also is.

Lemma 4A3. Let E and F be two Euclidean, or two projective, arrangements. The intersection mapping

$$i_{EF}\colon\ L(E) \times L(F) \to L(E \cup F)$$

by $i_{EF}(s, t) = s \cap t$ is well-defined and an isomorphism if and only if $E \cup F$ is the direct sum $E \oplus F$.

If it is the direct sum, then we have

$$\lambda(s \cap t) = \lambda s + \lambda t \quad \text{and} \quad [s, t] = E^d$$

for $s \in L(E)$ and $t \in L(F)$; also

$$(-1)^{r(E \oplus F)}\chi_{L(E \oplus F)}(y) = (-1)^{rE}\chi_{LE}(y) \cdot (-1)^{rF}\chi_{LF}(y).$$

The lemma for arrangements in P^d follows if it is true in Euclidean space, since an arrangement in P^d can be pulled back to one in R^{d+1} with the same rank, lattice, etc. (See Section 4B for details.) So let us assume that E and F are Euclidean.

The key to Lemma 4A3 is the first part. For

$$\chi_{LE\times LF}(y) = \sum_{s\in LE} \sum_{t\in LF} \mu(0, (s, t))\, y^{rE+rF-\lambda s-\lambda t}$$

$$= \chi_{LE}(y)\,\chi_{LF}(y)$$

by the multiplicative property of μ and the rank formula

$$\lambda(s \cap t) = \lambda(s, t) = \lambda s + \lambda t.$$

This rank formula, along with the modular law, also proves that $[s, t]$ is the whole space.

Suppose we have a direct sum $E \oplus F$. Look back to the coordinate system of Lemma 4A2. Each $s \in LE$ is a non-empty intersection of hyperplanes with only y-coordinates; each $t \in LF$ is a non-empty intersection of hyperplanes with only z-coordinates. Therefore $s \cap t$ is not empty, so i_{LE} is well-defined. Furthermore $s \cap t$ itself determines s, via the equation

$$s = \{(x, y, z)\colon \text{ for some } z\text{-coordinate } z', \ (x, y, z') \in s \cap t\}.$$

So if $s \cap t = s' \cap t'$, then s and s' must be equal; similarly $t = t'$. Thus i_{EF} is an isomorphism.

Conversely suppose that i_{EF} is an isomorphism. Then for any $s \in \operatorname{vert}(E)$[4] and $t \in \operatorname{vert}(F)$, $i_{EF}(s, t)$ exists in $L(E \cup F)$, so that $s \cap t \neq \phi$ and

$$r(E \cup F) \geqslant \lambda(s \cap t) = \lambda(s, t) = \lambda s + \lambda t = rE + rF.$$

The conclusion follows by the remark after (1.5). ■

Lemma 4A4. If $h \notin E$ and $r(E+h) = rE$, then

$$(-1)^{r(E+h)}\chi_{L(E+h)}(y) = (-1)^{rE}\chi_{LE}(y) + (-1)^{rE_h}\chi_{LE_h}(y).$$

---

[4] (Editor's note) A *relative vertex* of a Euclidean arrangement E is a minimum-dimensional non-empty intersection of its cuts. We write

$$\operatorname{vert}(E) = \text{the set of relative vertices of } E$$
$$= \{v \in LE\colon \lambda v = rE\}.$$

See p. 25 of the complete paper.

This is the most difficult of the four lemmas. I will not prove it from scratch, but will use Brylawski's theorem [DCG, §4] that the magnitude of the Möbius function of a geometry is a Tutte-Grothendieck invariant. In terms of arrangements, if $t \in LE$ and $T = E(t)$, and if $t \subseteq h \notin E$, then

$$|\mu_{L(T+h)}(0, t)| = |\mu_{LT}(0, t)| + |\mu_{L(T+h)}(h, t)|.$$

All these Möbius functions remain the same if T is replaced by E, since they only depend upon the intervals $[0, t]$, etc., of the semilattices. Now let's add up characteristic polynomials, using (1.6):

$$\begin{aligned}(-1)^{rE}\chi_{LE}(-y) + (-1)^{rE_h}\chi_{LE_h}(-y) \\ = \sum_{t \in LE:\ t \not\subseteq h} y^{rE-\lambda t}|\mu_{LE}(0, t)| \\ + \sum_{t \in LE:\ t \subseteq h} y^{rE-\lambda t}(|\mu_{LE}(0, t)| + |\mu_{L(E+h)}(h, t)|) \\ + \sum_{t \in L(E+h)-LE} y^{rE-\lambda t}|\mu_{L(E+h)}(h, t)|. \qquad (*)\end{aligned}$$

Tutte-Grothendieck invariance converts the middle summands to $y^{rE-\lambda t}|\mu_{L(E+h)}(0, t)|$. For the first summands, since $h \not\supseteq t$, $[0, t]$ is not enlarged so μ is not affected by adding h to E. To take care of the last summands, look at the interval $[0, t]_{L(E+h)}$ for $t \in L(E+h) - LE$. Writing $T = E(t)$, the arrangement $T + h$ is a direct sum $T \oplus h$. Thus $[0, t]_{L(E+h)} = L(T \oplus h) \simeq LT \times L(h) = [0, t^*]_{L(E+h)} \times [0, h] \simeq [h, t] \times [0, h]$, where $t^* = \bigcap T \neq t$, $t = t^* \cap h$. Now by the multiplicativity of μ, we have $\mu_{L(E+h)}(0, t) = \mu(h, t)\mu(0, h) = -\mu_{L(E+h)}(h, t)$. Therefore

$$\begin{aligned}(*) = \sum_{t \in L(E+h)} y^{rE-\lambda t}|\mu_{L(E+h)}(0, t)| \\ = (-1)^{r(E+h)}\chi_{L(E+h)}(-y).\end{aligned}$$

■

By the theory of Tutte-Grothendieck invariance, Theorem A is verified now if $c(E)$ and $(-1)^{rE}\chi_{LE}(-1)$ agree if $E = \phi$ or $E = \{h\}$. To check these is child's play. Just notice that $L(\phi) = \{0\}$ and $L(\{h\}) = \{0, h\}$, whence $\chi_{L(\phi)}(-1) = 1$ and $\chi_{L(\{h\})}(-1) = -2$. ■

4B. Proof of the projective case.

Theorem B is actually a variant formulation of the central case of Theorem A. One way to construct P^d is from the real linear space R^{d+1}. A k-dimensional projective subspace t (where $-1 \leqslant k \leqslant d$) corresponds to a $(k+1)$-dimensional linear subspace $\bar{t}$. An arrangement in P^d, $A = \{h_1, \ldots, h_n\}$, pulls back to a linear arrangement $\bar{A} = \{\bar{h}_1, \ldots, \bar{h}_n\}$ in R^{d+1}. $\bar{A}$ can be regarded as a $(d+1)$-dimensional Euclidean arrangement with at least one common point – the origin – so that $L(A) \simeq L(\bar{A})$. Hence

$$f_{\bar{A}}(x) = (-1)^{r\bar{A}} M_{L\bar{A}}(-x, -1) = (-1)^{rA} M_{LA}(-x, -1).$$

Provided that A is not empty, each region of A corresponds to two regions of $\bar{A}$, one the reflection in the origin of the other. The same is true of all the faces of A, save $\cap A$, which corresponds only to $\cap \bar{A}$. Thus

$$f_A(x) = \tfrac{1}{2}\,[x^{rA} + f_{\bar{A}}(x)]. \qquad \blacksquare$$

It is perfectly possible to go the other way; having somehow proved Theorem B (which is no harder than Theorem A), to derive Theorem A from it. Look at the projectivization E_P of the Euclidean arrangement E. The face numbers are related by

$$f_k(E) = f_k(E_P) - f_k(E'_\infty), \qquad \text{for } 0 \leqslant k \leqslant d.$$

In polynomial form,

$$f_E(x) = f_{E_P}(x) - x f_{E'_\infty}(x). \tag{4.2}$$

With the formula (5.3)[5] this yields a quick proof of the central case of Theorem A. Every t' in $L(E')$ (when E is central) is spanned in P^d by its finite part, so the mapping $t' \mapsto t' \cap E^d \in L(E)$ is a lattice isomorphism. Moreover $L(E') \simeq L(E_P)/\infty = L(E'_\infty)$. From (4.2) and (5.3),

---

[5] (Editor's note) Formula (5.3) is

$$f_{A \oplus h}(x) = (x+2) f_A(x) - x^{rA} = (-1)^{rA} M_{LA}(-x, -1) + x f_A(x).$$

See p. 56 of the complete paper.

$$
\begin{aligned}
f_E(x) &= f_{E'\oplus\infty}(x) - x f_{E'_\infty}(x) \\
&= (-1)^{rE'} M_{LE'}(-x, -1) + x f_{E'}(x) - x f_{E'_\infty}(x) \\
&= (-1)^{rE} M_{LE}(-x, -1).
\end{aligned}
$$

The trouble with this approach is that the non-central arrangements can't be handled half so easily. When E is non-central, E_P is not the direct sum of E' and ∞: in fact, $r(E_P) = r(E')$. As a result the use of (4.2) requires a close look at the relations among LE, LE', and LE_P. That is why I preferred to do Theorems A and B the other way round.

References

[LT3] Birkhoff, G.: *Lattice Theory,* Third Edition, Amer. Math. Soc. Colloq. Publ., Vol. 25, Amer. Math. Soc., Providence, R.I. 1967. MR 37 #2638.

[DCG] Brylawski, T.: A decomposition for combinatorial geometries, *Trans. Amer. Math. Soc. 171*(1972), 235-282. MR 46 #8869.

[P] Buck, R.C.: Partition of space, *Amer. Math. Monthly 50*(1943), 541-544. MR 5, 105.

[MI] Crapo, H.H.: Möbius inversion in lattices, *Arch. Math. (Basel) 19*(1968), 595-607. MR 39 #6791.

[CG] Crapo, H.H. and Rota, G.-C.: *On the Foundations of Combinatorial Theory: Combinatorial Geometries* (Prelim. Ed.), M.I.T. Press, Cambridge, Mass., 1970. MR 45 #74.

[CP] Grünbaum, B.: *Convex Polytopes,* Interscience, New York, 1967. MR 37 #2085. (See Chapter 18.)

[AH] ———, Arrangements of hyperplanes, *Proc. Second Louisiana Conf. on Combinatorics, Graph Theory, and Computing* (R.C. Mullin et al., eds.), Baton Rouge, 1971.

[AS] ———, *Arrangements and Spreads,* CBMS Regional Conference Series in Mathematics, No. 10, Amer. Math. Soc., Providence, R.I., 1972. MR 46 #6148.

[N] Harding, E.F.: The number of partitions of a set of N points in k dimensions induced by hyperplanes, *Proc. Edinburgh Math. Soc. (2)15*(1966/67), 285-289. MR 37 #4702.

[F] Roberts, S.: On the figures formed by the intercepts of a system of straight lines in a plane, and on analogous relations in space of three dimensions, *Proc. London Math. Soc. 19*(1888), 405-422.

[FCT] Rota, G.-C.: On the foundations of combinatorial theory, I. Theory of Möbius functions, *Z. Wahrscheinlichkeitstheorie und Verw. Gebiete 2*(1964), 340-368. MR 30 #4688.

[VK] Schläfli, L.: Theorie der vielfachen Kontinuität, reprinted in his *Gesammelte mathematische Abhandlungen,* Band I, Birkhäuser, Basel, 1950. (Written in 1850-52.)

[AHH] Smith, J.: *Arranging Hyperplanes in the Home,* Pan, London, 1926.

[T] Steiner, J.: Einige Gesetze über die Theilung der Ebene und des Raumes, *J. Reine Angew. Math. 1*(1826), 349-364.

[PH] Winder, R.O.: Partitions of N-space by hyperplanes, *SIAM J. Appl. Math. 14*(1966), 811-818. MR 34 #8281.

Chapter V
Recent advances

1. Regular matroids

In the final chapter of this anthology, we reprint two recent papers which show that the possibilities in matroid theory may not be as wide as has been thought.

The first of these papers is Seymour's "Decomposition of regular matroids." In this paper, Seymour proves that every regular matroid (i.e. matroid representable over every field) can be constructed by piecing together graphic and cographic matroids and copies of a ten-element matroid R_{10} . Here, "piecing together" means taking 1-sums (i.e. direct sums or disjoint unions), 2-sums (roughly speaking, joining two matroids at a point and then removing that point) and 3-sums (roughly speaking, joining two matroids at a line and then removing that line). Since regular matroids are essentially totally unimodular matrices, this result provides a surprising solution to the problem of characterizing totally unimodular matrices.

Seymour's theorem can also be regarded as a decomposition theorem. The easiest decomposition theorem is the following theorem characterizing series-parallel networks.

> *Theorem 1 (Duffin [65]):* Let G be a graph not containing a subdivision of K_4, the complete graph on four vertices. Then G can be constructed by taking disjoint unions, series extensions and parallel extensions of single edges.

A much more intricate decomposition theorem is due to Wagner. First, we need a definition. A graph G is said to be *a simplex sum* of the graphs G_1 and G_2 if G_1 and G_2 are edge subgraphs of G, $G = G_1 \cup G_2$, and $G_1 \cap G_2$ is a complete graph.

> *Theorem 2 (Wagner [60]):* Let G be a graph not containing K_5 as a subcontraction. Then G is a subgraph of a graph constructed by taking simplex sums of planar graphs and copies of the graph W.

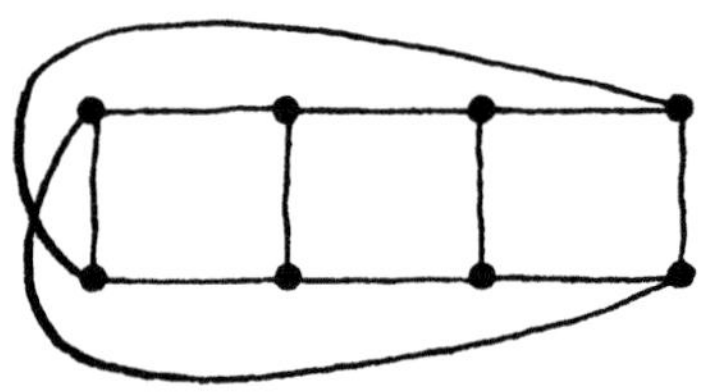

Figure 1. The Wagner graph W.

From this theorem, the equivalence of Hadwiger's conjecture for $n=5$ and the four-color conjecture can easily deduced. (Hadwiger's conjecture states that if a graph cannot be colored with $n-1$ colors, then it contains K_n as a subcontraction.) For further details and other decomposition theorems of this type, see Chap. 10 of Ore [67]. Seymour's theorem can be put in this form by observing that a matroid is regular if and only if it does not contain the four-point line, the Fano plane or its dual as minors (Tutte [II. 2]).

The proof of Seymour's theorem is quite involved. A plan of the proof is given at the end of §1. Many of the concepts and techniques – particularly the matroid version of Kelmans' theorem (§5) and the notion of splitters (§7) – are of independent interest.

Several related papers are listed below.

The matroid R_{10}: Bixby [77].

Applications: Seymour [81, 81′].

Grafts and graphs: Seymour [80].

Network flows: Seymour [77, 81].

Chromatic number of certain binary matroids: Walton and Welsh [80].

Further decomposition theories: Truemper [85, pre].

References

Bixby, R.E.: Kuratowski's and Wagner's theorems for matroids, *J. Combin. Theory Ser. B 22*(1977), 31-53.

Bixby, R.E.: A simple theorem on 3-connectivity, *Linear Algebra Appl. 45*(1982), 123-126.

Duffin, R.J.: Topology of series-parallel networks, *J. Math. Anal. Appl. 10* (1965), 303-318.

ORE, O.: *The four-color problem,* Academic Press, New York and London, 1967.
SEYMOUR, P.D.: The matroids with the max-flow min-cut property, *J. Combin. Theory Ser. B 23*(1977), 189-222.
SEYMOUR, P.D.: On Tutte's characterization of graphic matroids, *Combinatorics 79, Part I, Ann. Discrete Math. 8*(1980), 83-90.
SEYMOUR, P.D.: Some applications of matroid decomposition, *Algebraic Methods in Graph Theory, Vol. I, II (Proc. Conf., Szeged, 1978),* pp. 713-726, Colloq. Math. Soc. János Bolyai 25, North-Holland, Amsterdam, 1981.
SEYMOUR, P.D.: On Tutte's extension of the four-colour problem, *J. Combin. Theory Ser. B 31*(1981), 82-94.
SEYMOUR, P.D.: Matroids and multicommodity flows, *Europ. J. Combin. 2* (1981), 257-290.
TRUEMPER, K.: A decomposition theory for matroids. I: General results, *J. Combin. Theory Ser. B 39*(1985), 43-76.
TRUEMPER, K.: A decomposition theory for matroids. II: Minimal violation matroids, *J. Combin. Theory Ser. B 39*(1985), 282-297.
TRUEMPER, K.: A decomposition theory for matroids. III: Decomposition conditions, preprint.
WAGNER, K.: Bemerkungen zu Hadwigers Vermutung, *Math. Ann. 141* (1960), 433-451.
WALTON, P.N. and WELSH, D.J.A.: On the chromatic number of binary matroids, *Mathematika 27*(1980), 1-9.

2. Hereditary classes of matroids

Matroid theory originated as a common generalization of projective geometry and graph theory. In the last paper reprinted in this anthology, "Varieties of combinatorial geometries," by Kahn and Kung, it is shown that, in a certain sense, matroid theory is the intersection of these two subjects.

The motivation of this paper comes, in part, from the notion of a variety in universal algebra. For simplicity, we consider only varieties of groups. A collection of groups is said to be a *variety* if it is closed under subgroup, homomorphic image and direct product. For example, the collection of all abelian groups form a variety. There are two related basic results about varieties (Birkhoff [35]). The first states that free groups exist in a variety.

Theorem (Birkhoff): If $\mathcal{V}$ is a variety, then for every cardinal α there exists a group F_α in $\mathcal{V}$ such that every group in $\mathcal{V}$ which can be generated by α elements is a homomorphic image of F_α.

This theorem yields the second result, which states that varieties can be defined by equations.

Theorem (Birkhoff): For every variety $\mathcal{V}$, there exists a set of equations $E(\mathcal{V})$ such that a group G is in $\mathcal{V}$ if and only if G satisfies all the equations in $E(\mathcal{V})$.

As an example, the free group F_α in the variety of abelian groups is the direct product $\mathbb{Z}^\alpha$ of α copies of the integers under addition. The equation satisfied by this variety is (of course) $xy = yx$.

The direct analogue of a variety in matroid theory is a hereditary class. A collection $\mathcal{H}$ of matroids is said to be a *hereditary class* if it satisfies:

HC0. If G is a matroid in $\mathcal{H}$ and H is a matroid whose lattice of closed sets is isomorphic to the lattice of closed sets of G, then H is also in $\mathcal{H}$.

HC1. If G is in $\mathcal{H}$ and H is a minor of G, then H is in $\mathcal{H}$.

HC2. If G and H are in $\mathcal{H}$, then $G \oplus H$ is also in $\mathcal{H}$.

The rather convoluted condition HC0 indicates that it is more convenient to work with combinatorial geometries (or simple matroids) and we shall do so. An analogue of a free group for hereditary classes of geometries is a universal model. A geometry T_n in a hereditary class $\mathcal{H}$ is said to be a *universal model of rank n* if T_n is of rank n and every geometry in $\mathcal{H}$ of rank n is a subgeometry of T_n. (Universal models are more "cofree" objects than "free" objects. Geometrically, a universal model of rank n can be thought of as an "ambient space" in which the geometries of rank n in $\mathcal{H}$ are embedded.) Not every hereditary class has universal models – for example, the class of regular matroids or the class of series-parallel networks does not have universal models for all n. Thus, to capture the notion of a variety in universal algebra, we need to postulate the existence of universal models. In this spirit, we define a *variety* of combinatorial geometries to be a hereditary class of geometries with a universal model T_n for every non-negative integer n. The main result in the paper of Kahn and Kung is a classification of varieties of finite geometries.

This result (Theorem 14, §9) states that apart from two degenerate cases (matchstick and origami geometries), the only varieties of finite geometries are

- free geometries or Boolean algebras,
- geometries coordinatizable over a fixed finite field, and
- voltage graphic geometries with voltages in a fixed finite group.

The proof of this theorem starts with the basic observation (Lemma 1, §3) that if G is a geometry in a variety $\mathcal{T}$ and T is a subgeometry of G isomorphic to a universal model of $\mathcal{T}$, then T is a modular flat in G. The modularity of T allows points to be projected onto it using the projection argument (Lemma 3, §3). The two defining properties HC1 and HC2 of a hereditary class now play complementary rôles: HC1 allows points to be projected onto copies of a universal model T_n by contraction (if there are sufficient points) and HC2 ensures that there exist enough copies of T_n in universal models of higher rank. The proof can best be summarized by the following schematic diagram.

Is T_2 connected? —no→ Free geometries (§4)

↓ yes

Is T_3 connected? —no→ Matchstick geometries (§4)

↓ yes

Does T_n split for some n? (§5) —yes→ Origami geometries (§8)

↓ no

Is T_3 exactly a frame? —yes→ Group labelled voltage graphic geometries (§7)

↓ no

Geometries coordinatizable over $GF(q)$ (§6)

The problem of finding the right definition of varieties for infinite matroids and classifying them is still untouched. Its solution should lead to new ideas about infinite matroids.

We end with a brief survey of related work. Complete classes of matroids (i.e. hereditary classes closed under duality and matroid induction by graphs) are defined in Ingleton [77]. Sims proved in [77] that the class of matroids excluding $M(K_4)$, the cycle matroid of the complete graph on four vertices, is a complete class. Some of the techniques used in classifying varieties are used in Kahn and Kung [pre] to classify modularly complemented geometric lattices. Further work on hereditary classes can be found in Kung [pre].

References

BIRKHOFF, G.: On the structure of abstract algebras, *Proc. Cambridge Philos. Soc. 31*(1935), 433-454.

INGLETON, A.W.: Transversal matroids and related structures, *Higher Combinatorics* (M. Aigner, ed.), pp. 117-131, Reidel, Dordrecht, 1977.

KAHN, J. and KUNG, J.P.S.: A classification of modularly complemented lattices, *Europ. J. Combin., to appear.*

KUNG, J.P.S.: Numerically regular hereditary classes of combinatorial geometries, *Geometriae Dedicata, to appear.*

SIMS, J.A.: A complete class of matroids, *Quart. J. Math. Oxford Ser. (2) 28* (1977), 449-451.

Reprinted from *American Journal of Mathematics*,
volume 57, pages 800 through 804,
by permission of the Johns Hopkins University Press.

Decomposition of Regular Matroids

P. D. SEYMOUR

Merton College, Oxford, England, and University of Waterloo
Waterloo, Ontario, Canada

Communicated by the Editors

Received March 1, 1979

It is proved that every regular matroid may be constructed by piecing together graphic and cographic matroids and copies of a certain 10-element matroid.

1. INTRODUCTION

We shall assume familiarity with matroid theory; for an introduction, see Welsh [16].

Let M_1, M_2 be binary matroids with element sets S_1, S_2, respectively, where S_1, S_2 may intersect. We define a new binary matroid $M_1 \Delta M_2$ to be the matroid with element set $S_1 \Delta S_2$ and with cycles all subsets of $S_1 \Delta S_2$ of the form $C_1 \Delta C_2$, where C_i is a cycle of M_i $(i = 1, 2)$. (For sets S_1, S_2, $S_1 \Delta S_2$ denotes $(S_1 - S_2) \cup (S_2 - S_1)$. A *cycle* of a binary matroid is a subset of the elements expressible as a disjoint union of circuits. It is easy to see that if C, C' are cycles then $C \Delta C'$ is a cycle.)

We are only concerned with three special cases of this operation, as follows.

(i) When $S_1 \cap S_2 = \varnothing$ and $|S_1|, |S_2| < |S_1 \Delta S_2|$ (that is, $S_1, S_2 \neq \varnothing$) $M_1 \Delta M_2$ is a 1-*sum* of M_1 and M_2.

(ii) When $|S_1 \cap S_2| = 1$, and $S_1 \cap S_2 = \{z\}$, say, and z is not a loop or coloop of M_1 or M_2, and $|S_1|, |S_2| < |S_1 \Delta S_2|$ (that is, $|S_1|, |S_2| \geqslant 3$), $M_1 \Delta M_2$ is a 2-*sum* of M_1 and M_2.

(iii) When $|S_1 \cap S_2| = 3$, and $S_1 \cap S_2 = Z$, say, and Z is a circuit of M_1 and M_2, and Z includes no cocircuit of either M_1 or M_2, and $|S_1|$, $|S_2| < |S_1 \Delta S_2|$ (that is, $|S_1|, |S_2| \geqslant 7$), $M_1 \Delta M_2$ is a 3-*sum* of M_1 and M_2.

In each case M_1 and M_2 are called the *parts* of the sum.

It is helpful to visualize these in terms of graphs. Let G be a connected graph, and let $Y \subseteq V(G)$ be a minimal cut-set (that is, the deletion of Y from G produces a disconnected graph, but no proper subset of Y has this property). Choose non-empty subsets T_1, T_2 of $V(G)$, such that (T_1, Y, T_2) is a partition of $V(G)$, and no edge joins a vertex of T_1 to a vertex of T_2. Add a set Z of new edges, one joining every pair of distinct vertices in Y. Let G_1, G_2 be subgraphs so that $V(G_i) = Y \cup T_i$ $(i = 1, 2)$, $E(G_1) \cup E(G_2) = E(G) \cup Z$, and $E(G_1) \cap E(G_2) = Z$. Then if $|Y| = k$ and $1 \leqslant k \leqslant 3$, the polygon matroid $\mathscr{M}(G)$ of G is the k-sum of $\mathscr{M}(G_1)$ and $\mathscr{M}(G_2)$, provided that these both have fewer elements than $\mathscr{M}(G)$.

A matroid is *regular* if it is binary and has no minor isomorphic to the Fano matroid F_7 or its dual F_7^*. (M^* will always denote the dual of M.) Tutte proved that a matroid is regular if and only if it can be represented by the columns of a totally unimodular matrix [11], but we do not assume this result. The object of this paper is to show that every regular matroid may be obtained by means of 1-, 2-, and 3-sums, starting from graphic and cographic matroids and copies of a certain 10-element matroid which we call R_{10}. (A matroid is *graphic* if it is the polygon matroid of some graph.) Brylawski [4] proved the converse, that taking the 1-, 2-, or 3-sum of two regular matroids produces a regular matroid.

This theorem has several applications in combinatorics. Edmonds used it in an algorithm to recognize totally unimodular matrices (combining it with the work of Bixby and Cunningham, and Cunningham and Edmonds); combined with a theorem of Wagner it reduces Tutte's "tangential 2-block" conjecture to a conjecture about graphs; combined with Brylawski's theorem above it implies Tutte's characterization of regular matroids; combined with the Kuratowski–Wagner theorem it implies Tutte's characterization of graphic matroids (however, the proof of our result uses this theorem); the method of proof may be adapted to give a "short" proof of Tutte's characterization of graphic matroids, and a very short proof of a matroid max-flow min-cut theorem of the author; and there are some applications to "multicommodity flows" in matroids. Some of these will appear in subsequent papers [8–10].

The operations of taking 1-, 2-, and 3-sums, even in the extended sense of the next section, are not new. A 1-sum is simply a "direct sum" or "disjoint union", and 2-sums have also been investigated (for example, [2, 3]); 3-sums are rather less well-known, but are a special case of Brylawski's "modular flat" construction [4], and also occur in [5]. Our theorem too is not the first of its kind. There is a famous decomposition theorem in graph theory, due to Wagner [15], which is closely analogous to ours. Put into matroid language, it asserts that all (and only) graphic matroids without minors isomorphic to $\mathscr{M}(K_5)$ may be obtained by means of 1-, 2-, and 3-sums, starting from polygon matroids of planar graphs and copies of two special graphic matroids. (Actually, in Wagner's paper, it is only necessary to use one "special" graph, because he uses a different definition of 3-sum. But the adaptation is straightforward.) The close resemblance between Wagner's theorem and ours may help to make least the form of ours more plausible.

The proof of the theorem is in three steps.

(A) We define the (10-element, regular) matroid R_{10}, and show that every regular matroid may be obtained by 1- and 2-sums from regular matroids without R_{10} minors and copies of R_{10}. To do this we show that every regular matroid with an R_{10} minor is 2-separable, in Tutte's sense, except R_{10} itself. This step of the proof is essentially easy. (However, some of the lemmas are stated in greater generality than is needed here, because they have other applications.)

(B) We define the (12-element, regular) matroid R_{12}, and show that every regular matroid may be obtained by 1-, 2-, and 3-sums from regular matroids without R_{12} minors. To do this we partition the elements of R_{12} into two sets A, B of size 6, and show that every regular matroid with an R_{12} minor has a 3-separation (X, Y) with $A \subseteq X$, $B \subseteq Y$. This requires an analysis of the minimal matroids with an R_{12} minor, in which the connectivity between A and B is larger than it is in R_{12}.

(C) We show that every 3-connected regular matroid which is neither graphic nor cographic has an R_{10} or R_{12} minor. This is lengthy, but the matroid theory involved is simple—the difficulties are purely graph-theoretic.

Let us fix some notation and terminology. $E(M)$ is the set of elements of a matroid M, and if $E(M) = S$, M is said to be *on* S. For $Z \subseteq E(M)$, $M \times Z$ is the *restriction* of M to Z, that is, the matroid on Z with circuits those circuits of M included in Z. For convenience, $M \backslash Z$ denotes $M \times (E(M) - Z)$, and M/Z denotes $(M^* \backslash Z)^*$. $M \backslash Z$ and M/Z are the results of *deleting* and *contracting* Z, respectively. We abbreviate $M \backslash \{x\}$ by $M \backslash x$, etc. r_M is the rank function of M, and $r_M(E(M))$ is abbreviated $r(M)$. We shall also omit the subscript M when there is no risk of confusion. A *loop* is an element x such that $\{x\}$ is a circuit. Distinct x, y are *parallel* if $\{x, y\}$ is a circuit, and x and y are then both said to be *parallel elements*. The prefix "co-" dualizes a term: thus, coloop, cocircuit. Distinct x, y are *in series* if they are coparallel. A *parallel class* of M is a maximal subset $X \subseteq E(M)$ which contains no loops, so that every two distinct members of X are parallel. The parallel classes of M therefore form a partition of the non-loops of M. A *series class* is defined similarly. $X \subseteq E(M)$ is said to *span* $Y \subseteq E(M)$ if Y is a subset of the closure of X. $M \cong N$ denotes that M is isomorphic to N.

2. Sums and Separability

There is obviously a connection of some sort between being a k-sum (for $k = 1$, 2, or 3) and being k-separable in Tutte's sense. In this section we make the connection precise.

Before we do so, however, we shall extend the definition of 1- and 2-sums to non-binary matroids. Let M_1, M_2 be matroids on sets S_1, S_2, respectively. If $S_1 \cap S_2 = \emptyset$, and yet $S_1, S_2 \neq \emptyset$, we define the *1-sum* $M_1 \Delta M_2$ of M_1 and M_2 to be the matroid on $S_1 \cup S_2$ which has as circuits those sets which are circuits of M_1 or circuits of M_2. Clearly this extends the definition of a 1-sum for binary matroids.

Secondly, suppose that M_1, M_2 are matroids on S_1, S_2, and that $S_1 \cap S_2 = \{z\}$. Suppose also that z is not a loop or coloop of M_1 or M_2, and $|S_1|, |S_2| \geqslant 3$. Then the *2-sum* $M_1 \Delta M_2$ of M_1, M_2 is the matroid on $S_1 \Delta S_2$ with circuits the following sets:

(i) all sets $X_1 \subseteq S_1 - \{z\}$ which are circuits of M_1,
(ii) all sets $X_2 \subseteq S_2 - \{z\}$ which are circuits of M_2,
(iii) all sets $X_1 \cup X_2$, where $X_i \subseteq S_i - \{z\}$ and $X_i \cup \{z\}$ is a circuit of M_i $(i = 1, 2)$.

It is easy to see that this extends the definition of a 2-sum for binary matroids.

However, there appears to be no simple way to extend the definition of a 3-sum to non-binary matroids. There would, in any case, be no use for such a definition in this paper. (Brylawski's method [4] seems the least complicated.)

If (X_1, X_2) is a partition of S, we say that (X_1, X_2) is a *k-separation* of M if $|X_1|, |X_2| \geqslant k$ and

$$r(X_1) + r(X_2) \leqslant r(M) + k - 1.$$

It is easy to see that no matroid has a k-separation for $k < 1$. A k-separation is *exact* if equality holds in the displayed inequality. (Thus every 1-separation is exact.) M is *k-separable* if it has a k-separation. M is *k-connected* if it has no k'-separation for any $k' < k$.

The connection between 1-sums and 1-separation is particularly simple, as follows.

(2.1) *If (X_1, X_2) is a 1-separation of M then M is the 1-sum of $M \times X_1$ and $M \times X_2$; and conversely, if M is the 1-sum of M_1 and M_2 then $(E(M_1), E(M_2))$ is a 1-separation of M, and M_1, M_2 are isomorphic to (indeed, are) minors of M.*

The proof is trivial. A similar result holds for 2-sums, but the proof takes several steps.

(2.2) *Suppose that M is the 2-sum of M_1 and M_2. Then $(E(M_1) - E(M_2), E(M_2) - E(M_1))$ is an exact 2-separation of M.*

Proof. Put $E(M_1) \cap E(M_2) = \{z\}$, and $X_i = E(M_i) - \{z\}$ $(i = 1, 2)$. z is not a coloop of M_1, and so there is a set $B_1 \subseteq X_1$ independent in M_1, with $|B_1| = r(M_1) = r_M(X_1)$. Also, z is not a loop of M_2, and so there is a set $B_2 \subseteq X_2 \cup \{z\}$ with $z \in B_2$, independent in M_2, and with $|B_2| = r_M(X_2)$. From the definition of the circuits of $M_1 \Delta M_2$, we see that $(B_1 \cup B_2) - \{z\}$ includes no circuits of M, and so

$$r(M) \geqslant r_M(X_1) + r_M(X_2) - 1.$$

Moreover, $|X_1|, |X_2| \geqslant 2$, since $|E(M_i)| \geqslant 3$ $(i = 1, 2)$ and so (X_1, X_2) is a 2-separation. It is not a 1-separation of M, because (since z is neither a loop nor coloop of M_1 or M_2) M has a circuit intersecting both X_1 and X_2. Thus it is an exact 2-separation, as required.

(2.3) *Let (X_1, X_2) be a 2-separation of M. If C_1, C_2 are circuits of M and both intersect both X_1 and X_2, then $C_1 \cap X_1 \not\subset C_2 \cap X_1$.*

Proof. Suppose that $C_1 \cap X_1 \subset C_2 \cap X_1$. Choose $x_1 \in C_1 \cap X_1$, $x_2 \in (C_2 - C_1) \cap X_1$. Now $C_2 \cap X_1$ is independent in M, and so there is an independent set $B_1 \subseteq X_1$ with $C_2 \cap X_1 \subseteq B_1$, and with $|B_1| = r(X_1)$. Choose an independent set $B_2 \subseteq X_2$ with $|B_2| = r(X_2)$. Then B_2 spans X_2 in M, and so $(B_1 \cup B_2) - \{x_1, x_2\}$ spans x_1, x_2 and hence $E(M)$. Thus $r(M) \leqslant |B_1| + |B_2| - 2 = r(X_1) + r(X_2) - 2$, contrary to the hypothesis that (X_1, X_2) is a 2-separation.

(2.4) *Let (X_1, X_2) be a 2-separation of M, and let $Y_i \subseteq X_i$ $(i = 1, 2)$ be non-empty. Suppose that for $i = 1, 2$, there is a circuit C_i with $C_i \cap X_1 \neq \emptyset$, $C_i \cap X_2 \neq \emptyset$, $C_i \cap X_i = Y_i$. Then $Y_1 \cup Y_2$ is a circuit of M.*

Proof. Choose C_1, C_2 with these properties so that $C_1 \cup C_2$ is minimal. We prove that $C_1 = C_2$. First, suppose that $(C_1 \cup C_2) \cap X_1$ is not independent, and let C be a circuit with $C \subseteq (C_1 \cup C_2) \cap X_1$. Now $C \not\subseteq C_1$, and so there exists $x \in (C_2 - C_1) \cap C$. Choose $y \in C_2 \cap X_2$. There is a circuit $C_2' \subseteq (C_2 \cup C) - \{x\}$ with $y \in C_2'$. Then $C_2' \cap X_2 \subseteq C_2 \cap X_2$ and so $C_2' \cap X_1 \neq \emptyset$; but $y \in C_2'$ and so $C_2' \cap X_2 \neq \emptyset$. By (2.3), $C_2' \cap X_2 = C_2 \cap X_2 = Y_2$, and yet $C_1 \cup C_2' \subseteq (C_1 \cup C_2) - \{x\}$. This contradicts the minimality of $C_1 \cup C_2$. Thus $(C_1 \cup C_2) \cap X_1$ is independent, and similarly so is $(C_1 \cup C_2) \cap X_2$. For $i = 1, 2$, choose $B_i \subseteq X_i$, independent, with $(C_1 \cup C_2) \cap X_i \subseteq B_i$ and with $|B_i| = r(X_i)$. Now if $C_1 \cap X_2 = Y_2$ the result is true. We assume therefore that $C_1 \cap X_2 \neq Y_2$, and so by (2.3), $C_1 \cap X_2 \not\subseteq Y_2$. Choose $x_2 \in (C_1 \cap X_2) - Y_2$, and, similarly, choose $x_1 \in (C_2 \cap X_1) - Y_1$. Then $x_i \in B_i$ $(i = 1, 2)$, and $B_1 \cup B_2 - \{x_1, x_2\}$ spans x_1, x_2 and hence $E(M)$. Thus

$$r(M) \leqslant |B_1| + |B_2| - 2 = r(X_1) + r(X_2) - 2,$$

contrary to the hypothesis that (X_1, X_2) is a 2-separation.

(2.5) *Let* (X_1, X_2) *be an exact* 2*-separation of* M. *Then there are matroids* M_1, M_2 *on* $X_1 \cup \{z\}$, $X_2 \cup \{z\}$, *respectively* (*where* z *is a new element*), *so that* M *is the* 2*-sum of* M_1 *and* M_2.

Proof. Define the matroid M_1 on $X_1 \cup \{z\}$ to have as circuits those circuits of M which do not intersect X_2, together with all sets $Y \cup \{z\}$, where $\emptyset \neq Y \subseteq X_1$ is such that there is a circuit C of M with $C \cap X_1 = Y$, $C \cap X_2 \neq \emptyset$. The result follows easily from (2.4).

We summarize the results in the following.

(2.6) *If* (X_1, X_2) *is an exact* 2*-separation of* M *then there are matroids* M_1, M_2 *on* $X_1 \cup \{z\}$, $X_2 \cup \{z\}$, *respectively* (*where* z *is a new element*), *such that* M *is the* 2*-sum of* M_1 *and* M_2. *Conversely, if* M *is the* 2*-sum of* M_1 *and* M_2 *then* $(E(M_1) - E(M_2), E(M_2) - E(M_1))$ *is an exact* 2*-separation of* M, *and* M_1, M_2 *are isomorphic to minors of* M.

Proof. Because of (2.2) and (2.5), it only remains to show that if M is the 2-sum of M_1 and M_2 then M_1, M_2 are isomorphic to minors of M. Suppose then that $E(M_1) \cap E(M_2) = \{z\}$, and put $X_i = E(M_i) - \{z\}$ $(i = 1, 2)$. Then (X_1, X_2) is an exact 2-separation of M by (2.2) and so there is a circuit C with $C \cap X_1 \neq \emptyset$, $C \cap X_2 \neq \emptyset$. Choose $y \in C \cap X_2$, and put $M_1' = M \times (X_1 \cup C)/((C \cap X_2) - \{y\})$. Then M' is isomorphic to M_1, by (2.4). Similarly M_2 is isomorphic to a minor of M, as required.

For $k = 3$ the situation is a little more complicated. First, we are confined to binary matroids. Even there, being expressible as a 3-sum is not the same as having an exact 3-separation; indeed, we shall see that it is equivalent to having an exact 3-separation (X_1, X_2) such that $|X_1|, |X_2| \geq 4$. This is a technicality, but there is a second and more awkward difference—the parts of a 3-sum are not always isomorphic to minors of the whole. The most we can say is that the parts are isomorphic to minors of the whole, provided that the whole is 3-connected, and even this is non-trivial to prove; we defer the proof to Section 4.

We begin with a lemma.

(2.7) *Suppose that* M_1, M_2 *are binary matroids on* S_1, S_2, *respectively, and* M *is the* 3*-sum of* M_1 *and* M_2. *If* $Y_1 \cup Y_2$ *is a circuit of* M, *where* $\emptyset \neq Y_i \subseteq S_i - (S_1 \cap S_2)$ $(i = 1, 2)$, *then there exists* $z \in S_1 \cap S_2$ *such that for* $i = 1$, 2, $Y_i \cup \{z\}$ *is a circuit of* M_i.

Proof. Put $Z = S_1 \cap S_2$. Choose cycles C_1, C_2 of M_1, M_2, respectively, so that $Y_1 \cup Y_2 = C_1 \Delta C_2$. Then $C_1 \cap Z = C_2 \cap Z$, and moreover Z is a circuit of both M_1 and M_2; and so by replacing C_1, C_2 by $C_1 \Delta Z$, $C_2 \Delta Z$ if necessary, we may assume that $|C_1 \cap Z| \leq 1$. Now $C_1 \not\subseteq Y_1$, because $Y_2 \neq \emptyset$ and $Y_1 \cup Y_2$ is a circuit of M. Thus $C_1 \cap Z \neq \emptyset$. Put $C_1 \cap Z = \{z\}$;

then $C_2 \cap Z = \{z\}$. Now every circuit of M_i included in C_i contains z, because $Y_1 \cup Y_2$ is a circuit of M; yet C_i is a cycle of M_i, and so $C_i = Y_i \cup \{z\}$ is a circuit of M_i ($i = 1, 2$), as required.

When $A_1, ..., A_k$ are subsets of S and $Z \subseteq S$ is expressible as $\Delta(A_i : i \in I)$ for some $I \subseteq \{1, ..., k\}$, we say that $A_1, ..., A_k$ *surround* Z. When M is a binary matroid on S and $Z \subseteq S$, $\rho(Z)$ denotes the maximum number of circuits of M included in Z, with the property that no one of them is surrounded by the others. $\rho^*(Z)$ denotes the same quantity for M^*. (It is easy to see that $\rho(Z)$ is also the maximum number of cycles of M included in Z with the property that no one is surrounded by the others.)

It is well known that

(2.8) *If M is binary and $Z \subseteq S$, $r_M(Z) + \rho(Z) = |Z|$.*

(2.9) *If (X_1, X_2) is an exact 3-separation of a binary matroid M, with $|X_1|, |X_2| \geqslant 4$, then there are binary matroids M_1, M_2 on $X_1 \cup Z$, $X_2 \cup Z$, respectively (where Z contains three new elements), such that M is the 3-sum of M_1 and M_2. Conversely if M is the 3-sum of M_1 and M_2 then $(E(M_1) - E(M_2), E(M_2) - E(M_1))$ is an exact 3-separation of M, and $|E(M_1) - E(M_2)|, |E(M_2) - E(M_1)| \geqslant 4$.*

Proof. Suppose that (X_1, X_2) is an exact 3-separation of the binary matroid M, and $|X_1|, |X_2| \geqslant 4$. Let $\mathscr{C}_i$ be the set of circuits of M included in X_i ($i = 1, 2$). Now

$$\rho(X_1) + \rho(X_2) = \rho(X_1 \cup X_2) - 2,$$

and so there are two circuits C_1, C_2 of M so that $\mathscr{C}_1, \mathscr{C}_2$ and C_1, C_2 together surround all cycles of M. Take three new elements z_1, z_2, z_3 and put $Z = \{z_1, z_2, z_3\}$. For $i = 1$, 2, let M_i be the binary matroid on $X_i \cup Z$ with cycles the sets surrounded by $\mathscr{C}_i$ together with $(C_1 \cap X_i) \cup \{z_1\}$, $(C_2 \cap X_i) \cup \{z_2\}$ and Z. We claim that M is the 3-sum of M_1 and M_2. For certainly $M = M_1 \Delta M_2$, and for $i = 1$, 2, $(C_1 \cap X_i) \cup \{z_1\}$, $(C_2 \cap X_i) \cup \{z_2\}$ and Z are cycles of M_i, and so Z includes no cocircuits of M_i. It remains to check that Z is a circuit of M_i, and that M_i has fewer elements than M ($i = 1, 2$). The second assertion is true because $|X_1|, |X_2| \geqslant 4$ and $|Z| = 3$. If the first is false, then one of $\{z_1\}$, $\{z_2\}$, $\{z_1, z_2\}$ is a cycle of M_1 or M_2 (M_1, say), and so one of $C_1 \cap X_1$, $C_2 \cap X_1$, $(C_1 \Delta C_2) \cap X_1$ is surrounded by $\mathscr{C}_1$. But if Y is a cycle of M and $Y \cap X_1$ is surrounded by $\mathscr{C}_1$, then $Y \cap X_2$ is surrounded by $\mathscr{C}_2$, because $Y \cap X_2 \subseteq X_2$ and $Y \cap X_2 = Y \Delta (Y \cap X_1)$; and so Y is surrounded by $\mathscr{C}_1 \cup \mathscr{C}_2$. Thus one of C_1, C_2, $C_1 \Delta C_2$ is surrounded by $\mathscr{C}_1 \cup \mathscr{C}_2$, contrary to

$$\rho(X_1) + \rho(X_2) = \rho(X_1 \cup X_2) - 2.$$

Conversely, suppose that M is the 3-sum of M_1 and M_2. Put $E(M_1) \cap E(M_2) = Z = \{z_1, z_2, z_3\}$, and $E(M_i) - Z = X_i$ $(i = 1, 2)$. Certainly $|X_1|, |X_2| \geq 4$, because $|E(M_i)| < |E(M)|$ $(i = 1, 2)$. We must prove that

$$\rho(X_1) + \rho(X_2) = \rho(X_1 \cup X_2) - 2.$$

Let $\mathscr{C}_i$ be the set of circuits of M included in X_i $(i = 1, 2)$. Now Z includes no cocircuits of M_i, and so for each $z_j \in Z$ there is a circuit C_j^i of M_i with $C_j^i \cap Z = \{z_j\}$. Put $C_j = C_j^1 \Delta C_j^2$ so that C_j is a cycle of M $(j = 1, 2, 3)$. Now by (2.7), if C is any circuit of M such that $C \cap X_1, C \cap X_2 \neq \varnothing$, then $(C \cap X_i) \cup \{z_j\}$ is a circuit of M_i $(i = 1, 2)$, for some j; and then $C \Delta C_j$ is surrounded by $\mathscr{C}_1 \cup \mathscr{C}_2$. Thus $\mathscr{C}_1 \cup \mathscr{C}_2$ together with C_1, C_2 and C_3 surround all circuits of M. Moreover, $(C_1 \Delta C_2 \Delta C_3) \cap X_1 = C_2^1 \Delta C_2^1 \Delta C_3^1 \Delta Z$ and so it is surrounded by $\mathscr{C}_1$; and similarly $(C_1 \Delta C_2 \Delta C_3) \cap X_2$ is surrounded by $\mathscr{C}_2$. Thus $C_1 \Delta C_2 \Delta C_3$ is surrounded by $\mathscr{C}_1 \cup \mathscr{C}_2$. It follows that

$$\rho(X_1) + \rho(X_2) \geq \rho(X_1 \cup X_2) - 2.$$

Suppose that strict inequality holds here; then $\mathscr{C}_1 \cup \mathscr{C}_2$ surrounds one of C_1, C_2, $C_1 \Delta C_2$. Thus $\mathscr{C}_1$ surround one of $C_1 \cap X_1$, $C_2 \cap X_1$, $(C_1 \Delta C_2) \cap X_1$, and so one of these sets is a cycle of M_1. But $(C_1 \cap X_1) \cup \{z_1\}$, $(C_2 \cap X_1) \cup \{z_2\}$, $((C_1 \Delta C_2) \cap X_1) \cup \{z_1, z_2\}$ are cycles of M_1, and so one of $\{z_1\}$, $\{z_2\}$, $\{z_1, z_2\}$ is a cycle of M_1. This is contrary to the hypothesis that M is the 3-sum of M_1 and M_2, and completes the proof.

(2.10) *Let M be a matroid. Then*
(a) *the following are equivalent:*
 (i) *M is 1-separable.*
 (ii) *M is expressible as a 1-sum.*
(b) *The following are equivalent:*
 (i) *M is 1- or 2-separable.*
 (ii) *M is expressible as a 1- or 2-sum.*
(c) *If M is binary, the following are equivalent:*
 (i) *M is 1- or 2-separable, or has a 3-separation (X_1, X_2) with $|X_1|$, $|X_2| \geq 4$.*
 (ii) *M is expressible as a 1-, 2-, or 3-sum.*

Proof. That (ii) implies (i) follows from the previous results. To see that (i) implies (ii), observe that if M has a k-separation then it has an exact k'-separation for some $k' \leq k$.

3. Arcs and Fundamentals

If M is a matroid on S and $Z \subseteq S$, a *Z-arc* is a minimal non-empty subset $A \subseteq S - Z$ such that there is a circuit C with $C - Z = A$ and $C \cap Z \neq \varnothing$. A *Z-fundamental* (*for* A) is a circuit C such that $C - Z = A$, where A is a Z-arc.

We observe that all Z-arcs are non-empty and independent in M; and that no Z-arc is a proper subset of another.

(3.1) *If C is a circuit with $C \cap Z \neq \varnothing$, then $C - Z$ is expressible as a union of Z-arcs.*

Proof. We proceed by induction on $|C - Z|$. If $C \subseteq Z$ the result is trivial. We assume that $C - Z \neq \varnothing$. But $C \cap Z \neq \varnothing$, and so $C - Z$ includes a Z-arc A. Choose a Z-fundamental F for A. Choose $x \in A$. Now for each $y \in C - (A \cup Z)$ there is a circuit $C_y \subseteq (C \cup F) - \{x\}$ with $y \in C_y$, and $C_y \cap Z \neq \varnothing$, because $C_y \not\subseteq C - \{x\}$. By induction $C_y - Z$ is a union of Z-arcs; but

$$C - Z = A \cup \bigcup (C_y - Z \colon y \in C - (A \cup Z))$$

and the result follows.

(3.2) *If M is a matroid on S and $x, y \in S$ are distinct but are contained in the same circuits of M, then either x, y are in series or they are both coloops.*

The proof is elementary.

(3.3) *If A is a Z-arc and $x, y \in A$ are distinct, then x, y are in series in $M \times (A \cup Z)$.*

Proof. Because A is a Z-arc, there is no circuit of $M \times (Z \cup A)$ containing just one of x, y. The result follows from (3.2). (x, y cannot be coloops because they are contained in a circuit.)

When A is a Z-arc and $P \subseteq Z$, we say $A \to P$ if there is a Z-fundamental for A included in $A \cup P$. For its negation we write $A \not\to P$.

(3.4) *Suppose that A is a Z-arc, $P \subseteq Z$, $A \not\to P$, and $x_0 \in A$. If C is a circuit with $x_0 \in C$ and $C \cap Z \subseteq P$, then there is a Z-arc A' such that $A' \subseteq (A \cup C) - \{x_0\}$ and such that $A' \not\to P$.*

Proof. If possible, choose a counterexample in which $C - (A \cup Z)$ is minimal. Certainly $C \not\subseteq A \cup Z$, because A is a Z-arc and $A \not\to P$. Choose $y \in C - (A \cup Z)$. Let F be a Z-fundamental for A, and choose a circuit C_1

with $y \in C_1$ and $C_1 \subseteq (C \cup F) - \{x_0\}$. If there is a Z-arc $A' \subseteq C_1 - Z$ with $y \in A'$ then $A' \to P$, for otherwise the theorem is true; and if there is no such Z-arc, then by (3.1) $C_1 \cap Z = \emptyset$. Thus in either case there is a circuit $C_2 \subseteq C \cup A \cup P$ with $x_0 \notin C_2$ and $y \in C_2$. Choose a circuit $C_3 \subseteq (C \cup C_2) - \{y\}$ with $x_0 \in C_3$. Then $C_3 \cap Z \subseteq P$, and $C_3 - (A \cup Z) \subseteq (C - (A \cup Z)) - \{y\}$. This contradicts the minimality of $C - (A \cup Z)$, as required.

(3.5) *If $P \subseteq Z$ and A_1, A_2 are Z-arcs with $A_1 \cap A_2 \neq \emptyset$ and $A_2 \not\to P$, then for every $x \in A_1$ there is a Z-arc A with $x \in A \subseteq A_1 \cup A_2$ such that $A \not\to P$.*

Proof. Keeping A_1 fixed, choose the Z-arc A_2 with $A_1 \cap A_2 \neq \emptyset$ and $A_2 \not\to P$, such that $A_1 \cup A_2$ is minimal. By setting $A = A_2$ and $x_0 \in A_2 - A_1$, we see from (3.4) that every circuit included in $A_1 \cup A_2 \cup P$ is included in $A_1 \cup P$.

Now if $x \in A_1 \cap A_2$ the result is true, and we assume that $x \in A_1 - A_2$. Choose $y \in A_1 \cap A_2$. Let F_1, F_2 be Z-fundamentals for A_1, A_2, respectively. Choose a circuit C with $x \in C$ and $C \subseteq (F_1 \cup F_2) - \{y\}$. Then by (3.1) we can choose a circuit C_1 with $x \in C_1 - Z \subseteq C - Z$ so that either C_1 is a Z-fundamental or $C_1 \cap Z = \emptyset$. But $y \notin C_1$ and so $C_1 - Z \not\subseteq A_1$, because A_1 is a Z-arc; thus $(C_1 - Z) \cap (A_2 - A_1) \neq \emptyset$. From the result of the previous paragraph, $C_1 - Z$ is not a circuit, and so is a Z-arc and C_1 is a Z-fundamental; and moreover, $C_1 - Z \not\to P$, for the same reason. The result follows.

(3.6) *If A_1, A_2 are Z-arcs with $A_1 \cap A_2 \neq \emptyset$, and $P_1, P_2 \subseteq Z$ are such that $A_1 \not\to P_1$, $A_2 \not\to P_2$, then there is a Z-arc $A \subseteq A_1 \cup A_2$ such that $A \not\to P_1$, P_2.*

Proof. If possible, choose a counterexample with $A_1 \cup A_2$ minimal. Then certainly we have $A_1 \to P_2$, $A_2 \to P_1$. Let F_2 be a Z-fundamental for A_2 with $F_2 \subseteq A_2 \cup P_1$. Choose $x_0 \in A_1 \cap A_2$. Then by (3.4), taking $C = F_2$, we see that there is a Z-arc A_1' with $A_1' \subseteq (A_1 \cup A_2) - \{x_0\}$ such that $A_1' \not\to P_2$. Similarly, there is a Z-arc A_2' with $A_2' \subseteq (A_1 \cup A_2) - \{x_0\}$ such that $A_2' \not\to P_1$. Now $A_1' \not\subseteq A_1$; choose $y \in A_1' \cap (A_2 - A_1)$. In particular, $A_1' \cap A_2 \neq \emptyset$, and similarly $A_2' \cap A_1 \neq \emptyset$. By minimality of $A_1 \cup A_2$ we have $A_1 \cup A_2' = A_1 \cup A_2$, and hence $y \in A_2$. Thus $A_1' \cap A_2' \neq \emptyset$; but

$$A_1' \cup A_2' \subseteq (A_1 \cup A_2) - \{x_0\},$$

contrary to the minimality of $A_1 \cup A_2$.

(3.7) *Let M be a connected matroid on S, and let $Z \subseteq S$ be non-empty. Let (X_1, X_2) be a partition of S, so that no Z-arc intersects both X_1 and X_2,* *and such that for* $i = 1, 2$, *if* $A \subseteq X_i$ *is a Z-arc then* X_i *includes a Z-fundamental for A. Then*

$$r(X_1) + r(X_2) - r(S) = r(X_1 \cap Z) + r(X_2 \cap Z) - r(Z).$$

Proof. We proceed by induction on $|S - Z|$. If this is zero the result is trivial, and we assume not. Choose $X \subseteq S$ with $Z \subseteq X$ and $X \neq S$, maximal such that Z intersects every component of $M \times X$. Then $X \neq \varnothing$ and so there is a circuit C intersecting both X and $S - X$, and so by maximality of X, $S - X \subseteq C$. Thus $S - X$ is independent. Moreover, if $x, y \in S - X$ are distinct, then a circuit of M containing x must intersect X and so must contain y; and so by (3.2), x and y are in series in M. There is a circuit intersecting $S - X$ and Z, and so by (3.1) there is a Z-arc A including $S - X$. Now $A \subseteq X_1$ or X_2 (X_1, say), and so $S - X \subseteq X_1$. Let F be a Z-fundamental for A with $F \subseteq X_1$. Then $S - X \subseteq F$.

By induction, applied to the components of $M \times X$,

$$r(X_1 \cap X) + r(X_2 \cap X) - r(X) = r(X_1 \cap Z) + r(X_2 \cap Z) - r(Z).$$

But $r(X_1) = r(X_1 \cap X) + |S - X| - 1$ since the elements of $S - X$ are in series in M and $F \subseteq X_1$, and

$$r(S) = r(X) + |S - X| - 1$$

for the same reason. The result follows.

(3.8) *Let M be a matroid on S, let* $Z \subseteq S$, *and let* (P_1, P_2) *be a partition of Z. Then either there is a Z-arc A such that* $A \nrightarrow P_1$, $A \nrightarrow P_2$, *or there is a partition* (X_1, X_2) *of S such that* $X_i \cap Z = P_i$ $(i = 1, 2)$ *and*

$$r(X_1) + r(X_2) - r(S) = r(P_1) + r(P_2) - r(Z).$$

Proof. If $Z = \varnothing$ we may take $X_1 = S$, $X_2 = \varnothing$. Assume then that $Z \neq \varnothing$. If M is not connected we may treat its components separately, for if the theorem is true for each component it is true for M. Assume then that M is connected. Define

$$X_1 = P_1 \cup \bigcup (A \subseteq S - X\colon A \text{ is a } Z\text{-arc and } A \nrightarrow P_2).$$

Put $X_2 = S - X_1$. We claim that no Z-arc intersects both X_1 and X_2. For if A is a Z-arc and $A \cap X_1 \neq \varnothing$, then by (3.5), $A \subseteq X_1$. Moreover, if A is a Z-arc and $A \subseteq X_2$ then $A \rightarrow P_2$ by definition of X_1. There are now two possibilities:

(i) there is a Z-arc $A_1 \subseteq X_1$ such that $A_1 \nrightarrow P_1$. Pick a Z-arc A_2 with $A_1 \cap A_2 \neq \varnothing$ and $A_2 \nrightarrow P_2$; and then from (3.6) there is a Z-arc A such that $A \nrightarrow P_1$, $A \nrightarrow P_2$.

(ii) there is no such Z-arc $A_1 \subseteq X_1$. Then $A_1 \to P_1$ for every Z-arc A_1 included in X_1, and so the hypotheses of (3.7) are satisfied. The result follows.

4. Parts of a 3-Sum

Our first application of the results of the previous section is to prove the following theorem, postponed from Section 2.

(4.1) *If M is binary and is the 3-sum of M_1 and M_2, and M is 3-connected, then M_1, M_2 are isomorphic to minors of M.*

The main step in its proof is the following lemma.

(4.2) *Let Z be an independent cocircuit of a binary matroid M, with $|Z| = 3$. Then either there is a partition (R_1, R_2, R_3) of $E(M) - Z$ such that for each i,*

$$r(R_i) + r(E(M) - R_i) \leqslant r(M) + 1$$

or M has a minor M' such that:

(i) $|E(M')| = 6$,
(ii) *$Z \subseteq E(M')$ and Z is a cocircuit of M',*
(iii) *each $z \in Z$ is in a 2-element cocircuit of M'.*

Proof. We may assume that M is connected, for if the theorem is true for the component of M containing Z then it is true for M. Let $Z = \{z_1, z_2, z_3\}$ and put $P_i = Z - \{z_i\}$ $(i = 1, 2, 3)$. For every circuit C of M, $C \cap Z = \varnothing$ or $= P_i$ for some i, because Z is a cocircuit and M is binary. In particular, for every Z-arc A, $A \to P_i$ for some i. In fact $A \to P_i$ for a unique i, for if $A \to P_1$, $A \to P_2$, say, then $A \cup P_1$, $A \cup P_2$ are circuits, and so $P_1 \mathbin{\Delta} P_2 = P_3$ is a cycle of M (since M is binary), which is impossible since Z is independent. Define $v(A) = i$, where $A \to P_i$.

Suppose first that there are two Z-arcs A_1, A_2 with $v(A_1) \neq v(A_2)$ and $A_1 \cap A_2 \neq \varnothing$. Choose such a pair A_1, A_2 with $A_1 \cup A_2$ minimal. Let $A_1 \to P_1$, $A_2 \to P_2$, say. Put

$$\mathscr{C} = \{\varnothing, A_1 \cup P_1, A_2 \cup P_2, (A_1 \mathbin{\Delta} A_2) \cup P_3\}.$$

Then every member of $\mathscr{C}$ is a cycle of M. We claim that these are the only cycles included in $Z \cup A_1 \cup A_2$. For suppose that C_1 is another. Then for some $C' \in \mathscr{C}$, $C_1 \mathbin{\Delta} C' \subseteq A_1 \cup A_2$; and $C_1 \mathbin{\Delta} C' \notin \mathscr{C}$, because $C_1 \mathbin{\Delta} C' \neq \varnothing$ (since $C_1 \neq C'$). $C_1 \mathbin{\Delta} C'$ includes a circuit C, say; and then $C \subseteq A_1 \cup A_2$, and so $C \notin \mathscr{C}$. Now $C \not\subseteq A_2$, and so $C \cap (A_1 - A_2) \neq \varnothing$; choose $x_0 \in C \cap (A_1 - A_2)$. By (3.4) there is a Z-arc $A \subseteq (A_1 \cup A_2) - \{x_0\}$ such that $A \nrightarrow P_2$ (since $A_1 \nrightarrow P_2$). But $A \cap A_2 \neq \varnothing$, since $A \not\subseteq A_1 - A_2$, and yet

$$A \cup A_2 \subseteq (A_1 \cup A_2) - \{x_0\}.$$

This contradicts the minimality of $A_1 \cup A_2$. Thus the only cycles of M included in $A_1 \cup A_2 \cup Z$ are the members of $\mathscr{C}$. Choose $x_1 \in A_1 - A_2$, $x_2 \in A_2 - A_1$, $x_3 \in A_1 \cap A_2$, and put $X = \{x_1, x_2, x_3\}$. Put

$$M' = M \times (A_1 \cup A_2 \cup Z)/((A_1 \cup A_2) - X).$$

Then M' satisfies the theorem.

We may therefore assume that for any two Z-arcs A_1, A_2, if $v(A_1) \neq v(A_2)$ then $A_1 \cap A_2 = \varnothing$. Put

$$R_i = \bigcup (A \subseteq E(M) - Z\colon A \text{ is a } Z\text{-arc and } A \rightarrow P_i) \qquad (i = 1, 2, 3).$$

By (3.1), every element in $E(M) - Z$ is in a Z-arc, since M is connected, and so (R_1, R_2, R_3) is a partition of $E(M) - Z$.

Take a new element e and let M_1 be the binary matroid on $E(M) \cup \{e\}$ such that $M_1 \backslash e = M$ and such that $\{e, z_1, z_2\}$ is a circuit of M_1. Put $Z_1 = Z \cup \{e\}$, and put $X_1 = R_1 \cup \{e\}$, $X_2 = R_2 \cup R_3 \cup Z$. Then (X_1, X_2) is a partition of $E(M_1)$. Now let A be a Z_1-arc of M_1. Let F be a Z_1-fundamental for A (of M). Then either F or $F \Delta \{e, z_1, z_2\}$ is a circuit of M, and so A is a Z-arc of M. Hence $A \subseteq R_i$ for some i, and so A does not intersect both X_1 and X_2. If $A \subseteq R_2 \cup R_3$ then $A \cup P_2$ or $A \cup P_3$ is a Z_1-fundamental for A (of M_1), and if $A \subseteq R_1$ then $A \cup \{e\}$ is a Z_1-fundamental.

Thus by (3.7),

$$\begin{aligned} r_{M_1}(X_1) + r_{M_1}(X_2) - r(M_1) &= r_{M_1}(\{e\}) + r_{M_1}(Z) - r_{M_1}(Z_1) \\ &= 1 + 3 - 3 \\ &= 1. \end{aligned}$$

But $r_{M_1}(X_1) \geqslant r_M(R_1)$, $r_{M_1}(X_2) = r_M(E(M) - R_1)$, and $r(M_1) = r(M)$; thus $r(R_1) + r(E(M) - R_1) \leqslant r(M) + 1$.

Inequalities for R_2, R_3 are proved similarly. This completes the proof.

(4.3) *Suppose that M is the 3-sum of binary matroids M_1 and M_2, and that M is 3-connected. If (Y_1, Y_2) is a 2-separation of M_1 then for some i, $Y_i = \{x, z\}$, where $x \in E(M_1) - E(M_2)$, $z \in E(M_1) \cap E(M_2)$, and x, z are parallel in M_1.*

Proof. Put $Z = E(M_1) \cap E(M_2)$ and $X_i = E(M_i) - Z$ $(i = 1, 2)$, as usual. Now $Z \subseteq Y_1 \cup Y_2$ and $|Z| = 3$; we assume that $|Y_1 \cap Z| \geqslant 2$. If $Z \subseteq Y_1$ it is easy to see that $((Y_1 - Z) \cup X_2, Y_2)$ is a 2-separation of M, which is impossible. We deduce that $Y_1 \cap Z = Z - \{z\}$, say, and that $z \in Y_2$; and furthermore that $(Y_1 \cup \{z\}, Y_2 - \{z\})$ is not a 2-separation of M_1, for the same reason. But $r_{M_1}(Y_1 \cup \{z\}) = r_{M_1}(Y_1)$ since Z is a circuit of M_1, and so $|Y_2 - \{z\}| \leqslant 1$. Equality holds, since $|Y_2| \geqslant 2$; put $Y_2 - \{z\} = \{x\}$. Now Y_2 includes a circuit or cocircuit of M_1, because if not then $r_{M_1}(Y_2) = |Y_2| = 2$, and $r_{M_1}(Y_1) = r_{M_1}(Y_1 \cup Y_2)$, contrary to $r_{M_1}(Y_1) + r_{M_1}(Y_2) \leqslant r(M_1) + 1$. Thus $\{x, z\}$ includes a circuit or cocircuit of M_1. Now Z is a circuit of M_1 and so z is not a loop or coloop of M_1. And x is not a loop or coloop of M_1 because M is 3-connected and so x is not a loop or coloop of M. Thus $\{x, z\}$ is a circuit or cocircuit of M_1. It is not a cocircuit because $|\{x, z\} \cap Z| = 1$ and Z is a circuit. Hence x, z are parallel in M_1, as required.

(4.4) *If M_1 and M_2 are binary matroids on S_1, S_2, respectively, and Z_1, Z_2 are disjoint subsets of $S_1 - S_2$, then*

$$(M_1 \Delta M_2) \backslash Z_1 / Z_2 = (M_1 \backslash Z_1 / Z_2) \Delta M_2.$$

Proof. C is a cycle of $M \backslash Z_1 / Z_2$ if and only if $C \cup X$ is a cycle of M for some $X \subseteq Z_2$. The result follows.

Proof of (4.1). Suppose then that M is the 3-sum of M_1 and M_2, and M is 3-connected. Put $E(M_i) = S_i$ $(i = 1, 2)$, and $S_1 \cap S_2 = Z$. Then Z is an independent cocircuit of M_1^*, and so we may apply (4.2) to it. Suppose that (R_1, R_2, R_3) is a partition of $E(M_1^*) - Z$ such that for each i,

$$r_{M_1^*}(R_i) + r_{M_1^*}(S_1 - R_i) \leqslant r(M_1^*) + 1.$$

Now $|E(M_1^*)| \geqslant 7$ and so $|R_i| \geqslant 2$ for some i; $|R_1| \geqslant 2$, say. Then $(R_1, S_1 - R_1)$ is a 2-separation of M_1^* and hence of M_1, but it does not satisfy (4.3), a contradiction.

We deduce from (4.2) that M_1 has a six-element minor M_1', in which Z is a circuit and each $z \in Z$ is parallel to some other element of M_1'. Choose F_1, F_2 so that $M_1' \backslash F_1 / F_2$. Then

$$\begin{aligned} M \backslash F_1 / F_2 &= (M_1 \Delta M_2) \backslash F_1 / F_2 \\ &= (M_1 \backslash F_1 / F_2) \Delta M_2 \qquad \text{by (4.4)} \\ &= M_1' \Delta M_2 \\ &\cong M_2, \end{aligned}$$

and so M has a minor isomorphic to M_2, and similarly for M_1, as required.

The reader may be puzzled by the fact that although M is regular if and only if M^* is, the decomposition of M provided by our main theorem is not invariant under duality; for if M is the 3-sum of M_1 and M_2 then M^* need not (indeed, cannot) be the 3-sum of M_1^* and M_2^*. It can be shown that if $M=M_1 \Delta M_2$ then $M^*=M_1^* \Delta M_2^*$, and one may wonder why we do not define a 3-sum to be $M_1 \Delta M_2$ when $|S_1 \cap S_2|=2$, for then it would be invariant under duality. Indeed, if M is the 3-sum of M_1 and M_2, and $z \in E(M_1) \cap E(M_2)$, then $M=(M_1\backslash z) \Delta (M_2\backslash z)$, and the third common element of the matroids M_1, M_2 would appear to be irrelevant. The reason is that although our theorem could still be true (a fortiori) with this broader sense of 3-sum, the converse would become false; under this new definition one could produce the Fano matroid by a 3-sum, starting with, say, the polygon matroids of K_4 and the graph obtained from K_3 by replacing two edges by parallel pairs of edges.

There is an alternative definition of 3-sum which is invariant under duality and which would make our theorem and its converse both true. When M is a binary matroid, we say that distinct $e_1, e_2 \in E(M)$ are *adjacent* in M if M has no minor M' isomorphic to $\mathcal{M}(K_4)$ in which e_1, e_2 both appear and in which they correspond to non-adjacent edges of K_4. Clearly e_1, e_2 are adjacent in M if and only if they are adjacent in M^*. Then when M_1, M_2 have element sets S_1, S_2 and $S_1 \cap S_2=\{e_1, e_2\}$, we could say that $M_1 \Delta M_2$ is a 3-sum of M_1 and M_2 if e_1, e_2 are adjacent in both M_1 and M_2. This definition has all the desirable qualities we mentioned, and some more; however, it is rather complicated, and we prefer the version given. In passing, it can be shown that if M is regular and $e_1, e_2 \in E(M)$ are distinct, then e_1, e_2 are adjacent if and only if the matroid M' is regular, where M' has one extra element e_3, $M'\backslash e_3=M$, and $\{e_1, e_2, e_3\}$ is a cycle of M'.

5. Kelmans' Theorem

Kelmans [7] recently proved a result concerning the structure of 3-connected graphs. We need a matroid generalization of it, which we shall prove in this section.

We say that M is a *subdivision* of N if $E(M)$ can be partitioned into non-empty sets $P_1,..., P_k$, where $k=|E(N)|$, so that every pair of elements in the same P_i are in series, and so that if $y_i \in P_i$ $(1 \leqslant i \leqslant k)$ then

$$M/(E(M)-\{y_1,..., y_k\}) \cong N.$$

We say that M is *cyclically* 3-*connected* if it is a subdivision of a 3-connected matroid. It is an easy exercise to prove the following.

(5.1) *If M is connected, the following are equivalent*:

(i) *M is cyclically* 3-*connected*,

(ii) *for every partition* (X_1, X_2) *of* $E(M)$ *with* $r(X_1) + r(X_2) \leqslant r(M) + 1$, *one of* X_1, X_2 *is independent,*

(iii) *for every such partition* (X_1, X_2) *either all the members of* X_1 *or all the members of* X_2 *are pairwise in series.*

When $Z \subseteq E(M)$ and $A \subseteq E(M) - Z$ is a Z-arc such that for every series class P of $M \times Z$, $A \nrightarrow P$, we say A is an *adjoinable* Z-arc. The reason for the name is the following theorem.

(5.2) *Let* $Z \subseteq E(M)$ *be such that* $M \times Z$ *is cyclically* 3-*connected, and let* A *be an adjoinable* Z-*arc. Then* $M \times (Z \cup A)$ *is cyclically* 3-*connected.*

Proof. Let (X_1, X_2) be a partition of $Z \cup A$ such that

$$r(X_1) + r(X_2) \leqslant r(Z \cup A) + 1.$$

Suppose that both X_1 and X_2 include circuits of $M \times (Z \cup A)$. Now it is easy to see that

$$r(X_1 \cap Z) + r(X_2 \cap Z) - r(Z) \leqslant r(X_1) + r(X_2) - r(X_1 \cup X_2)$$

(this also follows from (8.1)), and so

$$r(X_1 \cap Z) + r(X_2 \cap Z) \leqslant r(Z) + 1.$$

Since $M \times Z$ is cyclically 3-connected, the elements in one of $X_1 \cap Z$, $X_2 \cap Z$ ($X_1 \cap Z$, say) are in series in $M \times Z$ by (5.1). Thus $X_1 \cap Z \subseteq P$ for some series class P of $M \times Z$. Let C be a circuit of M included in X_1. Then $C \subseteq A \cup P$. C is not a Z-fundamental, because $A \nrightarrow P$, and so $C \subseteq P$. But P is a series class of $M \times Z$, and $M \times Z$ is connected, and so $P = C = Z$. This contradicts $A \nrightarrow P$.

Thus for every such partition (X_1, X_2), one of X_1, X_2 is independent, and so by (5.1), M is cyclically 3-connected, as required.

It follows from (5.2) that if $Z_1 \subseteq Z_2 \subseteq \cdots \subseteq Z_r = E(M)$, and $M \times Z_1$ is cyclically 3-connected, and for each i $(1 \leqslant i \leqslant r-1)$ $Z_{i+1} - Z_i$ is an adjoinable Z_i-arc, then M is cyclically 3-connected. We would like a converse to this. The most appealing one would be that $\varnothing \neq Z \subseteq E(M)$ and M, $M \times Z$ are both cyclically 3-connected then we can find such a sequence $Z_1, \ldots, Z_r$ with $Z = Z_1$ and $Z_r = E(M)$. But this unfortunately is not true. For example, take $M = \mathcal{M}(K_5)$, and let Z be the set of edges in a K_4 subgraph; then there is no adjoinable Z-arc. For graphs, there are at least two ways to avoid this difficulty, both at the cost of some extra complication. Tutte [12] proved that a sequence $Z_1, \ldots, Z_r$ with prescribed Z_1, Z_r could always be found with the property that each $Z_{i+1} - Z_i$ is either an adjoinable Z_i-arc or a slightly less simple structure. This theorem of Tutte can be generalized to all cyclically 3-connected matroids (indeed, an earlier version of this paper proved and used such a generalization) but it becomes cumbersome. The following generalization of Kelmans' theorem is both easier to prove and for our purposes easier to apply.

(5.3) *Let $\emptyset \neq Z \subseteq E(M)$, and let M and $M \times Z$ both be cyclically 3-connected. Suppose that $M \times Z$ has at least two series classes. Then for some $r \geqslant 1$ there is a sequence $Z_1 \subseteq \cdots \subseteq Z_r = E(M)$ such that for each i $(1 \leqslant i \leqslant r-1)$, $Z_{i+1} - Z_i$ is an adjoinable Z_i-arc, and such that $M \times Z$ and $M \times Z_1$ are both subdivisions of the same 3-connected matroid.*

To prove this we use the following.

(5.4) *Let M be 3-connected, and let $Z \subset E(M)$ be such that $M \times Z$ is connected and $M \times Z$ has at least two series classes. Choose $Z_1 \subseteq E(M)$ with $|Z_1|$ maximum such that $M \times Z_1$ is a subdivision of $M \times Z$. Then there is an adjoinable Z_1-arc.*

Proof. Let $M \times Z$ have k series classes. Then $k \geqslant 2$. But no connected matroid has exactly two series classes, and so $k \geqslant 3$. M has at least $k+1$ series classes, since $Z \neq E(M)$. But $M \times Z_1$ is a subdivision of $M \times Z$, and so has k series classes; thus $Z_1 \neq E(M)$. Let the series classes of $M \times Z_1$ be $P_1, \ldots, P_k$. Let A_1 be any Z_1-arc (this must exist since M is connected and $Z_1 \neq E(M)$). We may assume that A_1 is not adjoinable, and so $A_1 \to P_1$, say. Suppose for a contradiction that $|P_1| = 1$, and $P_1 = \{p\}$, say. Then $A_1 \cup \{p\}$ is a circuit of M. Choose $a \in A_1$; and then a, p are parallel in $(M \times (Z_1 \cup A))/(A - \{a\}) = M_1$, say, and

$$\begin{aligned} M_1\backslash a &= (M \times (Z_1 \cup A))\backslash\{a\}/(A - \{a\}) \\ &= (M \times (Z_1 \cup A))\backslash\{a\}\backslash(A - \{a\}) \end{aligned}$$

since all elements in $A - \{a\}$ are coloops of $(M \times (Z_1 \cup A))\backslash\{a\}$. Thus

$$M_1\backslash a = M \times Z_1,$$

and so $M_1\backslash p \cong M \times Z_1$. Hence $M \times ((Z_1 - \{p\}) \cup A)$ is a subdivision of $M \times Z_1$. From the maximality of $|Z_1|$ we have $|A| = 1$ and $A = \{a\}$; but then $\{a, p\}$ is a circuit of M, which is impossible since M is 3-connected and $|E(M)| \geqslant k+1 \geqslant 4$.

It follows that $|P_1| \geqslant 2$. Now P_1 is a series class of $M \times Z_1$, and so

$$r(P_1) + r(Z_1 - P_1) \leqslant r(Z_1) + 1.$$

However, M is 3-connected, and so there is no partition (X_1, X_2) of $E(M)$ such that $X_1 \cap Z_1 = P_1$ and such that

$$r(X_1) + r(X_2) \leqslant r(M) + 1,$$

for if so then $|X_1| \geqslant |P_1| \geqslant 2$ and $|X_2| \geqslant k - 1 \geqslant 2$ and so (X_1, X_2) is a 2-separation.

From (3.8) there is a Z_1-arc A such that $A \nrightarrow P_1$, $A \nrightarrow (Z_1 - P_1)$. Then A is adjoinable, as required.

Proof of (5.3). By contracting all elements except one from every series class of M, we see that it suffices to prove (5.3) when M is 3-connected. The proof in that case is by induction on $|E(M) - Z|$. The result is trivial if this number is zero; assume then that $Z \neq E(M)$. By (5.4) there exists $Z' \subseteq E(M)$ such that $M \times Z'$ is a subdivision of $M \times Z$, and such that there is an adjoinable Z'-arc A, say. By (5.2), $M \times (Z' \cup A)$ is cyclically 3-connected, and $|Z' \cup A| > |Z|$, and so by induction there is a sequence $Z_2 \subseteq \cdots \subseteq Z_r = E(M)$ such that for $2 \leqslant i \leqslant r - 1$ each $Z_{i+1} - Z_i$ is an adjoinable Z_i-arc, and such that $M \times Z_2$ and $M \times (Z' \cup A)$ are subdivisions of the same 3-connected matroid N, say. The series classes of $M \times Z_2$ and of $M \times (Z' \cup A)$ are in 1–1 correspondence with the elements of N and so with each other in an obvious way. Let A' be the series class of $M \times Z_2$ corresponding to A. Put $Z_1 = Z_2 - A'$; and then A' is an adjoinable Z_1-arc, and $M \times Z_1$, $M \times Z$ are both subdivisions of the same 3-connected matroid, and so the sequence $Z_1 \subseteq \cdots \subseteq Z_r$ satisfies the theorem.

Actually, in our applications (5.4) is more convenient to use than (5.3)—we have proved (5.3) merely to show the connection with Kelmans' theorem, which is (5.3) restricted to graphic matroids.

6. A "Wheels and Whirls" Theorem

For $n \geqslant 2$, the *wheel* W_n is the binary matroid on $\{1,..., 2n\}$ which has cycles all sets surrounded by $\{2n-1, 2n, 1\}$ and the sets $\{i, i+1, i+2\}$ ($1 \leqslant i \leqslant 2n-3$, i odd). It is the polygon matroid of the graph consisting of a circuit with n vertices and one extra vertex joined (by one edge) to every vertex of the circuit.

We observe that $\{2, 4, 6,..., 2n\}$ is a circuit of W_n. The *whirl* $\mathscr{W}_n$ is the matroid on the same set, in which the same sets are independent, except that $\{2, 4, 6,..., 2n\}$ is independent in $\mathscr{W}_n$.

These matroids may be familiar to the reader, because Tutte [13] proved that if M is a matroid which is not 2-separable and M is not isomorphic to a wheel or whirl, then for some $x \in E(M)$, one of $M \backslash x$, M/x is not 2-separable. However, there is no connection between that theorem and the result we are concerned with here, as far as I can see.

First we prove the following lemma.

(6.1) *Let M be a connected matroid on S, and let $z_1,\ldots,z_{2n} \in S$ be distinct, where $n \geqslant 2$. Let $Z_i = \{z_i, z_{i+1}, z_{i+2}\}$ $(1 \leqslant i \leqslant 2n)$, reading subscripts modulo $2n$. Suppose that for i odd, Z_i is a circuit, and for i even, Z_i is a cocircuit. Then $S = \{z_1,\ldots,z_{2n}\}$ and M is isomorphic to W_n or $\mathscr{W}_n$.*

Proof. Put $P = \{z_i : i \text{ even}\}$ and $Q = \{z_i : i \text{ odd}\}$. For i odd, Z_i is a circuit and $Z_i - Q = \{z_{i+1}\}$; thus Q spans P, and $r(P \cup Q) = r(Q) \leqslant n$. Similarly, if r^* denotes r_{M^*}, $r^*(P \cup Q) \leqslant n$. But for any subset X of a matroid, $r(X) + r^*(X) \geqslant |X|$, with equality only if X is a separator; yet M is assumed to be connected. It follows that $S = P \cup Q$, and that $r(M) = n$.

Let $\mathscr{C}$ contain all sets $X \subseteq S$ such that $|X \cap Q| = 2$ $(X \cap Q = \{z_{j_1}, z_{j_2}\}$, say, where $j_2 > j_1)$ and such that $X \cap P$ is either $P \cap \{z_i : j_1 < i < j_2\}$ or $P - \{z_i : j_1 < i < j_2\}$. It is easy to see that if some nonempty $X \subseteq S$ is such that $|X \cap Z_i| \neq 1$ for $1 \leqslant i \leqslant 2n$, i even, then X includes either P or a member of $\mathscr{C}$; and so in particular, every circuit of M includes either P or a member of $\mathscr{C}$.

On the other hand, every member of $\mathscr{C}$ is a circuit. To show this, it suffices (by symmetry) to show that

$$\{z_1, z_{2r+1}\} \cup \{z_2, z_4,\ldots, r_{2r}\}$$

is a circuit $(r = 1,\ldots, n-1)$, and we prove this by induction on r. It is true by hypothesis if $r = 1$. If $r > 1$, then by induction,

$$\{z_1, z_{2r-1}\} \cup \{z_2, z_4,\ldots, z_{2r-2}\}$$

is a circuit, and so is $\{z_{2r-1}, z_{2r}, z_{2r+1}\}$; and so

$$\{z_1, z_{2r+1}\} \cup \{z_2, z_4,\ldots, z_{2r}\}$$

includes a circuit. But this circuit must include P or a member of $\mathscr{C}$, and so the inclusion is an equality, as required. Thus all members of $\mathscr{C}$ are circuits of M, and every other circuit includes P. If P is a circuit then $M \cong W_n$. If not, then P spans S, because $r(M) = n$, and every set $P \cup \{z\}$ $(z \in Q)$ is a circuit. Then $M \cong \mathscr{W}_n$, as required.

The main result of this section is the following.

(6.2) *Let M be a matroid on S, and let $x, y \in S$ be distinct, and let N be another matroid. Suppose that $M \backslash x / y \cong N$, and that N is connected, and has no loops, coloops, or series or parallel elements. Then one of the following is true:*

(i) *x is a loop, coloop, or parallel element of M;*
(ii) *y is a loop, coloop, or series element of M;*

(iii) *for some* x', $y' \in S$, $M\backslash x'/y' \cong N$ *but* x' *is not a loop, coloop, or parallel element of* M/y';

(iv) *for some* x', $y' \in S$, $M\backslash x'/y' \cong N$ *but* y' *is not a loop, coloop, or series element of* $M\backslash x'$;

(v) *there exists* $z \in S - \{x, y\}$ *so that* $\{x, y, z\}$ *is both a circuit and a cocircuit of* M;

(vi) *for some* $n \geqslant 2$, $N \cong W_n$ *or* $\mathcal{W}_n$ *and* $M \cong W_{n+1}$ *or* $\mathcal{W}_{n+1}$, *respectively.*

Proof. We assume that (i)–(iv) are all false. It follows that M has no loops, coloops, or series or parallel elements. For suppose that C is a circuit of M with $|C| \leqslant 2$. C does not include a circuit of $M\backslash x/y$ because N has no circuits of size $\leqslant 2$, and so $x \in C$. Thus x is a loop or parallel element of M and (i) is true, contrary to our assumption. Similarly, M has no coloops or series elements. It follows that M is connected (because $M\backslash x/y$ is connected, and $\{x, y\}$ includes no circuits or cocircuits).

We define inductively a sequence $x_1, y_1, x_2, y_2, \ldots$ of elements of M with the following properties:

(i) for all $i \geqslant 1$, $M\backslash x_i/y_i \cong N$ and $M/y_i\backslash x_{i+1} \cong N$,

(ii) for $i \geqslant 1$, $\{x_i, y_i, x_{i+1}\}$ is a circuit, and $\{y_i, x_{i+1}, y_{i+1}\}$ is a cocircuit.

To do so, we define $x_1 = x$ and $y_1 = y$. Inductively, having defined $x_1, y_1, \ldots, x_i, y_i$, we have that $M\backslash x_i/y_i \cong N$ and so by the falsity of (iii), x_i is a loop, coloop, or parallel element of M/y_i. M has no loops, coloops, or parallel elements, and so for some element $x_{i+1} \neq x_i, y_i$, $\{x_i\, y_i, x_{i+1}\}$ is a circuit of M. We see that $M\backslash x_i/y_i \cong M/y_i\backslash x_{i+1}$, because x_i, x_{i+1} are parallel in M/y_i; and so $M/y_i\backslash x_{i+1} \cong N$. Now by the falsity of (iv), y_i is a series element of $M\backslash x_{i+1}$, and there is an element $y_{i+1} \neq y_i$, x_{i+1} so that $\{y_i, x_{i+1}, y_{i+1}\}$ is a cocircuit of M. And again, $M\backslash x_{i+1}/y_{i+1} \cong M/y_i\backslash x_{i+1}$, because y_i, y_{i+1} are in series in $M\backslash x_{i+1}$, and so $M\backslash x_{i+1}/y_{i+1} \cong N$. This completes the inductive definition.

For $i \geqslant 1$, define $z_{2i-1} = x_i$, $z_{2i} = y_i$. Define $Z_i = \{z_i, z_{i+1}, z_{i+2}\}$. Then for i odd, Z_i is a circuit, and for i even it is a cocircuit. Since M is finite, there is repetition in the sequence $z_1, z_2, \ldots$. Choose $i < j$ with j minimum so that $z_i = z_j$. Then $j \geqslant i + 3$, since z_i, z_{i+1}, z_{i+2} are distinct by the construction.

Suppose first that $j = i + 3$. If $i = 1$ then $\{z_1, z_2, z_3\} = Z_1$ is both a circuit and a cocircuit of M, and so (v) is true.

Secondly, suppose that $j = i + 3$ and $i = 2$. By the falsity of (iv), y_1 is a series element of $M\backslash x_1$, and so there is a cocircuit $\{z_1, z_2, z\}$ for some $z \in S - \{z_1, z_2\}$. But Z_3 is a circuit and $Z_3 = \{z_3, z_4, z_2\}$, and so $z \in \{z_3, z_4\}$. Thus if Y denotes $\{z_1, z_2, z_3, z_4\}$, we have $r(Y) = 2$ and $r^*(Y) = 2$, and so Y is a separator. Hence $Y = S$, which is impossible because $|S| = |E(N)| + 2 \geqslant 6$ (since N has no loops, coloops, or series or parallel elements).

Thirdly, suppose that $j = i + 3$ and $i \geqslant 3$. One of Z_{i-2}, Z_{j-2} is a circuit and one is a cocircuit, and yet they have intersection $\{z_i\}$, by the minimality of j. This is impossible, and so we may assume that $j \geqslant i + 4$.

Suppose that $j - i$ is odd. Then one of Z_i, Z_{j-2} is a circuit and one is a cocircuit, but their intersection is $\{z_i\}$, a contradiction. Thus $j - i$ is even.

Suppose that $i \geqslant 2$. Then one of Z_{i-1}, Z_{j-2} is a circuit and one is a cocircuit, and we obtain a contradiction as before.

Hence $i = 1$, and for some $n \geqslant 2$, $x_1 = x_{n+1}$ and $x_1, y_1, \ldots, x_n, y_n$ are all distinct. To apply Lemma (6.1), it suffices to prove that $\{y_n, x_1, y_1\}$ is a cocircuit.

Suppose now that $n = 2$. Then $\{x_1, y_1, x_2\}$ and $\{x_2, y_2, x_1\}$ are circuits, and $\{y_1, x_2, y_2\}$ is a cocircuit, and $\{y_2, x_1, y_3\}$ is a cocircuit. Therefore $|\{y_2, x_1, y_3\} \cap \{x_1, y_1, x_2\}| \neq 1$, and so $y_3 = y_1$ or x_2. In either case, $r^*(\{x_1, y_1, x_2, y_2\}) = 2$ and as before $Y = S$, a contradiction.

Thus $n \geqslant 3$. Then $\{y_n, x_1, y_{n+1}\}$ is a cocircuit, but $\{x_1, y_1, x_2\}$ and $\{x_2, y_2, x_3\}$ are circuits, and so $y_{n+1} = y_1$ and $\{y_n, x_1, y_1\}$ is a cocircuit. Thus the hypotheses of (6.1) are satisfied, and we conclude that M (and hence N) is isomorphic to a wheel or a whirl. Then (vi) holds.

7. Splitters

Let $\mathscr{F}$ be a class of matroids, closed under minors and under isomorphism. $N \in \mathscr{F}$ is said to be a *splitter* for $\mathscr{F}$ if every $M \in \mathscr{F}$ with a minor isomorphic to N is 1- or 2-separable unless $M \cong N$.

Such objects occur in graph theory, but only in relatively out-of-the-way places. The class of graphic matroids, for example, has no splitters. In a non-graphic context, however, they appear to be more central. We shall see that the class of regular matroids does have a splitter, the matroid we call R_{10}.

(7.1) *Suppose that $N \in \mathscr{F}$, and that for every $M \in \mathscr{F}$, if $M/x \cong N$ then x is a loop, coloop, or series element of M. Then for every $M \in \mathscr{F}$, if M has a minor isomorphic to N, then there exists $Z \subseteq E(M)$ such that $M \times Z$ is a subdivision of N.*

Proof. M has a minor isomorphic to N. Choose disjoint subsets X, $Y \subseteq E(M)$ such that $M \backslash X / Y \cong N$, with Y minimal. For $y \in Y$, let M_y be $(M \backslash X)/(Y - \{y\})$. Then $M_y / y \cong N$, and so by hypothesis y is a loop, coloop or series element of M_y. If y is a loop or coloop then $M_y / y = M_y \backslash y$, and so $M_y \backslash (X \cup \{y\})/(Y - \{y\}) \cong N$, contrary to the minimality of Y. Thus each $y \in Y$ is a series element of M_y. Let the elements of $M \backslash X / Y$ be $z_1, \ldots, z_k$. Let L_i be the subset of $\{z_1, \ldots, z_k\} \cup Y$ containing z_i and those $y \in Y$ for which in

M_y, y is in series with z_i but with no z_j for $j < i$. Then $(L_1,..., L_k)$ is a partition of $E(M\backslash X)$ and $M\backslash X$ is a subdivision of N, as required.

For example, we have the following, which will be used later in the paper.

(7.2) *If M is regular and has an $\mathcal{M}(K_{3,3})$ minor, then there exists $Z \subseteq E(M)$ such that $M \times Z$ is a subdivision of $\mathcal{M}(K_{3,3})$.*

Proof. We merely verify the hypotheses of (7.1), taking $\mathcal{F}$ to be the class of regular matroids. A mechanical way to do so is to take a representation of $\mathcal{M}^*(K_{3,3})$ over $GF(2)$, and examine the matroids obtained from this by adding one new vector in all possible ways. Such a matroid M will have an element e so that $M\backslash e \cong \mathcal{M}^*(K_{3,3})$, and all such matroids which are binary will be constructed (except the one in which e is a coloop). Their duals are the matroids which need to be checked. For details, see the Appendix.

(7.3) *Suppose that $N \in \mathcal{F}$, and is non-null and connected, and that the following statements are true:*

(i) *all circuits and cocircuits of N have at least three elements;*

(i) *for every $M \in \mathcal{F}$, if $M\backslash x \cong N$ then x is a loop, coloop, or parallel element of M;*

(iii) *for every $M \in \mathcal{F}$, if $M/x \cong N$ then x is a loop, coloop, or series element of M;*

(iv) *if $N \cong W_n$ or $\mathcal{W}_n$ for some $n \geqslant 2$, then $W_{n+1} \notin \mathcal{F}$ or $\mathcal{W}_{n+1} \notin \mathcal{F}$, respectively.*

Then N is a splitter for $\mathcal{F}$.

The proof is in two steps.

Step 1. *If $M \in \mathcal{F}$ and $Z \subseteq E(M)$, and $M \times Z$ is a subdivision of N, then for every Z-arc A there is a series class P of $M \times Z$ such that $A \to P$.*

Proof. If possible, take a counterexample M, Z, A with $|E(M)|$ minimum. Then obviously $E(M) = Z \cup A$, and $|A| = 1$. (For all the elements in A are in series, and so if $|A| > 1$ we may produce a smaller counterexample by contracting.) Let $A = \{a\}$, say. Let the series classes of $M\backslash a$ be $P_1,..., P_k$, where $k = |E(N)|$, so that $(P_1,..., P_k)$ is a partition of Z.

Now if $|P_i| = 1$ for all i, then $M\backslash a \cong N$, and yet a is not a loop, coloop, or parallel element of M, contrary to hypothesis. Thus for some i, $|P_i| > 1$. We assume $|P_i| > 1$. Choose $x_1, x_1' \in P_1$. Now $(M/x_1) \times (Z - \{x_1\})$ is a subdivision of N, and A is a $(Z - \{x_1\})$-arc of M/x_1 (since $A \cup \{x_1\}$ includes no circuits of M); and so by induction there is a circuit of M/x_1 containing a and included in $\{a\} \cup P_{j_1}$, for some j_1. This is therefore not a circuit of M, and so there is a circuit C_1 of M with x_1, $a \in C_1$ and $C_1 - \{x_1, a\} \subseteq P_{j_1}$, for some $j_1 \neq 1$. Similarly there is a circuit C'_1 with x'_1, $a \in C'_1$ and $C'_1 - \{x'_1, a\} \subseteq P_{j'_1}$ for some $j'_1 \neq 1$. But $(C_1 \cup C'_1) - \{a\}$ includes a circuit C, and $C \subseteq Z$ and so is a union of at least three P_i's (because N has no circuits with $\leqslant 2$ elements). Thus $|P_1| = 2$, and $P_{j_1} \subseteq C_1$. Suppose now that for some $i > 1$, $|P_i| \geqslant 2$, $|P_2| \geqslant 2$, say. Then we have similarly $|P_2| = 2$. Choose $x_2 \in P_2$; then there is a circuit C_2 with x_2, $a \in C_2$ and $C_2 - \{x_2, a\} \subseteq P_{j_2}$ for some $j_2 \neq 2$. Certainly $C_1 \neq C_2$ (because even if $j_1 = 2$ and $j_2 = 1$, $C_1 \neq C_2$ since $P_{j_1} \subseteq C_1$), and so there is a circuit $C \subseteq (C_1 \cup C_2) - \{a\}$. But then C is a union of at least three P_i's, and yet $C_1 \cup C_2$ only includes two P_i's, a contradiction.

This proves that $|P_1| = 2$, $|P_i| = 1$ $(2 \leqslant i \leqslant k)$. Thus $M \backslash a / x_1 \cong N$, and we may apply (6.2). We discover from (6.2) that one of the following holds.

(i) a is a loop, coloop, or parallel element of M. This is impossible since $\{a\}$ is a Z-arc of M and $\{a\} \nrightarrow P$ for any series class P of $M \times Z$.

(ii) x_1 is a loop, coloop, or series element of M. But $x_1 \in C_1$, and so it is not a loop or coloop, and if it is in series with y, say, then $y \in C_1$. But $y \neq a$, and so $y \in Z$, and so $y \in P_1$ since P_1 is a series class of $M \backslash a$. This is impossible since $C_1 \cap P_1 = \{x_1\}$.

(iii) For some $x', y' \in E(M)$, $M \backslash x' / y' \cong N$ but x' is not a loop, coloop, or parallel element of M / y'. This is impossible by hypothesis (ii) of (7.3).

(iv) For some $x', y' \in E(M)$, $M \backslash x' / y' \cong N$ but y' is not a loop, coloop, or series element of $M \backslash x'$. This is impossible by hypothesis (iii) of (7.3).

(v) There exists $z \neq a$, x_1 such that $\{a, x_1, z\}$ is both a circuit and a cocircuit of M. Thus $z \in Z$, and z, x_1 are in series in $M \backslash a$; and so $z \in P_1$. This is impossible because $\{a\} \nrightarrow P_1$.

(vi) For some $n \geqslant 2$, $N \cong W_n$ or $\mathcal{W}_n$, and $M \cong W_{n+1}$ or $\mathcal{W}_{n+1}$, respectively. This contradicts hypothesis (iv) of (7.3), and completes the proof of Step 1.

Step 2. *If $M \in \mathcal{F}$ is* 3*-connected and has a minor isomorphic to N, then $M \cong N$, that is, N is a splitter for $\mathcal{F}$.*

Proof. $M \in \mathcal{F}$ has a minor isomorphic to N, and so by (7.1) there exists $Z \subseteq E(M)$ such that $M \times Z$ is a subdivision of N. Suppose for a contradiction that we can choose Z thus with $Z \neq E(M)$. Then by (5.4) we can choose Z thus so that in addition there is an adjoinable Z-arc A. But this contradicts Step 1. So $Z = E(M)$, and M is a subdivision of N. But M is 3-connected, and $|E(M)| \geqslant |E(N)| \geqslant 4$, and so M has no pairs of elements in series. It follows that $M \cong N$, as required.

For the result of this paper we need only two applications of (7.3) (more will appear in subsequent papers, however).

The matroid R_{10} was, as far as I know, first introduced by Bixby [1]. (He calls matroids isomorphic to it matroids of *Type R*.) We define R_{10} to be the linear independence matroid of the ten 5-tuples over $GF(2)$ with three 1's and two 0's. It may be verified that another, less symmetric but more convenient, representation of R_{10} is the following: take the nine 6-tuples over $GF(2)$ with two 1's and four 0's which represent $\mathcal{M}(K_{3,3})$, and add the 6-tuple of all 1's.

It may be verified that R_{10} is regular, and is isomorphic to its dual (although not self-dual). Its automorphism group is doubly transitive. $R_{10}\backslash e \cong \mathcal{M}(K_{3,3})$ and $R_{10}/e \cong \mathcal{M}^*(K_{3,3})$ for any element e. A totally unimodular representation of R_{10} is given in Fig. 1.

$$\begin{bmatrix} 1 & 0 & 0 & 0 & 0 & -1 & 1 & 0 & 0 & 1 \\ 0 & 1 & 0 & 0 & 0 & 1 & -1 & 1 & 0 & 0 \\ 0 & 0 & 1 & 0 & 0 & 0 & 1 & -1 & 1 & 0 \\ 0 & 0 & 0 & 1 & 0 & 0 & 0 & 1 & -1 & 1 \\ 0 & 0 & 0 & 0 & 1 & 1 & 0 & 0 & 1 & -1 \end{bmatrix}$$

R_{10}

FIGURE 1

(7.4) *R_{10} is a splitter for the class of regular matroids.*

Proof. It is necessary only to check that R_{10} satisfies the hypotheses of (7.3). To verify this, take a 5-dimensional representation of R_{10} over $GF(2)$, and examine the result of adding any other vector to this representation. It will be found that either the vector added is the zero vector, or is one already in the representation, or the enlarged binary matroid produced will not be regular. For details, see the Appendix. This verifies (7.3)(ii), and (7.3)(iii) follows from this, because R_{10} is isomorphic to its dual.

(7.5) *$\mathcal{M}(K_5)$ is a splitter for the class of regular matroids without minors isomorphic to $\mathcal{M}(K_{3,3})$.*

Proof. Again, we verify that the hypotheses of (7.3) hold; with the difference that we must check (7.3)(ii) and (7.2)(iii) separately. (See the Appendix.)

There is another application of (7.3) which is not needed for this paper, but which we state here for convenient reference in subsequent papers.

(7.6) ***F_7 is a splitter for the class of binary matroids without F_7^* minors.***

The proof is again case-checking. (See the Appendix.)

8. Connectivity

If $A, B \subseteq S$ are disjoint and M is a matroid on S, we define $k_M(A, B)$ to be the minimum over all partitions (X, Y) of S with $A \subseteq X$, $B \subseteq Y$, of

$$r(X) + r(Y) - r(M).$$

$k_M(A, B)$ should be thought of as the "connectivity" between A and B in M. It is easy to see that

$$0 \leqslant k_M(A, B) \leqslant |A|, |B|$$

and that $k_M(A, B) = k_{M^*}(A, B)$.

(8.1) *If (X, Y) is a partition of $E(M)$, and $M' = M\backslash Z_1/Z_2$, where $Z_1 \cap Z_2 = \emptyset$ and $Z_1 \cup Z_2 = Z$, say, then*

$$k_{M'}(X - Z, Y - Z) \leqslant k_M(X, Y);$$

that is,

$$r_{M'}(X - Z) + r_{M'}(Y - Z) - r(M') \leqslant r_M(X) + r_M(Y) - r(M).$$

Proof. Using duality, it suffices to prove the result when $Z_1 = \{z\}$, say and $Z_2 = \emptyset$. By symmetry we may assume that $z \in X$. We must prove that

$$r_M(X - \{z\}) + r_M(Y) - r(M') \leqslant r_M(X) + r_M(Y) - r(M).$$

But $r_M(X - \{z\}) \leqslant r_M(X)$, and if equality occurs, then z is not a coloop of M and so $r(M') = r(M)$.

(8.2) *If $A, B \subseteq E(M)$ are disjoint, and M' is obtained from M by deleting and contracting elements from $E(M) - (A \cup B)$, then*

$$k_{M'}(A, B) \leqslant k_M(A, B).$$

Proof. Choose a partition (X, Y) of $E(M)$ with $A \subseteq X$, $B \subseteq Y$, so that $k_M(X, Y) = k_M(A, B)$. If $Z = E(M) - E(M')$, we have from (8.1) that $k_{M'}(X - Z, Y - Z) \leqslant k_M(X, Y)$, and so $k_{M'}(A, B) \leqslant k_M(A, B)$, as required.

For the remainder of this section, we shall assume that M is a matroid on a set S, and that N is a minor of M; that (A, B) is a partition of $E(N)$, and $k_N(A, B) = k$; that $k_M(A, B) > k$; and that for every $x \in S - (A \cup B)$, and for $M' = M\backslash x$, M/x, either N is not a minor of M' or $k_{M'}(A, B) = k$. We shall analyse the structure of M with respect to N. (We should perhaps stress that N is a minor of M; no isomorphism is involved.)

(8.3) *For each* $x \in S-(A \cup B)$ *either* $k_{M\backslash x}(A, B) > k$ *or* $k_{M/x}(A, B) > k$.

Proof. Suppose that $k_{M\backslash x}(A, B) \leqslant k$. Then there exists a partition (X_1, Y_1) of $S-\{x\}$ with $A \subseteq X_1$, $B \subseteq Y_1$, and

$$r_M(X_1)+r_M(Y_1)-r(M\backslash x) \leqslant k < r_M(X_1 \cup \{x\})+r_M(Y_1)-r(M).$$

Thus $r_M(X_1) \neq r_M(X_1 \cup \{x\})$, so x is not a loop, and $r(M\backslash x)=r(M)$, so x is not a coloop. We have

$$r_M(X_1)+r_M(Y_1)-r(M) \leqslant k.$$

Now suppose that $k_{M/x}(A, B) \leqslant k$. Then there exist X_2, $Y_2 \subseteq S$ such that $X_2 \cup Y_2 = S$, $X_2 \cap Y_2 = \{x\}$, $A \subseteq X_2$, $B \subseteq Y_2$, and

$$r_M(X_2)-1+r_M(Y_2)-1-(r(M)-1) \leqslant k;$$

that is, $r_M(X_2)+r_M(Y_2)-r(M) \leqslant k+1$. Adding, we obtain

$$r_M(X_1)+r_M(X_2)+r_M(Y_1)+r_M(Y_2)-2r(M) \leqslant 2k+1,$$

and hence, using submodularity of the rank function,

$$r_M(X_1 \cup X_2)+r_M(Y_1 \cap Y_2)+r_M(X_1 \cap X_2) + r_M(Y_1 \cup Y_2)-2r(M) \leqslant 2k+1.$$

Now every element of S is in precisely two of X_1, X_2, Y_1, Y_2, and so $(X_1 \cup X_2, Y_1 \cap Y_2)$ and $(X_1 \cap X_2, Y_1 \cup Y_2)$ are both partitions of S; and $A \subseteq X_1 \cup X_2$, $X_1 \cap X_2$, and $B \subseteq Y_1 \cap Y_2$, $Y_1 \cup Y_2$. Thus

$$r_M(X_1 \cup X_2)+r_M(Y_1 \cap Y_2)-r(M) \geqslant k+1$$

and

$$r_M(X_1 \cap X_2)+r_M(Y_1 \cup Y_2)-r(M) \geqslant k+1.$$

Adding, we obtain a contradiction, as required.

(8.4) *For each* $x \in S-(A \cup B)$, *either* N *is not a minor of* $M\backslash x$ *or* N *is not a minor of* M/x.

Proof. This follows from (8.3) and our hypothesis about M, N.

Thus N may be expressed as $M\backslash P/Q$, where (P, Q) is a partition of $S-(A \cup B)$, in a unique way.

(8.5) *There is no circuit* C *with* $C \subseteq P \cup Q$, $|C-Q| \leqslant 1$, *and there is no* *cocircuit D with $D \subseteq P \cup Q$, $|D - P| \leqslant 1$. In particular, Q is independent, and $A \cup B \cup Q$ spans P.*

Proof. Suppose that C is a circuit with $C \subseteq Q$. Choose $q \in C$; then q is a loop of $M \backslash P/(Q - \{q\})$, and so $M \backslash P/(Q - \{q\}) \backslash \{q\} = N$, contrary to (8.4). Secondly, suppose that $C \subseteq P \cup Q$ is a circuit with $C \cap P = \{p\}$. Then p is a loop of $M \backslash (P - \{p\})/Q$, and so M/p has N as a minor, contrary to (8.4). The second assertion of the theorem follows by duality.

For $z \in P \cup Q$, let M_z be $M \backslash z$ if $z \in P$, and be M/z if $z \in Q$. Now N is a minor of M_z, and so by hypothesis $k_{M_z}(A, B) = k$. Thus for each $z \in P \cup Q$ there is a partition (X_z, Y_z) of $S - z$ such that $A \subseteq X_z$, $B \subseteq Y_z$, and

$$r_{M_z}(X_z) + r_{M_z}(Y_z) = r(M_z) + k.$$

(8.6) *For distinct $z, z' \in P \cup Q$, either $z \in X_{z'}$ or $z' \in X_z$ but not both. In particular, for each $z \in P \cup Q$, (X_z, Y_z) is uniquely defined.*

Proof. The second statement follows from the first immediately. Suppose the first is false; then by exchanging A and B if necessary, we may assume that $z \in X_{z'}$, $z' \in X_z$. By duality we may assume that $z \in P$. Now $(X_z \cup X_{z'}, Y_z \cap Y_{z'})$ is a partition of S, and so

$$r_M(X_z \cup X_{z'}) + r_M(Y_z \cap Y_{z'}) - r(M) \geqslant k + 1 \qquad (*)$$

since $k_M(A, B) > k$.

There are now two cases, $z' \in P$, and $z' \in Q$. Suppose first that $z' \in P$. Then $M_z = M \backslash z$ and $M_{z'} = M \backslash z'$. By (8.5), z, z' are not coloops of M, and so $r(M_z) = r(M_{z'}) = r(M)$. Thus

$$r_M(X_z) + r_M(Y_z) - r(M) \leqslant k$$

and

$$r_M(X_{z'}) + r_M(Y_{z'}) - r(M) \leqslant k.$$

Adding, and using submodularity and (*), we obtain

$$r_M(X_z \cap X_{z'}) + r_M(Y_z \cup Y_{z'}) - r(M) \leqslant k - 1.$$

Let M' be $M \backslash \{z, z'\}$. By (8.5), $r(M') = r(M)$, and since M' has N as a minor, and since $(X_z \cap X_{z'}, Y_z \cup Y_{z'})$ is a partition of the elements of M', we have a contradiction to (8.2).

Now suppose that $z' \in Q$. Then $M_z = M \backslash z$ and $M_{z'} = M/z'$. By (8.5), z is not a coloop of M, and z' is not a loop, and so $r(M_z) = r(M)$, $r(M_{z'}) = r(M) - 1$. Thus

$$r_M(X_z) + r_M(Y_z) - r(M) \leqslant k,$$

and

$$r_M(X_{z'} \cup \{z'\}) - 1 + r_M(Y_{z'} \cup \{z'\}) - 1 - (r(M) - 1) \leqslant k,$$

that is,

$$r_M(X_{z'} \cup \{z'\}) + r_M(Y_{z'} \cup \{z'\}) - r(M) \leqslant k + 1.$$

Adding, and using submodularity and (*), we obtain

$$r_M((X_z \cap X_{z'}) \cup \{z'\}) + r_M((Y_z \cup Y_{z'}) \cup \{z'\}) - r(M) \leqslant k.$$

Let M' be $M\backslash z/z'$. By (8.5), $r(M') = r(M) - 1$, and since M' has N as a minor, since $(X_z \cap X_{z'}, Y_z \cup Y_{z'})$ is a partition of the elements of M', and since $r_{M'}(Z) = r_M(Z \cup \{z'\}) - 1$ for $Z \subseteq S - \{z, z'\}$, we have a contradiction to (8.2).

(8.7) *For distinct $z, z' \in P \cup Q$, if $z \in X_{z'}$ then $X_z \subseteq X_{z'}$. In particular, $P \cup Q$ may be ordered as $z_1, \ldots, z_n$, so that for each i,*

$$X_{z_i} = A \cup \{z_1, \ldots, z_{i-1}\},$$
$$Y_{z_i} = \{z_{i+1}, \ldots, z_n\} \cup B.$$

Proof. The second statement follows from the first and (8.6), because we would have $z \in X_{z'}$ if and only if $|X_z| < |X_{z'}|$. To prove the first, we may assume from duality that $z \in P$. There are two cases, $z' \in P$ and $z' \in Q$.

Suppose first that $z' \in P$. As in (8.6), we have

$$r_M(X_z) + r_M(Y_z) - r(M) \leqslant k$$

and

$$r_M(X_{z'}) + r_M(Y_{z'}) - r(M) \leqslant k.$$

Now $(X_z \cup X_{z'}, Y_z \cap Y_{z'})$ is a partition of $E(M\backslash z')$, and so

$$r_M(X_z \cup X_{z'}) + r_M(Y_z \cap Y_{z'}) - r(M) \geqslant k.$$

Using these three inequalities and submodularity, we obtain

$$r_M(X_z \cap X_{z'}) + r_M(Y_z \cup Y_{z'}) - r(M) \leqslant k.$$

But $(X_z \cap X_{z'}, Y_z \cup Y_{z'})$ is a partition of $E(M\backslash z)$, and so, by the uniqueness of (X_z, Y_z) asserted in (8.6), we have $X_z \cap X_{z'} = X_z$, that is, $X_z \subseteq X_{z'}$, as required.

Now suppose that $z' \in Q$. Then as in (8.6) we have

$$r_M(X_z) + r_M(Y_z) - r(M) \leqslant k$$

and

$$r_M(X_{z'} \cup \{z'\}) + r_M(Y_{z'} \cup \{z'\}) - r(M) \leqslant k + 1.$$

But $(X_z \cap X_{z'}, Y_z \cup Y_{z'})$ is a partition of the elements of $M \backslash z$, and so

$$r_M(X_z \cap X_{z'}) + r_M(Y_z \cup Y_{z'}) - r(M) \geqslant k.$$

Using these three inequalities and submodularity, we obtain

$$r_M((X_z \cup X_{z'}) \cup \{z'\}) + r_M((Y_z \cap Y_{z'}) \cup \{z'\}) - r(M) \leqslant k + 1,$$

that is,

$$r_{M_{z'}}(X_z \cup X_{z'}) + r_{M_{z'}}(Y_z \cap Y_{z'}) - r(M_{z'}) \leqslant k.$$

Thus, by the uniqueness of $(X_{z'}, Y_{z'})$, we have $X_z \cup X_{z'} = X_{z'}$, that is, $X_z \subseteq X_{z'}$, as required.

(8.8) *$z_1, \ldots, z_n$ are alternately members of P and members of Q.*

Proof. Suppose this is false; then by duality, we may assume that z_i, $z_{i+1} \in P$. By (8.7),

$$r_M(A \cup \{z_1, \ldots, z_{i-1}\}) + r_M(\{z_{i+1}, \ldots, z_n\} \cup B) - r(M) \leqslant k$$

and

$$r_M(A \cup \{z_1, \ldots, z_i\}) + r_M(\{z_{i+2}, \ldots, z_n\} \cup B) - r(M) \leqslant k.$$

Now

$$r_M(A \cup \{z_1, \ldots, z_i\}) + r_M(\{z_{i+1}, \ldots, z_n\} \cup B) - r(M) > k$$

since $k_M(A, B) > k$, and so

$$r_M(A \cup \{z_1, \ldots, z_{i-1}\}) + r_M(\{z_{i+2}, \ldots, z_n\} \cup B) - r(M) < k.$$

Let M' be $M \backslash \{z_i, z_{i+1}\}$. Then as before, $r(M') = r(M)$; and so $k_{M'}(A, B) < k$, contrary to (8.2).

(8.9) *For all $i > 1$, if $z_i \in P$ there is a circuit C of M with $z_{i-1}, z_i \in C$, and $C - \{z_{i-1}, z_i\} \subseteq (Q \cap \{z_j : j > i\}) \cup B$. If $z_i \in Q$ there is a cocircuit D with $z_{i-1}, z_i \in D$ and $D - \{z_{i-1}, z_i\} \subseteq (P \cap \{z_j : j > i\}) \cup B$.*

A similar result holds for all $i < n$ with A and B exchanged.

Proof. By duality we may assume that $z_i \in P$, so that $z_{i-1} \in Q$, by (8.7). Let M_0 be M/z_{i-1}. By (8.7),

$$r_{M_0}(A \cup \{z_j : 1 \leqslant j \leqslant i-2\}) + r_{M_0}(\{z_j : i \leqslant j \leqslant n\} \cup B) - r(M_0) \leqslant k.$$

Let M' be $M_0 \backslash (P - \{z_i\})$, and then by (8.1)

$$r_{M'}(A \cup (Q \cap \{z_j : 1 \leqslant j \leqslant i-2\})) \\ + r_{M'}((Q \cap \{z_j : i+1 \leqslant j \leqslant n\}) \cup \{z_i\} \cup B) - r(M') \leqslant k.$$

Let M'' be $M' \backslash z_i$. By (8.2), since N is a minor of M'', we have

$$r_{M''}(A \cup (Q \cap \{z_j : 1 \leqslant j \leqslant i-2\})) \\ + r_{M''}((Q \cap \{z_j : i+1 \leqslant j \leqslant n\}) \cup B) - r(M'') \geqslant k.$$

The first terms in these inequalities are equal; and by (8.5), $r(M'') = r(M')$; thus

$$r_{M'}((Q \cap \{z_j : i+1 \leqslant j \leqslant n\}) \cup \{z_i\} \cup B) \\ \leqslant r_{M''}((Q \cap \{z_j : i+1 \leqslant j \leqslant n\}) \cup B),$$

that is, there is a circuit C of M' and hence of M_0 with $z_i \in C$, and $C - \{z_i\} \subseteq (Q \cap \{z_j : i+1 \leqslant j \leqslant n\}) \cup B$. Now C is not a circuit of M, because by (8.7),

$$r_M(A \cup \{z_j : 1 \leqslant j \leqslant i-1\}) + r_M\{(z_j : i+1 \leqslant j \leqslant n\} \cup B) - r(M) \leqslant k$$

and yet

$$r_M(A \cup \{z_j : 1 \leqslant j \leqslant i-1\}) + r_M(\{z_j : i \leqslant j \leqslant n\} \cup B) - r(M) > k$$

since $k_M(A, B) > k$, and thus $\{z_j : i+1 \leqslant j \leqslant n\} \cup B$ does not span z_i. Hence $C \cup \{z_{i-1}\}$ is a circuit of M. This completes the proof.

(8.10) *If $z_1 \in P$, there is no circuit C with $z_1 \in C \subseteq P \cup Q \cup B$. If $z_1 \in Q$, there is no cocircuit D with $z_1 \in D \subseteq P \cup Q \cup B$. Similar results hold for z_n and A.*

Proof. If $z_1 \in P$, then by (8.7),

$$r_M(A) + r_M(\{z_2, \ldots, z_n\} \cup B) - r(M) \leqslant k,$$

but

$$r_M(A) + r_M(\{z_1, \ldots, z_n\} \cup B) - r(M) > k$$

since $k_M(A, B) > k$. Thus there is no circuit C with $z_1 \in C \subseteq P \cup Q \cup B$. The other results follow by duality and symmetry.

9. Separability because of R_{12}

Now we use the results of the previous section to prove that every regular matroid with an R_{12} minor is 3-separable. In an attempt to clarify the proof, the relevant properties of R_{12} have been abstracted, in the following theorem. (R_{12} itself is defined later in this section.)

(9.1) *Let $\mathscr{F}$ be a class of matroids, closed under minors and under isomorphism. Let $N \in \mathscr{F}$, and let (A, B) be a partition of $E(N)$, with $k_N(A, B) = k$. Suppose that N, $\mathscr{F}$ have the following properties:*

(i) *for each $x \in A$ there is a circuit C and a cocircuit D of N containing x, with $C, D \subseteq A$,*

(ii) *for each $M \in \mathscr{F}$, if $M \backslash x = N$ and x is not a coloop of M, there is a circuit C of M with $x \in C$ and $C - \{x\}$ included in one of A, B,*

(iii) *for each $M \in \mathscr{F}$, if $M/y = N$ and y is not a loop of M, there is a cocircuit D of M with $y \in D$ and $D - \{y\}$ included in one of A, B,*

(iv) *for each $M \in \mathscr{F}$, if $M \backslash x/y = N$, suppose that there is a cocircuit D of M with $\{x, y\} \subset D \subseteq B \cup \{x, y\}$; then either there is a circuit C of M with $x \in C \subseteq B \cup \{x, y\}$ or x is parallel to an element of A in M/y,*

(v) *for each $M \in \mathscr{F}$, if $M \backslash x/y = N$, suppose that there is a circuit C of M with $\{x, y\} \subset C \subseteq B \cup \{x, y\}$; then either there is a cocircuit D of M with $y \in D \subseteq B \cup \{x, y\}$, or y is in series with an element of A in $M \backslash x$.*

Then $k_M(A, B) = k$ for each $M \in \mathscr{F}$ with N as a minor.

(*Remark.* Statements (ii) and (iii) form a dual pair, as do (iv) and (v), and (i) is invariant under duality. Thus if N, $\mathscr{F}$ have these properties, then so do N^*, $\mathscr{F}^* = \{M^*: M \in \mathscr{F}\}$. However, there is no symmetry between A and B.)

Proof. We proceed by induction on $|E(M)|$. The result is clear if $|E(M)| = |E(N)|$, because then $M = N$, and so we assume that $|E(M)| > |E(N)|$. We assume for a contradiction that $k_M(A, B) > k$. Thus, by induction, M and N have the properties discussed in Section 8, and so (8.4)–(8.10) are true for them. We assume by duality that $z_1 \in P$, where P, Q, $z_1, \ldots, z_n$ are defined as before.

(1) *z_1 is parallel in $M \backslash (P - \{z_1\})/Q$ to some element $a \in A$.*

Put $M_1 = M \backslash (P - \{z_1\})/Q$. Then $M_1 \backslash z_1 = N$, and z_1 is not a coloop of M_1, by (8.5), and so by (ii) there is a circuit C_0 of M_1 with $z_1 \in C_0$ and $C_0 - \{z_1\}$ included in one of A, B. If $C_0 - \{z_1\} \subseteq B$, then there is a circuit C'

of M with $z_1 \in C'$ and $C' - \{z_1\} \subseteq Q \cup B$, contrary to (8.10). Thus $C_0 - \{z_1\} \subseteq A$, and there is a circuit C' of M with $z_1 \in C'$, $C' - \{z_1\} \subseteq Q \cup A$. By (8.10) applied to z_n and A, we deduce $n > 1$. By (8.9) with A, B exchanged, we see that there is a circuit C of M with $z_1, z_2 \in C$, $C - \{z_1, z_2\} \subseteq A$. By (8.9) again, there is a cocircuit D' of M with z_1, $z_2 \in D'$, $D' - \{z_1, z_2\} \subseteq P \cup B$. Let M_2 be $M \backslash (P - \{z_1\})/(Q - \{z_2\})$. Now $D' \cap (Q - \{z_2\}) = \varnothing$, and so $D' - (P - \{z_1\})$ is a union of cocircuits of M_2. Choose a cocircuit D of M_2 with $z_1 \in D \subseteq D' - (P - \{z_1\})$. C is a circuit of $M \backslash (P - \{z_1\})$, and D is a cocircuit of this, and so $|C \cap D| \neq 1$; thus $z_2 \in D$. Now $D \cap B \neq \varnothing$, because by (8.5), $P \cup \{z_2\}$ includes no cocircuits of M, and so $D \neq \{z_1, z_2\}$. But $M_2 \backslash z_1/z_2 = N$, and so by (iv), either there is a circuit C'' of M_2 with $z_1 \in C'' \subseteq B \cup \{z_1, z_2\}$ or z_1 is parallel to an element of A in M_2/z_2. The first alternative implies that there is a circuit C_1 of M with $z_1 \in C_1 \subseteq B \cup Q \cup P$, contrary to (8.10); and the second is the desired result.

We observe that N is isomorphic to $N' = M \backslash ((P - \{z_1\}) \cup \{a\})/Q$, because z_1 and a are parallel in $M \backslash (P - \{z_1\})/Q$, and an isomorphism is given by the map ϕ: $A \cup B \to ((A - \{a\}) \cup \{z_1\}) \cup B$, defined by

$$\phi(x) = x \qquad (x \neq a),$$
$$\phi(a) = z_1.$$

We observe that under this isomorphism, $\phi(A) = (A - \{a\}) \cup \{z_1\}$ and $\phi(B) = B$. Now N' is a minor of $M \backslash a$, and so by induction, $k_{M\backslash a}((A - \{a\}) \cup \{z_1\}, B) = k$; that is, there is a partition X, Y of $S - \{a\}$ with $(A - \{a\}) \cup \{z_1\} \subseteq X$, $B \subseteq Y$, and

$$r_M(X) + r_M(Y) - r(M \backslash a) = k.$$

a is not a coloop of M, because it is not a coloop of N by (i), and so $r(M \backslash a) = r(M)$. Thus

$$r_M(X) + r_M(Y) - r(M) = k.$$

However, $r_M(X \cup \{a\}) + r_M(Y) - r(M) > k$ since $k_M(A, B) > k$, and so $r_M(X \cup \{a\}) > r_M(X)$. In particular, there is no circuit C of M with $a \in C \subseteq A$. However, by (i), there is a circuit C of M with $a \in C$, $C - Q \subseteq A$, $C \cap P = \varnothing$. Thus $C \cap Q \neq \varnothing$; choose $q \in C \cap Q$. Then $q = z_i$ for some i. Now $i \neq 1$, because $z_1 \in P$ by hypothesis; and so by (8.9) there is a cocircuit D of M with $z_i \in D$, $D - \{z_i\} \subseteq P \cup B$. But then $|C \cap D| = 1$, which is impossible. This completes the proof of (9.1).

R_{12} is defined to be the linear independence matroid of the columns of the matrix of Fig. 2, which has entries over $GF(2)$. It will be seen that the matrix obtained by deleting the first six columns is symmetric, and so R_{12} is isomorphic to its dual. (However, it is not self-dual). R_{12} has just two circuits of cardinality 3, and they are disjoint. We define B to be the union of these two circuits, and A to be the set of the remaining six elements. Thus (A, B) is a partition of $E(R_{12})$.

$$\begin{bmatrix} 1&0&0&0&0&0&1&1&1&0&0&0\\ 0&1&0&0&0&0&1&1&0&1&0&0\\ 0&0&1&0&0&0&1&0&0&0&1&0\\ 0&0&0&1&0&0&0&1&0&0&0&1\\ 0&0&0&0&1&0&0&0&1&0&1&1\\ 0&0&0&0&0&1&0&0&0&1&1&1 \end{bmatrix}$$

R_{12}

FIGURE 2

(9.2) *If M is regular, and M has a minor isomorphic to R_{12}, then M has an exact 3-separation (X_1, X_2) with $|X_1|, |X_2| \geqslant 4$ (indeed, $\geqslant 6$).*

Proof. We take $\mathscr{F}$ to be the class of regular matroids, and N to be R_{12}; we take (A, B) to be the partition of $E(R_{12})$ defined as above, and $k = 2$. It suffices to verify that the hypotheses of (9.1) hold. Each of these is an assertion about the members of $\mathscr{F}$ with at most two more elements than N (that is, at most 14 elements) and so checking each is a finite problem, which in principal could be left to the reader. However, the following observations may considerably reduce the amount of work the reader needs to do.

Checking hypothesis (i) is straightforward, and (iii) follows from (ii) because R_{12} is isomorphic to its dual, and the isomorphism exchanges A and B. It remains to check (ii), (iv), and (v). First it should be verified that there are only two non-isomorphic matroids M in $\mathscr{F}$ such that for some element e, $M \backslash e \cong R_{12}$ and e is not a loop, coloop, or parallel element of M. Then simply checking these two matroids verifies (ii). To verify (iv) we observe that M/y must be one of our two matroids (in a counterexample) and so we just check the ways of reinserting y. To verify (v), we observe that $M \backslash x$ must be the dual of one of our two matroids, and proceed similarly. For full details, see the Appendix.

10. GRAFTS

Now we begin the third part of the proof—that every regular matroid which is not 2-separable is either graphic or cographic, or has a minor isomorphic to R_{10} or R_{12}. The proof of this is essentially graph theory, in the sense that although we need a mild application of (5.4), all the difficulties are graph-theoretic.

A graph is *simple* if it has no loops or multiple edges. $V(G)$ and $E(G)$ denote the sets of vertices and edges of G, respectively. When $u, v \in V(G)$ are distinct, u, v are said to be 3-*connected* if there are three paths linking u and v, vertex-disjoint except for u and v. G is 3-*connected* if $|V(G)| \geqslant 4$ and every pair of vertices is 3-connected; or equivalently, if $|V(G)| \geqslant 4$ and the result of deleting any two vertices is connected.

The following is due to Tutte [13].

(10.1) *If G is connected and $|E(G)| \geqslant 4$, then $\mathcal{M}(G)$ is 3-connected if and only if G is simple and 3-connected.*

If $e \in E(G)$, $G \backslash e$ and G/e are the graphs obtained from G by deleting and contracting e, respectively (in the sense of graph theory). Then for disjoint subsets F_1, F_2 of $E(G)$, we define $G \backslash F_1/F_2$ in the natural way.

Let G be a graph, and let $T \subseteq V(G)$. The pair (G, T) is called a *graft*. With every graft (G, T) we can associate a binary matroid on the set $E(G) \cup \{\Omega\}$, where Ω is a new element. To define $\mathcal{M}(G, T)$, we associate with each $e \in E(G) \cup \{\Omega\}$ a vector $v(e)$ over $GF(2)$, and then we say $X \subseteq E(G) \cup \{\Omega\}$ is a cycle of $\mathcal{M}(G, T)$ if and only if $\sum_{e \in X} v(e) = 0$. The vectors $v(e)$ are defined as follows.

Let V be a $|V(G)|$-dimensional vector space over $GF(2)$, and let $\{b_v : v \in V(G)\}$ be a basis of V. Then for $e \in E(G) \cup \{\Omega\}$,

$$v(e) = 0 \text{ if } e \in E(G) \text{ and is a loop,}$$

$$v(e) = b_u + b_v \text{ if } e \in E(G) \text{ and has distinct ends } u, v,$$

$$v(\Omega) = \sum_{v \in T} b_v.$$

It is easy to see that an equivalent definition of $\mathcal{M}(G, T)$ is as follows. A *T-join* of G is a subset X of $E(G)$ such that X includes no circuits of G and such that the vertices incident with an odd number of edges in X are precisely the vertices in T. Then $\mathcal{M}(G, T)$ is a matroid on $E(G) \cup \{\Omega\}$, where Ω is a new element, in which for $C \subseteq E(G) \cup \{\Omega\}$, C is a circuit of $\mathcal{M}(G, T)$ if and only if and only if either $\Omega \notin C$ and C is the edge-set of a circuit of G, or $\Omega \in C$ and $C - \{\Omega\}$ is a T-join.

We observe that $\mathcal{M}(G, T) \backslash \Omega = \mathcal{M}(G)$; that if $|T|$ is odd then Ω is a coloop of $\mathcal{M}(G, T)$ (because there are no T-joins); and that if $|T| = 0$ or 2 then $\mathcal{M}(G, T)$ is graphic. It follows that if $\mathcal{M}(G, T)$ is not graphic then $|T|$ is even and $|T| \geqslant 4$. (This is not a sufficient condition for $\mathcal{M}(G, T)$ to be non-graphic. For example, it may be shown that if G is outerplanar then $\mathcal{M}(G, T)$ is graphic for all $T \subseteq V(G)$.)

We observe also that if M is binary and $M \backslash e$ is graphic, and $M \backslash e \cong \mathcal{M}(G)$, say (where $V(G) \neq \emptyset$), then $M \cong \mathcal{M}(G, T)$ for some $T \subseteq V(G)$. To see this, observe that every representation of $M \backslash e$ over $GF(2)$ may be extended to a representation of M (unless e is a coloop, when the result is clear), and so in particular the representation $e \to v(e)$ ($e \in E(G)$) (defined as above) may be extended, using a vector v, say, to a representation of M. Choose $T \subseteq V(G)$ so that $v = v(\Omega)$ (again, defined as above), and then $M \cong \mathcal{M}(G, T)$.

Some important examples of grafts are given in Fig. 3. In the figure, the circles contain the vertices in T in each case. (The last graft has been labelled so that the element labelled i of $\mathcal{M}(G, T)$ corresponds to the ith column of the representation in Section 9.)

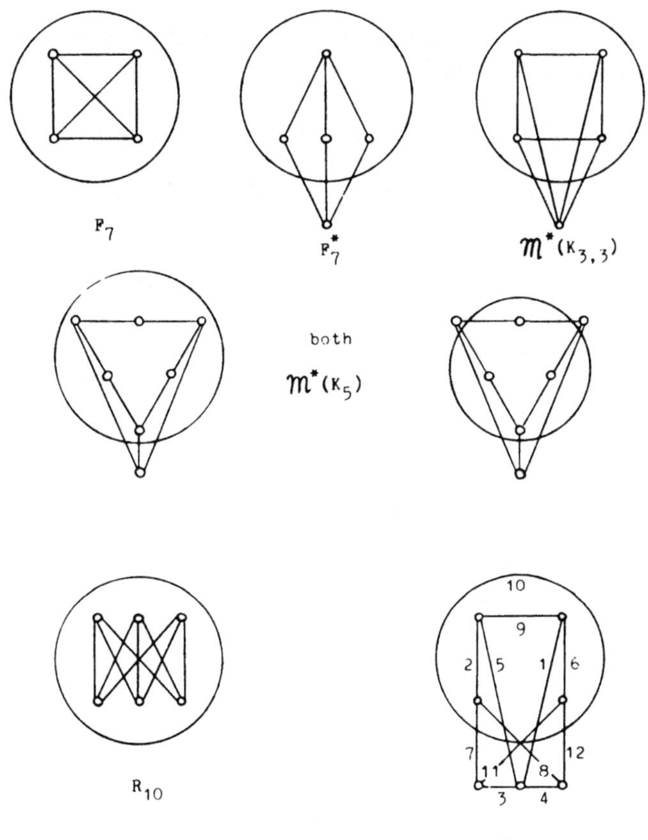

FIGURE 3

When $T \subseteq V(G)$ and e is an edge of G with ends u, v, say, let T/e be a subset of $V(G/e)$ defined as follows (here w is the vertex of G/e obtained by identifying u, v under contraction of e).

$$T/e = T \quad \text{if} \quad u, v \notin T \text{ or if } u = v,$$
$$T/e = (T - \{u\}) \cup \{w\} \quad \text{if} \quad u \in T, v \notin T,$$
$$T/e = (T - \{v\}) \cup \{w\} \quad \text{if} \quad u \notin T, v \in T,$$
$$T/e = T - \{u, v\} \quad \text{if} \quad u, v \in T \text{ and } u \neq v.$$

We observe that if $|T|$ is even then so is $|T/e|$, and indeed $|T/e| = |T|$ unless both $u, v \in T$. When $F \subseteq E(G)$ and $F = \{e_1, \ldots, e_k\}$, T/F is defined to be

$$(\cdots((T/e_1)/e_2)\cdots/e_k).$$

This definition is seen to be independent of the ordering of the members of F. Then it is easy to check the following.

(10.2) *If (G, T) is a graft and F_1, $F_2 \subseteq E(G)$ are disjoint, then*

$$\mathcal{M}(G, T)\backslash F_1/F_2 = \mathcal{M}(G\backslash F_1/F_2, T/F_2).$$

11. Some Graph Theory Lemmas

When G is simple and $e \in E(G)$, $G \circ e$ denotes the graph obtained as follows: let the ends of e be u, v; delete e; if u is cubic in G, contract one edge incident with it; and similarly for v.

(11.1) *If G is simple and 3-connected and $|V(G)| \geqslant 5$, and $e \in E(G)$, then one of G/e, $G \circ e$ is 3-connected.*

Proof. The result is true when $|V(G)| = 5$, as may easily be verified by checking cases. We therefore assume that $|V(G)| \geqslant 6$, so that $|V(G \circ e)| \geqslant 4$.

Let the ends of e be u, v. Assume that $G \circ e$ is not 3-connected. Then $G \circ e$ has a two-vertex cut-set. There are therefore vertices x_1, x_2 of G, so that $\{x_1, x_2\}$ is a cut-set of $G\backslash e$ (but not of G, and so every path linking u to v in G uses one of x_1, x_2, except for the path using e), and such that the neighbour set of u in G is not $\{x_1, x_2, v\}$, and the neighbour set of v is not $\{x_1, x_2, u\}$.

Assume that G/e is not 3-connected, and so there is a vertex y and two non-empty sets X_1, X_2 such that $(X_1, X_2, \{y\}, \{u, v\})$ is a partition of $V(G)$, and such that no vertex in X_1 is adjacent to any vertex in X_2.

Choose $z_1 \in X_1$. G is 3-connected, and so there are three paths linking z_1 with u, v, y, respectively, vertex-disjoint except for z_1. Thus there is a path linking u and v within $X_1 \cup \{u, v\}$, not using e, and so one of x_1, x_2 (x_1, say) is in X_1. Similarly $x_2 \in X_2$, and so $y \neq x_1$, x_2.

Let U_1, V_1 be the sets of vertices in $X_1 \cup \{y\}$ adjacent to u, v, respectively. Let G_1 be the restriction of G to $X_1 \cup \{y\}$. Then every path of G_1 linking U_1 to V_1 passes through x_1; and so there is a partition $(U, V, \{x_1\})$ of $X_1 \cup \{y\}$ with $U_1 \subseteq U \cup \{x_1\}$, $V_1 \subseteq V \cup \{x_1\}$, and with no vertex in U adjacent in G_1 to any in V. Now $y \in U \cup V$; we assume $y \in V$ without loss of generality. Then no vertex in U is adjacent in G to any vertex of G not in U except u, x_1. But G is 3-connected, and so $U = \emptyset$. Thus $U_1 = \emptyset$ or $\{x_1\}$; but $U_1 \neq \emptyset$ since $\{v, y\}$ is not a cut-set of G, and so $U_1 = \{x_1\}$. Thus x_1 is the only vertex in X_1 adjacent to u. Similarly, for one of u, v, x_2 is the only vertex in X_2 adjacent to it.

Now $|V(G)| \geqslant 6$, and so one of $|X_1|$, $|X_2| \neq 1$. We assume $|X_1| \geqslant 2$ without loss of generality. Choose $z_1 \in X_1 - \{x_1\}$, and choose three paths of G linking z_1 to x_1, v, y, respectively, vertex-disjoint except for z_1. We deduce that there is a path of G linking v to y avoiding x_1 and contained within $X_1 \cup \{v, y\}$. Hence u is not adjacent to y (for this would give us a u–v path avoiding x_1, x_2, and e). Now we recall that the neighbour set of u is not $\{x_1, x_2, v\}$, and so some $z_2 \in X_2 - \{x_2\}$ is adjacent to u. Thus x_2 is the only neighbour of v in X_2 (because it is clearly not the only neighbour of u), and $|X_2| \geqslant 2$. But then as before there is a path linking u to y avoiding x_2 and contained within $X_2 \cup \{u, y\}$; this, composed with the path within $X_1 \cup \{v, y\}$ linking v and y and avoiding x_1 gives us a u–v path avoiding x_1, x_2, and e, a contradiction.

G is said to be a *subdivision* of a graph H if G may be obtained from H by (repeatedly) picking an edge and replacing it by two adjacent edges in series and a vertex of valency 2. It is easy to see that if G is connected, and $\mathscr{M}(G)$ has at least four series classes, then $\mathscr{M}(G)$ is cyclically 3-connected if and only if G is a subdivision of a simple 3-connected graph.

A series class of $\mathscr{M}(G)$ which is independent in $\mathscr{M}(G)$ is called a *line* of a graph G. However, when G is a subdivision of a 3-connected graph, a line is the edge-set of a path subgraph of G, and it will be convenient to use the word "line" to refer to such a path as well. We shall use expressions such as "the line L passes through vertex v" and "vertex v is an end of L" without further explanation.

G is said to be a *minor* of a graph H if $G = H \backslash F_1 / F_2$ for disjoint subsets F_1, F_2 of $E(H)$. (We observe that this definition does not permit the removal of isolated vertices, and so if G has at least k components then so does every minor G. This is for technical convenience only.) It is easy to see that if G has a $K_{3,3}$ minor then it has a subgraph which is a subdivision of $K_{3,3}$.

(11.2) *Let G be a 3-connected graph with a $K_{3,3}$ minor, and let v_1, v_2, v_3 be distinct vertices of G, pairwise adjacent. Then G has the graph of Fig. 4 as a minor.*

The proof is in three steps, as follows.

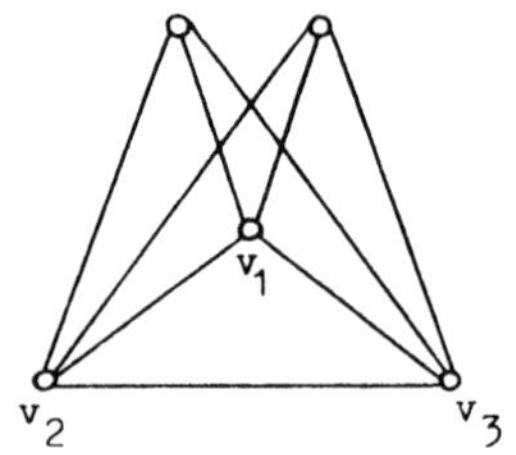

FIGURE 4

(11.3) *If* $|V(G)| = 6$ *then* (11.2) *is true.*

Proof. Check cases.

(11.4) *If G is a subdivision of $K_{3,3}$ and $v_1, v_2, v_3 \in V(G)$ are distinct, and do not all lie on the same line of G, then either $G \cong K_{3,3}$, or there is an edge e, not joining two of v_1, v_2, v_3, such that G/e is a subdivision of $K_{3,3}$ and v_1, v_2, v_3 do not all lie on the same line of G/e.*

Proof. We assume $G \not\cong K_{3,3}$, so that G has a vertex v of valency 2. Let e_1, e_2 be the edges incident with v, and let their respective other ends be u_1, u_2. Now if $v \neq v_1, v_2, v_3$ we may take $e = e_1$ and (11.4) holds. We assume therefore that $v = v_3$. Now u_1, u_2, v all lie on the same line of G and so $\{u_1, u_2\} \neq \{v_1, v_2\}$. We assume without loss of generality that $u_1 \neq v_1, v_2$. We may assume that (11.4) is not satisfied taking $e = e_1$, and so u_1 is an end of the line of G through v_1 and u_1, v_1, v_2 are all on the same line of G. Thus $u_2 \neq v_1, v_2$, and so similarly, u_2, v_1, v_2 are all on the same line of G. This is impossible.

Proof of (11.2). We use induction on $|E(G)|$. Evidently we may assume that G is simple. Let H be a subgraph of G which is a subdivision of $K_{3,3}$, with as few edges as possible. Suppose that $v_1 \notin V(H)$. Let e be an edge of G incident with v_1 but not with v_2 or v_3 (this exists since G is 3-connected). Now G/e, $G \circ e$ both have $K_{3,3}$ minors, and v_1, v_2, v_3 is a triangle of G/e; thus we may assume that G/e is not 3-connected. Hence by (11.1) $G \circ e$ is 3-connected, and we may assume that v_1, v_2, v_3 is not a triangle of $G \circ e$. Thus v_1 is cubic in G.

Now G/e is not 3-connected, and so there is a vertex y and there are non-empty sets X_1, X_2 such that $(X_1, X_2, \{u, v_1\}, \{y\})$ is a partition of $V(G)$ (where u is the other end of e), such that no vertex in X_1 is adjacent to any in X_2. Now $\{u, y\}$ is not a cut-set of G, and so v_1 is adjacent to a vertex in X_1 and a vertex in X_2. But this is impossible, since v_1 has neighbours v_2, v_3, u, and $u \notin X_1, X_2$, and v_2, v_3 are adjacent.

We deduce that $v_1 \in V(H)$, and similarly $v_2, v_3 \in V(H)$. Now v_1, v_2, v_3

are not all on the same line of H because H was chosen so that $|E(H)|$ was minimum. By (11.4) there exists $F_2' \subseteq E(H)$ so that $H/F_2' \cong K_{3,3}$ and no two of v_1, v_2, v_3 are identified under contraction of F_2'.

Choose $F_1 \subseteq E(G) - E(H)$ maximal so that the graph $G \backslash F_1$ is connected, and put $F_2 = (E(G) - F_1) - (E(H) - F_2')$. The result follows by applying (11.3) to $G \backslash F_1/F_2$. (This complicated choice of F_1, F_2 is to avoid isolated vertices in the minor.)

(11.5) *Let G be a graph, and let H be a subgraph with the same vertex-set. Suppose that H is a subdivision of $K_{3,3}$, and that every vertex v of valency 2 in H is adjacent in G to a vertex not in the line of H which passes through v. Then G is 3-connected.*

Proof. Suppose that x, y are distinct vertices of G, that $X, Y \subseteq V(G)$ are non-empty, and that $(X, Y, \{x, y\})$ is a partition of $V(G)$, and that no vertex in X is adjacent in G to any vertex in Y. Since $K_{3,3}$ is 3-connected, one of X, Y (X, say) contains no vertex with valency 3 in H, and indeed X is a subset of a line L of H, and contains neither of the ends of L. Thus x, y are also in L. Choose $v \in X$; and then v is not adjacent to any vertex not in L, contrary to hypothesis.

12. Extensions of 3-Connected Graphs

In this section we prove the following theorem.

(12.1) *Let M be a binary matroid, and let $M \backslash e \cong \mathscr{M}(G)$, where G is a 3-connected graph with a $K_{3,3}$ minor. Then either M is graphic or M has a minor isomorphic to F_7^*, R_{10}, or R_{12}.*

The proof will be by means of grafts. We have $M \backslash e \cong \mathscr{M}(G)$, and so there exists $T \subseteq V(G)$ so that $M \cong \mathscr{M}(G, T)$. If $|T| = 0$ or 2 or is odd then M is graphic. The result is therefore implied by the next theorem.

(12.2) *Let G be a 3-connected graph with a $K_{3,3}$ minor, and let $T \subseteq V(G)$, with $|T|$ even and $|T| \geqslant 4$. Then for some disjoint subsets F_1, F_2 of $E(G)$, $(G \backslash F_1/F_2, T/F_2)$ is one of the three grafts of Fig. 3 which correspond to F_7^*, R_{10}, and R_{12}.*

(We see that this is a purely graph-theoretic statement.)

Proof. If possible, let (G, T) be a counterexample, with $|E(G)|$ as small as possible. Clearly G is simple. Let H be a subgraph of G which is a subdivision of $K_{3,3}$, and choose H so that $|V(H)|$ is as small as possible. We show first that $V(H) = V(G)$. Let us suppose that this is not true.

(1) *Let e be an edge of G, with ends u, v and suppose that $v \notin V(H)$. If G/e is 3-connected then $u, v \in T$ and $|T| = 4$. If $G \circ e$ is 3-connected, then one of the following holds*:

(i) *u is cubic, $|T| = 4$, and T contains u and both its neighbours different from v,*

(ii) *v is cubic, $|T| = 4$, and T contains v and both its neighbours different from u,*

(iii) *u, v are cubic, $|T| = 6$, and T contains u, v and all their neighbours.*

For if G/e is 3-connected, then $u, v \in T$ and $|T| = 4$, because otherwise $(G/e, T/e)$ would be a smaller counterexample. Similarly if $G \circ e$ is 3-connected, one of (i), (ii), (iii) holds. (Recall that in the definition of $G \circ e$, it was not stipulated which edge incident with u should be contracted when u is cubic; thus we may choose this edge, if possible, so that it does not join two vertices in T.)

(2) *Each $v \in V(G) - V(H)$ is in T and is cubic.*

For if $v \notin T$ then by (1), $|T| = 4$, and T contains all neighbours of v and all their neighbours (except v), which is impossible from 3-connectedness. Thus $v \in T$. If v is not cubic, then by (1), $|T| = 4$ and T contains all neighbours of v, which is impossible by counting.

(3) $|T| = 4$, *provided that $V(H) \neq V(G)$.*

For suppose that $|T| \geqslant 6$. Then by (1), $|T| = 6$, and T contains $v \in V(G) - V(H)$, its neighbours, and all their neighbours, which is impossible by 3-connectedness.

Choose $v \in V(G) - V(H)$, and let u_1, u_2, u_3 be its neighbours.

(4) *T does not contain all of u_1, u_2, u_3.*

For if it does, then by (3), $T = \{v, u_1, u_2, u_3\}$. Choose $v' \notin T$. Choose three paths P_1, P_2, P_3 between v and v', vertex-disjoint except for v, v'. Then each P_i uses just one of u_1, u_2, u_3, u_i, say. Choose $F_1 \subseteq E(G)$, maximal such that F_1 contains no edge of any P_i, and such that the graph $(V(G), E(G) - F_1)$ is connected. Let $F_2 \subseteq E(G)$ contain those edges not in F_1 and not in any P_i, and also those edges in P_i not incident with u_i $(i = 1, 2, 3)$. Then $(G \backslash F_1/F_2, T/F_2)$ is the graft corresponding to F_7^*, a contradiction.

We assume that $u_3 \notin T$. Let e be the edge joining v and u_3.

(5) $u_1, u_2 \in T$.

For by applying (1) to u_3 and v, we see that T contains all neighbours of v different from u_3, and so $u_1, u_2 \in T$.

Now by (1) and (4), G/e is not 3-connected. Choose a vertex y and non-empty subsets X_1, X_2 of $V(G)$ so that $(X_1, X_2, \{u_3, v, y\})$ is a partition of $V(G)$ and such that no vertex in X_1 is adjacent to any in X_2. Now $\{u_3, y\}$ is not a cut-set, and so v is adjacent to a vertex in X_1 and a vertex in X_2; $u_1 \in X_1$, $u_2 \in X_2$, say.

Now $v \notin V(H)$, and so all vertices in H within one of $X_i \cup \{u_3, y\}$ $(i = 1, 2)$ $(X_2 \cup \{u_3, y\}$, say) are on the same line of H.

G is 3-connected, and so there are two paths linking u_2 to u_3, y, respectively, contained within $X_2 \cup \{u_3, y\}$ and vertex-disjoint except for u_2. Thus there are subsets Y_1, Y_2 of $X_2 \cup \{u_3, y\}$, such that $Y_1 \cap Y_2 = \varnothing$, $u_3 \in Y_1$, $y \in Y_2$, $u_2 \notin Y_1 \cup Y_2$, and for $i = 1, 2$, the restriction of G to Y_i is connected, and u_2 is adjacent to a vertex in Y_i. Choose Y_1, Y_2 with these properties so that $Y_1 \cup Y_2$ is maximal. Certainly $Y_1 \cup Y_2 \subseteq (X_2 - \{u_2\}) \cup \{u_3, y\}$, and we claim that equality holds. For if not, then some vertex $z \in X_2 - \{u_2\}$ is not in $Y_1 \cup Y_2$ but is adjacent to a vertex in $Y_1 \cup Y_2$ (in Y_1, say); then we may replace Y_1, Y_2 by $Y_1 \cup \{z\}$, Y_2, contrary to the maximality of $Y_1 \cup Y_2$. Hence (Y_1, Y_2) is a partition of $(X_2 - \{u_2\}) \cup \{u_3, y\}$.

Let F be the set of all edges with both ends in the same Y_i. Then G/F is the graph of Fig. 5. (Here y_1 and y_2 are the vertices obtained from Y_1, Y_2, respectively, after contraction of F. In the figure y_1 and y_2 may or may not be adjacent.)

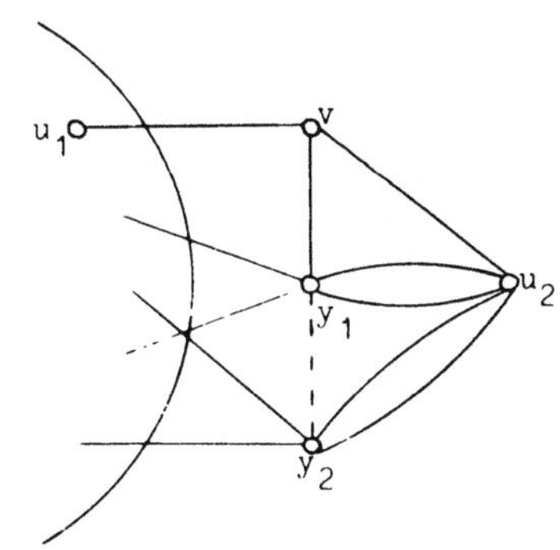

FIGURE 5

(6) *G/F is 3-connected.*

Let w be any vertex in X_1. Then there are three paths of G linking w with u_2, vertex-disjoint except for w and u_2; and so there are three paths Q_1, Q_2,

Q_3 of G/F, linking w with v, y_1, y_2, respectively, vertex-disjoint except for w. Thus w is 3-connected to u_2 in G/F. Also v, y_1, y_2 are 3-connected to u_2 because they are adjacent to it. Suppose therefore that (Z_1, Z_2, Z) is a partition of $V(G/F)$ with $|Z| \leqslant 2$ and Z_1, Z_2 non-empty, such that no vertex in Z_1 is adjacent to any in Z_2. Then we must have $u_2 \in Z$, because every other vertex is 3-connected to u_2. In particular, $Z - \{u_2\}$ is not a cut-set, and so u_2 has a neighbour in Z_1 and one in Z_2. Therefore, to show that G/F is 3-connected, it remains to show that the pairs v, y_1; v, y_2; y_1, y_2 are all 3-connected. v, y_1 are obviously 3-connected because they are adjacent. Now if $F = \varnothing$ we have $G/F = G$, and so G/F is 3-connected as required. We thus may assume that $F \neq \varnothing$, and hence one of $Y_1 - \{u_3\}$, $Y_2 - \{y\}$ is non-empty. If $Y_1 - \{u_3\} \neq \varnothing$ then because G is 3-connected, some vertex of $Y_1 - \{u_3\}$ is adjacent to a vertex not in $Y_1 - \{u_3\}$, different from u_3 and u_2 and hence in Y_2; thus y_1 and y_2 are adjacent in G/F. Similarly, y_1, y_2 are adjacent in G/F if $Y_2 - \{y\} \neq \varnothing$. Thus y_1, y_2 are 3-connected in G/F. It remains to check the pair v, y_2. Choose $w \in X_1$ and choose paths Q_1, Q_2, Q_3 as before; and then $Q_3 \cup Q_1$, and y_2, u_2, v, and y_2, y_1, v are three paths of G/F, linking y_2 to v, pairwise vertex-disjoint except for y_2 and v. Hence y_2 is 3-connected to v, and so G/F is 3-connected, as required.

(7) *$F = \varnothing$ and $X_2 = \{u_2\}$.*

For if not we have a contradiction to the minimality of G, since G/F has a $K_{3,3}$ minor and $|T/F| \geqslant 4$ (because $v, u_1, u_2 \in T/F$ and $|T/F|$ is even).

(8) *$y \in T$.*

For let f be the edge of G joining u_2 and y. Then $G \circ f$ is not 3-connected, because in it v is only adjacent to two vertices; and so by (11.1), G/f is 3-connected. Now G/f has a $K_{3,3}$ minor, and so by the minimality of G, $|T/f| \leqslant 2$ and so $y \in T$.

(9) *$V(H) = V(G)$.*

To complete the proof of (9), let G' be the graph obtained from G by deleting u_2 and v and adding three new edges joining the pairs (y, u_1), (y, u_3), (u_1, u_3), respectively. Clearly G' is 3-connected, and has a $K_{3,3}$ minor, and so by (11.2) there exist $F_1, F_2 \subseteq E(G')$, disjoint, so that $G' \backslash F_1/F_2$ is the graph of Fig. 4, where v_1, v_2, v_3 are relabelled as u_1, u_3, y. Moreover we can choose F_1, F_2 so that they contain none of the new edges; and then $(G \backslash F_1/F_2, T/F_2)$ is the graft of Fig. 3 corresponding to R_{12}, a contradiction.

We recall that H is a subgraph of G which is a subdivision of $K_{3,3}$, with $|V(H)|$ minimum. It follows that

(10) *Two vertices in the same line of H are adjacent in H if they are adjacent in G.*

For otherwise we could find a subdivision of $K_{3,3}$ with fewer vertices. We say that distinct vertices are *collinear* if they are in the same line of H.

(11) *For any edge e of H, in a line L of H, say, with ends u_1, u_2, say, one of the following is true*:

(i) *u_1, u_2 are both ends of L,*
(ii) *$u_1, u_2 \in T$ and $|T| = 4$,*
(iii) *One of u_1, u_2, say, u_1, is an end of L and there is a vertex v_e not on L which is cubic in G, adjacent to u_2 in G, and collinear with u_1.*
(We see that v_e necessarily has valency 2 in H.)

For suppose that u_2 is not an end of L. Then H/e is a subdivision of $K_{3,3}$, and so by the minimality of G, either $|T/e| = 2$ or G/e is not 3-connected. $|T/e| = 2$ gives alternative (ii); and G/e not 3-connected gives alternative (iii), by (11.5) and (10).

(12) $H \not\cong K_{3,3}$.

For if $T \subseteq V(K_{3,3})$ and $|T| = 4$ then we can make $(K_{3,3} \backslash F_1/F_2, T/F_2)$ the graft corresponding to F_7^* if we choose F_1, F_2 suitably. If $|T| = 6$ we have the graft corresponding to R_{10}; and by hypothesis $|T| \geqslant 4$ and is even.

Thus H has a vertex of valency 2. Let v be such a vertex, and let e_1, e_2 be edges of H incident with it, and let u_1, u_2 be their respective other ends. Let $e_3, \ldots, e_k$ be the other edges of G incident with v, and let $u_3, \ldots, u_k$ be their other ends. Let L be the line of H through v; then by (10), none of $u_3, \ldots, u_k$ is on L.

(13) *v has valency* 3 *in G* (*that is*, $k = 3$).

Now for each $i \geqslant 3$, $G \backslash e_i$ has a $K_{3,3}$ minor, and so by choice of G, $G \backslash e_i$ is not 3-connected. Suppose that $k \geqslant 4$. By (11.5) each u_i has valency 3 in G and 2 in H. Let x_i, y_i be the vertices adjacent to u_i in H. By applying (11) to x_i, u_i we see that $|T| = 4$ and $x_i, u_i \in T$ (case (iii) of (11) cannot occur, because v is the only vertex of G adjacent to u_i in G but not on the same line of H as u_i, and v is not cubic in G). Similarly, $y_i \in T$. Thus $\bigcup_{3 \leqslant i \leqslant k} \{x_i, u_i, y_i\} \subseteq T$, and yet $|T| = 4$. Let L' be the line of H containing x_3, u_3, y_3. Then for each $i \geqslant 3$, $|\{x_i, u_i, y_i\} \cup \{x_3, u_3, y_3\}| \leqslant 4$, and so at least two of x_i, u_i, y_i are on L'; thus all three are. Hence $u_3, \ldots, u_k$ are all on L'. Now $L' \neq L$ and so one of u_1, u_2 (u_1, say) is not on L'. We know that every vertex in T is on L', because $|T| \leqslant 4$ and $|\{x_3, u_3, y_3\} \cup$

$|\{x_4, u_4, y_4\}| \geq 4$, and so $u_1, v \notin T$. Apply (11) to u_1, v; we deduce that u_1 is an end of L and there is a vertex w of G, cubic in G, adjacent to v and collinear with u_1. But then w must be on L', which is impossible.

(14) $|T| = 4$, *and* T *contains* v *and at least one of* u_1, u_2.

For u_3 is the only vertex of G adjacent to v and not on L, by (13). u_3 is not collinear with both u_1 and u_2; we may therefore assume that u_3 is not collinear with u_1. Thus there is no vertex except u_2 adjacent to v and collinear with u_1. We apply (11) to v, u_1, and deduce that $|T| = 4$ and $v, u_1 \in T$.

(15) $u_1, u_2 \in T$.

For suppose that $u_2 \notin T$. By (11) applied to v, u_2 we deduce that u_2 is an end of L, and u_3 is cubic in G, adjacent to v, and collinear with u_2. Let x, y be the vertices different from v, adjacent to u_3, where either $y = u_2$ or y is between $u_2$and u_3 in H. Now u_3 has valency 2 in H, and so T contains u_3 and one of x, y by (14). Suppose that $y \notin T$. Then $u_3, x \in T$ and $T = \{v, u_1, u_3, x\}$. Delete all edges except those in H and the edge e_3; contract all edges in the line through u_3 except the two edges incident with u_3; contract all edges in L except e_1 and e_2; contract all edges in the third line of H passing through u_2; and contract all except one edge in every other line of H. This produces the graft corresponding to R_{12}, a contradiction.

Thus $y \in T$, and $T = \{v, u_1, u_3, y\}$. (So $y \neq u_2$, since $u_2 \notin T$.) Delete e_1 and all edges of G not in H, except e_3; and contract all remaining edges except e_2, e_3, the three edges of H incident with y or u_3, and an edge incident with u_2 on the third line of H passing through u_2. This produces the graft corresponding to F_7^*, a contradiction.

(16) *There are at most two vertices with valency* 2 *in* H *and all such vertices lie on* L.

For by (15), T contains any such vertex and both its neighbours in H and yet $|T| = 4$.

(17) *All vertices in* T *lie on* L.

For we know that $v, u_1, u_2 \in T$ and lie on L. Suppose that $v' \in T$ and does not lie on L. By (16), v is the only vertex of H of valency 2, and so G is the graph of Fig. 6. (G has no other edges, because any other edge could be deleted without destroying 3-connectedness.) If $v' = u_3$ or x_1, delete the edges (u_3, x_2), (u_3, u_1), (u_2, x_1) and contract (u_3, x_1), (x_2, x_3); this produces the graft corresponding to F_7^*, a contradiction. Similarly $v' \neq x_2$, and so $v' = x_3$. But then we have the graft corresponding to R_{12}, a contradiction.

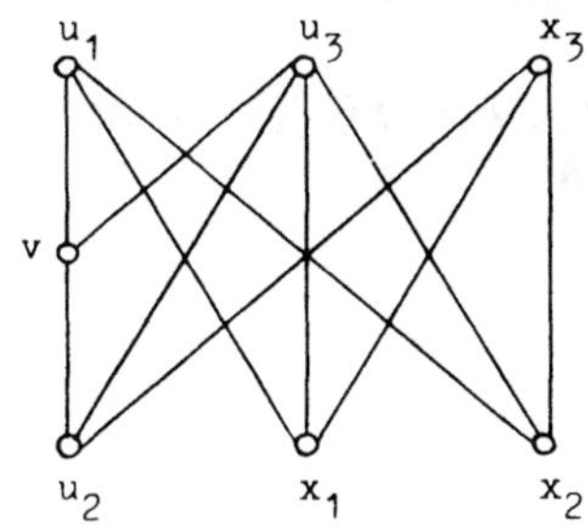

FIGURE 6

(18) *Conclusion.* Thus L contains four vertices, and H is the graph of Fig. 7, and $|T| = \{p, q, r, s\}$.

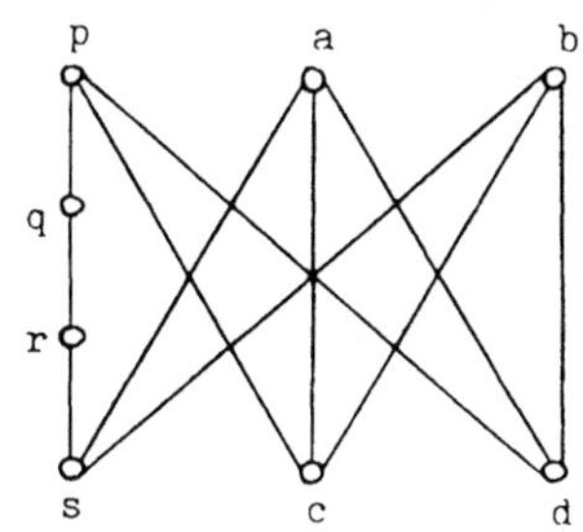

FIGURE 7

If q is adjacent to c in G, then adding the edge (q, c) to H and removing (p, c) gives an alternative choice of H which does not satisfy (17), which is impossible. Similarly q is not adjacent to d, and r is not adjacent to a or b. Thus, relabelling if necessary, we may assume that q is adjacent to a and r to c. (And G has no other edges, because they could be removed without destroying 3-connectedness.) But then deleting (r, s), (a, c), (a, d), (p, c) and contracting (b, d), (b, c), (a, s) gives the graft corresponding to F_7^*, a contradiction. This completes the proof of (12.2) and hence (12.1).

13. EXTENSIONS OF SUBDIVISIONS

We now prove the following.

(13.1) *Let G be a subdivision of a simple 3-connected graph H which has a $K_{3,3}$ minor, and let $T \subseteq V(G)$ be given. Then one of the following is true:*

(i) *$\mathcal{M}(G, T)$ is graphic,*

(ii) *there exists $F \subseteq E(G)$ such that $|L - F| = 1$ for each line L of G (so that $G/F \cong H$) and such that $\mathcal{M}(G/F, T/F)$ is non-graphic (that is, $|T/F| \geqslant 4$ and is even),*

(iii) *there exists F_1, $F_2 \subseteq E(G)$, disjoint, so that $(G \backslash F_1/F_2, T/F_2)$ is one of the grafts of Fig. 8.*

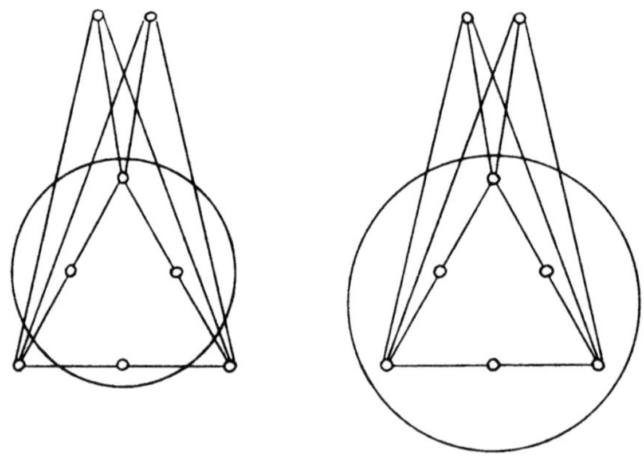

FIGURE 8

(13.1) may be deduced from the following lemma.

(13.2) *Let G be a subdivision of a simple 3-connected graph H, and let $T \subseteq V(G)$ be given. Then one of the following is true:*

(i) *$\mathcal{M}(G, T)$ is graphic,*

(ii) *there exists $F \subseteq E(G)$ such that $|L - F| = 1$ for each line L of G and such that $|T/F| \geqslant 4$ and is even,*

(iii) *there are three lines L_1, L_2, L_3 of G forming a circuit, with $T \subseteq V(L_1) \cup V(L_2) \cup V(L_3)$, and there exists $F \subseteq E(G)$ such that $|L_i - F| = 2$ $(i = 1, 2, 3)$, $|L - F| = 1$ for every other line L of G, and T/F contains all three of the vertices of G/F of valency 2.*

Proof of (13.1) *from* (13.2). Let G, T be as in (13.1). If (ii) of (13.2) holds then (ii) of (13.1) holds, by (10.3). Thus by (13.2), we may assume that alternative (iii) of (13.2) holds, and that $|T|$ is even. Let v_1, v_2, v_3 be the vertices in the corresponding triangle of H. By (11.2), there exists F_1', $F_2' \subseteq E(H)$, disjoint, so that $H \backslash F_1'/F_2'$ is the graph of Fig. 4. Let F_1'', F_2'' be the corresponding subsets (in the natural way) of $E(G/F)$. Put $F_1 = F_1''$,

$F_2 = F_2'' \cup F$. Then $(G \backslash F_1/F_2, T/F_2)$ is one of the grafts of Fig. 8, as required.

Proof of (13.2). If possible, choose a counterexample (G, T) with $|E(G)|$ minimum. Then clearly $|T| \geqslant 4$ and is even, for otherwise $\mathcal{M}(G, T)$ is graphic.

(1) *No two edges of G are in series in* $\mathcal{M}(G, T)$.

For if e_1, e_2 are in series in $\mathcal{M}(G, T)$, they are in series in G and hence are in the same line of G. We contract e_1. By the minimality of the counterexample we know that one of (i), (ii), (iii) holds for $(G/e_1, T/e_1)$. If (ii) or (iii) is true for $(G/e_1, T/e_1)$ then it is true for (G, T) which is impossible. Thus (i) holds and $\mathcal{M}(G, T)/e_1$ is graphic. But e_1 and e_2 are in series in $\mathcal{M}(G, T)$, and so $\mathcal{M}(G, T)$ is graphic, a contradiction.

(2) *Every vertex of valency* 2 *of G is in T. No two vertices of valency* 2 *are adjacent.*

For if v has valency 2 and $v \notin T$, it is easy to see that the edges incident with v are in series in $\mathcal{M}(G, T)$, contrary to (1). If v_1, v_2 both have valency 2 and are adjacent, then we know that $v_1, v_2 \in T$, and again it is easy to see that the two edges with just one end in $\{v_1, v_2\}$ are in series in $\mathcal{M}(G, T)$.

If v is a vertex of G of valency 2, let e, f be the edges incident with it, and let x, y be their other ends. Let G_v be the graph G with e, f exchanged. Thus $E(G_v) = E(G)$ and $G_v \cong G$. It is easy to check that $\mathcal{M}(G_v, T \Delta \{x, y\}) = \mathcal{M}(G, T)$.

(3) *If* $v_1, \ldots, v_k$ *are vertices of G of valency* 2, *and* v_i *is adjacent to* x_i, y_i $(i = 1, \ldots, k)$, *then*

$$|T \Delta \{x_1, y_1\} \Delta \cdots \Delta \{x_k, y_k\}| \geqslant 4.$$

For $\mathcal{M}(G', T') = \mathcal{M}(G, T)$, where $T' = T \Delta \{x_1, y_1\} \Delta \cdots \Delta \{x_k, y_k\}$ and G' is obtained from G by exchanging for each i the edges incident with v_i. But $\mathcal{M}(G, T)$ is not graphic, and so $|T'| \geqslant 4$.

(4) *There do not exist vertices* v_1, v_2, v_3 *of G of valency* 2 *and vertices* u_1, u_2, u_3 *of valency* $\geqslant 3$ *such that the following all hold*:

(a) u_1 *is adjacent to* v_1 *but not to* v_2, v_3,
(b) u_2 *is adjacent to* v_2 *but not to* v_3,
(c) u_3 *is adjacent to* v_3.

For suppose that they do exist. For $i = 1, 2, 3$, let e_i, f_i be the edges incident with v_i, with e_i incident with u_i. Choose $F' \subseteq E(G)$ so that F' contains none of $e_1, e_2, e_3, f_1, f_2, f_3$ and so that for every vertex $v \neq v_1, v_2, v_3$ of valency 2, F' contains just one edge incident with v. For $i = 1, 2, 3$, choose $g_i \in \{e_i, f_i\}$ as follows:

$$g_i = f_i \quad \text{if} \quad u_i \in T/(F' \cup \{g_j : 1 \leqslant j < i\}),$$
$$g_i = e_i \quad \text{otherwise.}$$

Put $F = F' \cup \{g_1, g_2, g_3\}$; then $u_1, u_2, u_3 \in T/F$ (because of hypotheses (a), (b), (c)), and so $|T/F| \geqslant 4$ (since it is even) and (ii) is satisfied, which is impossible.

(5) *G has at least three vertices of valency* 2.

Let $v_1, \ldots, v_k$ be the vertices of G of valency 2. For each i, let e_i, f_i be the edges incident with v_i, and let x_i, y_i be their other ends.

Now $k > 0$, since (ii) is not satisfied with $F = \varnothing$.

If $k = 1$, then $x_1 \in T$, since (ii) is not satisfied with $F = \{e_1\}$; and similarly $y_1 \in T$. By (3), $|T - \{x_1, y_1\}| \geqslant 4$, and so $|T/e_1| \geqslant 4$; thus (ii) is satisfied with $F = \{e_1\}$, a contradiction.

Now suppose that $k = 2$. By relabelling if necessary we may assume that $x_1 \neq x_2, y_2$, and $x_2 \neq x_1, y_1$, although possibly $y_1 = y_2$. If there exists $w \neq x_1, x_2, y_1, y_2, v_1, v_2$ with $w \in T$, we define g_i for $i = 1, 2$ as follows:

$$\text{if } x_i \notin T \quad \text{put} \quad g_i = e_i,$$
$$\text{if } x_i \in T \quad \text{put} \quad g_i = f_i.$$

Put $F = \{g_1, g_2\}$. Then $x_1, x_2, w \in T/F$ and so (ii) is satisfied, which is impossible. Thus $T \subseteq \{x_1, x_2, y_1, y_2, v_1, v_2\}$. Now if $x_1, x_2 \in T$ then by (3), $|T \Delta \{x_1, y_1\} \Delta \{x_2, y_2\}| \geqslant 4$, and so $y_1, y_2 \notin T$, and $y_1 \neq y_2$; but then (ii) is satisfied with $F = \{f_1, f_2\}$. We assume $x_1 \notin T$ without loss of generality. If $x_2 \notin T$, put $F = \{e_1, e_2\}$, and (ii) is satisfied. Thus $x_2 \in T$. If $y_2 \in T$, then $T \Delta \{x_2, y_2\} \subseteq \{v_1, v_2, y_1\}$ contrary to (3). Thus $y_2 \notin T$. But $|T| \geqslant 4$, and so $y_1 \neq y_2$, and $y_1 \in T$. Put $F = \{e_1, f_2\}$ and then (ii) is satisfied, a contradiction.

(6) *G has exactly three vertices of valency* 2, *and the corresponding edges of H form a triangle.*

For let L_1, L_2, L_3 be the lines of G passing through v_1, v_2, v_3, respectively. If $L_1 \cup L_2 \cup L_3$ does not include a circuit then u_1, u_2, u_3 can be chosen to contradict (4). Thus, since H is simple, $L_1 \cup L_2 \cup L_3$ is a circuit. This is true for every choice of v_1, v_2, v_3, and so $k = 3$.

(7) *Conclusion.* We may thus assume that $x_1 = y_2$, $x_2 = y_3$, $x_3 = y_1$. Now $v_1, v_2, v_3 \in T$, and so, by the falsity of (iii), there exists $w \in T$ with $w \neq x_1, x_2, x_3, v_1, v_2, v_3$. Choose $g_1 \in \{e_1, f_1\}$ so that $x_3 \in T/g_1$. Choose $g_2 \in \{e_2, f_2\}$ so that $x_1 \in T/\{g_1, g_2\}$. Put $g_3 = f_3$, and put $F = \{g_1, g_2, g_3\}$. Then $x_3, x_1, w \in T/F$, and so $|T/F| \geqslant 4$ and (ii) is satisfied, which is impossible. This completes the proof.

14. Proof of the Main Theorem

If (G, T) is one of the grafts of Fig. 8, then $\mathscr{M}(G, T)$ has an R_{12} minor, obtained by contracting the element Ω. Thus we may combine (12.1) and (13.1) to give the following.

(14.1) *Let M be a binary matroid, and let $M \backslash e \cong \mathscr{M}(G)$, where G is a subdivision of a graph H which is simple and 3-connected, and has a $K_{3,3}$ minor. Then either M is graphic or M has a minor isomorphic to one of F_7^*, R_{10}, and R_{12}.*

Proof. Choose $T \subseteq V(G)$ so that $M \cong \mathscr{M}(G, T)$. By (13.1), either M is graphic, or it has an R_{12} minor, or there exists $F \subseteq E(G)$ so that G/F is isomorphic to H and M/F is non-graphic. In the third case we apply (12.1) to M/F.

We use (14.1) for the third and final step in the proof of the following main theorem.

(14.2) *Let M be a 3-connected regular matroid. Then either M is graphic or cographic or M has a minor isomorphic to one of R_{10}, R_{12}.*

Proof. Let M be regular and 3-connected, and be neither graphic nor cographic. By Tutte's characterization of the graphic matroids [14], M has a minor isomorphic to one of $\mathscr{M}(K_{3,3})$, $\mathscr{M}(K_5)$. Suppose that M has no $\mathscr{M}(K_{3,3})$ minor. $M \ncong \mathscr{M}(K_5)$ since M is not graphic; but $\mathscr{M}(K_5)$ is a splitter for the class of regular matroids without $\mathscr{M}(K_{3,3})$ minors, by (7.5), and so M is 1- or 2-separable, a contradiction. Thus M has a minor isomorphic to $\mathscr{M}(K_{3,3})$.

By (7.2), there exists $Z_0 \subseteq E(M)$ such that $M \times Z_0$ is a subdivision of $\mathscr{M}(K_{3,3})$. We may therefore choose $Z \subseteq E(M)$ with $|Z|$ maximum such that $M \times Z$ is graphic, is cyclically 3-connected, and has an $\mathscr{M}(K_{3,3})$ minor. By (5.4), there is an adjoinable Z-arc A. By (5.2), $M \times (Z \cup A)$ is cyclically 3-connected and has an $\mathscr{M}(K_{3,3})$ minor, and so by the maximality of Z,

$M \times (Z \cup A)$ is not graphic. Choose $e \in A$, and choose a connected graph G and $T \subseteq V(G)$ such that $M \times Z \cong \mathcal{M}(G)$ and

$$(M \times (Z \cup A))/(A - \{e\}) \cong \mathcal{M}(G, T).$$

Now $M \times (Z \cup A)$ is a subdivision of $\mathcal{M}(G, T)$ and so $\mathcal{M}(G, T)$ is not graphic. But $\mathcal{M}(G)$ is cyclically 3-connected, and so (by the remark after (11.1)) G is a subdivision of a simple 3-connected graph; and G has a $K_{3,3}$ minor. By (14.1), $\mathcal{M}(G, T)$, and hence M, has a minor isomorphic to one of F_7^*, R_{10}, or R_{12}, but F_7^* is impossible since M is regular. This completes the proof.

Now we can prove the main result.

(14.3) *Every regular matroid M may be constructed by means of* 1-, 2-, *and* 3-*sums, starting with matroids each isomorphic to a minor of M and each either graphic or cographic or isomorphic to R_{10}.*

Proof. We use induction on $|E(M)|$. Let M be a regular matroid. If M is graphic or cographic or isomorphic to R_{10} the result is true. Suppose not. Then by (14.2), if M is 3-connected then it has an R_{10} or R_{12} minor. By (7.4) if M has an R_{10} minor, it is 2-separable; and by (9.2) if M has an R_{12} minor then it has a 3-separation (X, Y) with $|X|, |Y| \geqslant 4$. By (2.10), M is expressible as a 1-, 2-, or 3-sum. By (2.1) and (2.6), if M is expressible as a 1- or 2-sum then the parts of the sum are isomorphic to minors of M. If M is not expressible as a 1- or 2-sum then it is 3-connected, and so by (4.1) the parts of the 3-sum are isomorphic to minors of M. Thus M is expressible as the 1-, 2-, or 3-sum of two matroids M_1, M_2, which are both isomorphic to minors of M. Thus M_1 and M_2 are both regular; and so, since they both have fewer elements than M, they may both be obtained in the required way, by induction. Hence so may M.

15. APPENDIX: CASE ANALYSIS

Now we give the detailed case-checking postponed from previous sections. The matroids concerned are all binary, and we represent them by matrices with entries over $GF(2)$, each element corresponding to a column, in the usual way. In order to show that a given matroid has another as a minor, we give matrix representations of both, list the columns of the first corresponding to elements which are to be deleted and contracted, and give a bijection from the remaining columns to the columns of the second matrix. The reader is presumed to be familiar with the matrix operations which correspond to matroid deletion and contraction, and to be able to test if a bijection between the column sets of two matrices over $GF(2)$ gives an isomorphism of the matroids they represent; thus the reader may verify that the bijection given is a matroid isomorphism in each case.

In Fig. 9 we list some matrix representations. The case analysis for (7.2), (7.4), (7.5), and (7.6) is given in the table of Fig. 10. To explain this table, let us take, for example, the fifth line. This deals with $\mathcal{M}^*(K_{3,3})$. If we examine the matrix for $\mathcal{M}^*(K_{3,3})$ in Fig. 9, we see that from symmetry there are only two non-isomorphic ways to add a column to this matrix which is non-zero and not the same as a column already in the matrix. These two possible columns are $(1, 0, 1, 0)^T$ and $(1, 1, 1, 0)^T$. (T denotes transpose.) We deal with these two cases simultaneously, using the symbol $*$, which indicates that the entry may be either 0 or 1. From the fifth line of Fig. 10 we learn the following. Let N_1 be a matroid isomorphic to $\mathcal{M}^*(K_{3,3})$, on a set $(1, 2,..., 9)$, where element i is represented by the ith column of the matrix in Fig. 9. Let M be a matroid on $\{1, 2,..., 9, e\}$, represented by the columns of the same matrix together with one additional column $(1, 0, 1, 0)^T$ or $(1, 1, 1, 0)^T$, so that $M\backslash e = N_1$. Let $N_2 \cong F_7$, and be on a set $\{1, 2,..., 7\}$, where again element i corresponds to the ith column of the matrix for F_7 in Fig. 9. Then

$$M\backslash Z_1/Z_2 \cong N_2,$$

where $Z_1 = \{5, 6\}$, $Z_2 = \{2\}$, and an isomorphism ϕ from $M\backslash Z_1/Z_2$ to N_2 is obtained by mapping elements 1, 3, 4, 7, 8, e, 9 of $M\backslash Z_1/Z_2$ to elements 1, 2, 3, 4, 5, 6, 7 of N_2, respectively. We deduce that if M is a binary matroid and $M\backslash e \cong \mathcal{M}^*(K_{3,3})$, then either M has an F_7 minor, or e is a loop, coloop, or parallel element of M. By taking duals, it follows that if $N \cong \mathcal{M}(K_{3,3})$ and $\mathcal{F}$ is the class of regular matroids, then the hypotheses of (7.1) are satisfied, and so (7.2) is true. The table similarly lists all the 1-element extensions necessary to verify (7.4), (7.5), and (7.6).

The table also contains information about R_{12}, which we use to verify that R_{12} satisfies the hypotheses of (9.1). However, the argument here needs more explanation. We follow the outline given after (9.2). M_1 and M_2 denote the matroids represented by the matrices obtained from the R_{12} matrix of Fig. 9 by adding the columns $(1, 1, 0, 0, 1, 1)^T$ and $(0, 0, 1, 1, 0, 0)^T$, respectively. The "R_{12}" entries in the table show the following.

(15.1) *If a non-zero new column is added to the R_{12} matrix of figure 9, and if the new matroid obtained is regular, then either the new column is the same as an old column, or it is one of* $(1, 1, 0, 0, 0, 0)^T$, $(1, 1, 0, 0, 1, 1)^T$, $(0, 0, 0, 0, 1, 1)^T$, *or* $(0, 0, 1, 1, 0, 0)^T$. *The first three of these give isomorphic matroids, the isomorphisms preserving the sets A, B and the new element.*

Now (9.1)(i) is easily verified—recall that B is the union of the two 3-element circuits of R_{12}, and so $A = \{3, 4, 7, 8, 11, 12\}$ and

$$F_7 = \begin{bmatrix} 1 & 0 & 0 & 0 & 1 & 1 & 1 \\ 0 & 1 & 0 & 1 & 0 & 1 & 1 \\ 0 & 0 & 1 & 1 & 1 & 0 & 1 \end{bmatrix}$$

$$F_7^* = \begin{bmatrix} 1 & 0 & 0 & 0 & 0 & 1 & 1 \\ 0 & 1 & 0 & 0 & 1 & 0 & 1 \\ 0 & 0 & 1 & 0 & 1 & 1 & 0 \\ 0 & 0 & 0 & 1 & 1 & 1 & 1 \end{bmatrix}$$

$$\mathcal{M}^*(K_{3,3}) = \begin{bmatrix} 1 & 0 & 0 & 0 & 1 & 0 & 0 & 1 & 1 \\ 0 & 1 & 0 & 0 & 1 & 1 & 0 & 0 & 1 \\ 0 & 0 & 1 & 0 & 0 & 1 & 1 & 0 & 1 \\ 0 & 0 & 0 & 1 & 0 & 0 & 1 & 1 & 1 \end{bmatrix}$$

$$\mathcal{M}(K_5) = \begin{bmatrix} 1 & 0 & 0 & 0 & 1 & 1 & 1 & 0 & 0 & 0 \\ 0 & 1 & 0 & 0 & 1 & 0 & 0 & 1 & 1 & 0 \\ 0 & 0 & 1 & 0 & 0 & 1 & 0 & 1 & 0 & 1 \\ 0 & 0 & 0 & 1 & 0 & 0 & 1 & 0 & 1 & 1 \end{bmatrix}$$

$$\mathcal{M}^*(K_5) = \begin{bmatrix} 1 & 0 & 0 & 0 & 0 & 0 & 1 & 1 & 0 & 0 \\ 0 & 1 & 0 & 0 & 0 & 0 & 1 & 0 & 1 & 0 \\ 0 & 0 & 1 & 0 & 0 & 0 & 1 & 0 & 0 & 1 \\ 0 & 0 & 0 & 1 & 0 & 0 & 0 & 1 & 1 & 0 \\ 0 & 0 & 0 & 0 & 1 & 0 & 0 & 1 & 0 & 1 \\ 0 & 0 & 0 & 0 & 0 & 1 & 0 & 0 & 1 & 1 \end{bmatrix}$$

$$R_{10} = \begin{bmatrix} 1 & 0 & 0 & 0 & 0 & 1 & 1 & 0 & 0 & 1 \\ 0 & 1 & 0 & 0 & 0 & 1 & 1 & 1 & 0 & 0 \\ 0 & 0 & 1 & 0 & 0 & 0 & 1 & 1 & 1 & 0 \\ 0 & 0 & 0 & 1 & 0 & 0 & 0 & 1 & 1 & 1 \\ 0 & 0 & 0 & 0 & 1 & 1 & 0 & 0 & 1 & 1 \end{bmatrix}$$

$$R_{12} = \begin{bmatrix} 1 & 0 & 0 & 0 & 0 & 0 & 1 & 1 & 1 & 0 & 0 & 0 \\ 0 & 1 & 0 & 0 & 0 & 0 & 1 & 1 & 0 & 1 & 0 & 0 \\ 0 & 0 & 1 & 0 & 0 & 0 & 1 & 0 & 0 & 0 & 1 & 0 \\ 0 & 0 & 0 & 1 & 0 & 0 & 0 & 1 & 0 & 0 & 0 & 1 \\ 0 & 0 & 0 & 0 & 1 & 0 & 0 & 0 & 1 & 0 & 1 & 1 \\ 0 & 0 & 0 & 0 & 0 & 1 & 0 & 0 & 0 & 1 & 1 & 1 \end{bmatrix}$$

$$M_7^* = \begin{bmatrix} 1 & 1 & 1 & 0 & 0 & 0 & 1 & 0 & 0 & 0 & 0 & 0 & 0 \\ 1 & 1 & 0 & 1 & 0 & 0 & 0 & 1 & 0 & 0 & 0 & 0 & 0 \\ 1 & 0 & 0 & 0 & 1 & 0 & 0 & 0 & 1 & 0 & 0 & 0 & 0 \\ 0 & 1 & 0 & 0 & 0 & 1 & 0 & 0 & 0 & 1 & 0 & 0 & 0 \\ 0 & 0 & 1 & 0 & 1 & 1 & 0 & 0 & 0 & 0 & 1 & 0 & 0 \\ 0 & 0 & 0 & 1 & 1 & 1 & 0 & 0 & 0 & 0 & 0 & 1 & 0 \\ 0 & 0 & 1 & 1 & 0 & 0 & 0 & 0 & 0 & 0 & 0 & 0 & 1 \end{bmatrix}$$

FIGURE 9

N_1	e	Z_1	Z_2	N_2	ϕ
F_7	none				
F_7^*	$(1, 0, 0, 1)^T$	—	5	F_7	1, 2, 4, 3, e, 6, 7
	$(1, 1, 0, 0)^T$	—	3	F_7	1, 2, 4, 5, 6, e, 7
	$(1, 1, 1, *)^T$	—	4	F_7	1, 2, 3, 5, 6, 7, e
$\mathscr{M}^*(K_{3,3})$	$(1, *, 1, 0)^T$	5, 6	2	F_7	1, 3, 4, 7, 8, e, 9
$\mathscr{M}(K_5)$	$(1, 1, 1, *)^T$	7, 9, 10	4	F_7	1, 2, 3, 8, 6, 5, e
$\mathscr{M}^*(K_5)$	$(1, 1, 0, 0, 0, 0)^T$	—	4, 10	$\mathscr{M}^*(K_{3,3})$	1, 2, 6, 5, e, 9, 3, 8, 7
	$(1, 1, 0, 0, 1, 1)^T$	—	3, 4	$\mathscr{M}^*(K_{3,3})$	1, 2, 6, 5, 7, 9, 10, 8, e
	$(*, *, *, 1, 1, 1)^T$	7	1, 2, 3	F_7	4, 5, 6, 10, 9, 8, e
	$(0, *, 1, 1, 0, 0)^T$	7	1, 2, 10	F_7	4, 5, 6, 3, 9, 8, e
R_{10}	$(1, 0, 1, *, *,)^T$	9, 10	4, 5	F_7	1, 2, 3, 8, e, 6, 7
	$(1, 1, 0, 0, 0)^T$	1, 2	4, 8	F_7	3, 5, 7, 10, e, 9, 6
	$(1, 1, 1, 1, 1)^T$	7, 10	1, 3	F_7	2, 4, 5, 9, 6, 8, e
R_{12}	$(1, 0, 1, *, *, 0)^T$	6, 9, 12	4, 5, 11	F_7	1, 2, 3, 10, e, 8, 7
	$(1, 0, 0, *, *, 1)^T$	8, 9, 12	4, 5, 7	F_7	2, 3, 6, 11, 10, 1, e
	$(1, *, 1, *, 1, *)^T$	8, 10, 12	2, 4, 6	F_7	1, 3, 5, 11, 9, 7, e
	$(1, *, 1, 1, 0, *)^T$	5, 9, 10	2, 6, 12	F_7	1, 3, 4, 11, 8, 7, e
	$(1, 1, 0, 0, 1, 1)^T$	—	—	M_1	1,..., 12, e
	$(0, 0, 1, 1, 0, 0)^T$	—	—	M_2	1,..., 12, e
	$(1, 1, 0, 0, 0, 0)^T$	—	—	M_1	9, 10, 11, 12, 5, 6, 7, 8, 1, 2, 3, 4, e
M_2^*	$(1, 1, *, 0, 0, 0, 1)^T$	1, 2, 6, 12	5, 9, 10	F_7^*	
	$(1, 1, 0, 0, 1, 1, 1)^T$	1, 5, 6, 7	2, 9, 10	F_7^*	8, 7, 11, x, 3, 4, y
	$(0, 0, 1, *, 0, 0, 1)^T$	2, 3, 6, 9	1, 5, 10	F_7^*	12, 11, 8, x, 3, 4, y
					7, 12, x, 8, 4, y, 11

FIGURE 10

$B = \{1, 2, 5, 6, 9, 10\}$. Both $(1, 1, 0, 0, 1, 1)^T$ and $(0, 0, 1, 1, 0, 0)^T$ are spanned by columns corresponding to elements in A and so, from the table, (9.1)(ii) is true. (9.1)(iii) is true since $R_{12} \cong R_{12}^*$, and it remains to verify (9.1)(iv) and (v).

First we verify (9.1)(v), since it is easier. By duality we must show the following:

(15.2) *If M is regular and $M \backslash y/x \cong R_{12}$, and if there is a cocircuit D of M with $\{x, y\} \subset D \subseteq A \cup \{x, y\}$, then either there is a circuit C of M with $y \in C \subseteq A \cup \{x, y\}$, or y is parallel in M/x to an element of B.*

(This is equivalent to (9.1)(v) because $R_{12} \cong R_{12}^*$, and under such an isomorphism A and B are interchanged.) Now $\{x, y\} \subset D$, and so y is not a coloop of M/x. If y is a loop in M/x then either $\{y\}$ or $\{x, y\}$ is a circuit of M and (15.2) is true. If y, z are parallel in M/x (for some $z \neq x, y$) then (15.2) is true if $z \in B$, and if $z \in A$ then either $\{y, z\}$ or $\{x, y, z\}$ is a circuit of M and again (15.2) is true. Thus we may assume that y is not a loop, coloop, or parallel element of M/x. However, $M/x \backslash y \cong R_{12}$, and so $M/x \cong M_1$ or M_2. In either case y is spanned in M/x by A, and so (15.2) is true. This verifies (9.1)(v).

Finally, we verify (9.1)(iv). We must show the following:

(15.3) *If M is regular and $M \backslash x/y \cong R_{12}$, and if there is a cocircuit D of M with $\{x, y\} \subset D \subseteq B \cup \{x, y\}$, then either there is a circuit C of M with $x \in C \subseteq B \cup \{x, y\}$, or x is parallel in M/y to an element of A.*

As before, we may assume that x is not a loop, coloop, or parallel element of M/y, and so $M/y \cong M_1$ or M_2. If $M/y \cong M_1$, then x is spanned in M/y by B, and so (15.3) is true. We therefore assume that $M/y \cong M_2$, so that $M^* \backslash y \cong M_2^*$, and is represented by the last matrix of Fig. 9, columns 1–12 representing elements 1–12, and column 13 representing x. Add a column $c = (c_1, \ldots, c_7)^T$ corresponding to y, to make a matrix representing M^*. Now by hypothesis there is a circuit C of M^* with $\{x, y\} \subset C \subseteq B \cup \{x, y\}$, and $B = \{1, 2, 5, 6, 9, 10\}$; thus, $c_7 = 1$, $c_1 = c_2$, and $c_5 = c_6$; and not all $c_1, \ldots, c_6$ are zero. Moreover, $(M^*/x)\backslash y \cong R_{12}$, and so, by (15.1), either y is a loop, coloop, or parallel element of M^*/x, or $M^*/x \cong M_1$ or M_2. Thus, either $(c_1, \ldots, c_7)^T$ agrees with one of the columns of our matrix representing M_2^* in Fig. 9 in the first six rows, or (by (15.1)) it is one of $(1, 1, 0, 0, 0, 0, 1)^T$, $(1, 1, 0, 0, 1, 1, 1)^T$, $(0, 0, 0, 0, 1, 1, 1)^T$, $(0, 0, 1, 1, 0, 0, 1)^T$. Using symmetry and the results $c_1 = c_2$, etc., we may assume that c is one of $(1, 1, 0, 0, 0, 0, 1)^T$, $(1, 1, 0, 0, 1, 1, 1)^T$, $(0, 0, 1, 1, 0, 0, 1)^T$, $(1, 1, 1, 0, 0, 0, 1)^T$, $(0, 0, 1, 0, 0, 0, 1)^T$. But for each of these choices of c, M^* has an F_7^* minor as shown by the last lines of the table. This completes the proof.

ACKNOWLEDGMENTS

Several people have helped with parts of this paper. Kathie Cameron worked out the Appendix in matrix terms, compactly enough so that full details could be included. Dominic Welsh suggested that (7.3) could be extended to non-binary matroids. Klaus Truemper and Jack Edmonds helped me to understand the matrix and algorithmic implications of the theorem, and I am also grateful to Edmonds for several valuable discussions on other aspects of the paper, and his guidance on presentation. Theorem (3.8) was derived from a suggestion of László Lovász. And I would particularly like to thank the anonymous referee who meticuloush · ' the paper and uncovered a number of errors.

References

1. R. E. Bixby, Kuratowski's and Wagner's theorems for matroids, *J. Combinatorial Theory Ser. B* **22** (1977), 31–53.
2. R. E. Bixby and W. H. Cunningham, Matroids, graphs, and 3-connectivity, *in* "Graph Theory and Related Topics" (J. A. Bondy and U. S. R. Murty, Eds.), pp. 91–103, Academic Press, New York, 1979.
3. T. Brylawski, A combinatorial model for series–parallel networks, *Trans. Amer. Math. Soc.* **154** (1971), 1–22.
4. T. Brylawski, Modular constructions for combinatorial geometries, *Trans. Amer. Math. Soc.* **203** (1975), 1–44.
5. W. H. Cunningham, "A Combinatorial Decomposition Theory," Thesis (Research Report CORR 74–13), University of Waterloo, 1973.
6. W. H. Cunningham, On matroid connectivity, to appear.
7. A. K. Kelmans, A theory of 3-connected graphs, *in* "Proceedings, Trudi Colloquium on Algebraic Methods in Graph Theory," Szeged, Hungary, in press.
8. P. D. Seymour, Matroids and multicommodity flows, *European J. Combinatorics*, in press.
9. P. D. Seymour, On Tutte's characterization of graphic matroids, *in* "Combinatorics 1979," Annals of Discrete Mathematics, North-Holland, Amsterdam, in press.
10. P. D. Seymour, On Tutte's extension of the four-colour problem, *J. Combinatorial Theory Ser. B*, in press.
11. W. T. Tutte, A homotopy theorem for matroids, I, II, *Trans. Amer. Math. Soc.* **88** (1958), 144–174.
12. W. T. Tutte, "Connectivity in Graphs," Univ. of Toronto Press, Toronto, 1966.
13. W. T. Tutte, Connectivity in matroids, *Canad. J. Math.* **18** (1966), 1301–1324.
14. W. T. Tutte, Matroids and graphs, *Trans. Amer. Math. Soc.* **90** (1959), 527–552.
15. K. Wagner, "Beweis einer Abschwächung der Hadwiger–Vermutung," *Math. Ann.* **153** (1964), 139–141.
16. D. J. A. Welsh, "Matroid Theory," Academic Press, New York, 1976.

VARIETIES OF COMBINATORIAL GEOMETRIES

BY

J. KAHN[1] AND J. P. S. KUNG

ABSTRACT. A *hereditary class* of (finite combinatorial) geometries is a collection of geometries which is closed under taking minors and direct sums. A *sequence of universal models* for a hereditary class $\mathfrak{T}$ of geometries is a sequence (T_n) of geometries in $\mathfrak{T}$ with rank $T_n = n$, and satisfying the universal property: if G is a geometry in $\mathfrak{T}$ of rank n, then G is a subgeometry of T_n. A *variety* of geometries is a hereditary class with a sequence of universal models.

We prove that, apart from two degenerate cases, the only varieties of combinatorial geometries are

(1) the variety of free geometries,

(2) the variety of geometries coordinatizable over a fixed finite field, and

(3) the variety of voltage-graphic geometries with voltages in a fixed finite group.

1. Introduction. The notion of free objects in a variety is one of the most important and pervasive ideas in universal algebra. In this paper, we investigate how that notion can be interpreted in the context of the theory of combinatorial geometries (or matroids). No detailed knowledge of universal algebra is required for reading this paper. However, we do assume familiarity with the basic concepts of the theory of combinatorial geometries [**2**, **4** and **13**].

To fix our terminology, let G be a finite geometric lattice. Its maximum and minimum are denoted by $\hat{1}$ and $\hat{0}$. Let S be the set of points (or atoms) in G. The lattice structure of G induces the structure of a combinatorial geometry, also denoted by G, on S. The cardinality $|G|$ of the geometry G is the cardinality of the set S of points. All geometries appearing in this paper are assumed to be of finite cardinality.

Let T be a subset of S. The *deletion* of T from G is the geometry on the point set $S \setminus T$ obtained by restricting G to the subset $S \setminus T$. The *contraction* G/T of G by T is the geometry induced by the geometric lattice $[\overline{T}, \hat{1}]$ on the set S' of all flats in G covering $\overline{T}$ (the closure of T). Thus, in our convention, the contraction of a geometry is always a geometry—indeed, it is isomorphic to the simplification of the matroid G/T as it is usually defined. A geometry which can be obtained from G by deletions and contractions is called a *minor* of G. Minors are the subobjects in the category of combinatorial geometries and strong maps.

Finally, we note that we shall often use "is" for "is isomorphic to" when precision threatens to turn into pedantry.

Received by the editors July 7, 1980.

AMS (MOS) subject classifications (1970). Primary 05B35, 08A15.

[1] Supported in part by an AMS Fellowship and ONR Contract #N0014-76-C-0366.

2. Varieties of combinatorial geometries. A collection $\mathcal{T}$ of combinatorial geometries is a *hereditary class* if it is closed under taking minors and direct sums. More precisely, a hereditary class $\mathcal{T}$ satisfies the following axioms.

H1. If G is in $\mathcal{T}$ and H is isomorphic to a minor of G, then H is also in $\mathcal{T}$.

H2. If G and H are in $\mathcal{T}$, then so is their direct sum $G \oplus H$.

Hereditary classes of geometries have been discussed earlier, most notably in **[9]**, but little is known about them.

Formally, the definition of a hereditary class is a direct analog of the definition of a variety in universal algebra (see, for example, **[3]**). However, in universal algebra, a variety $\mathcal{V}$ of algebras is a very structured collection of objects. For example, a theorem of Birkhoff says that it can be defined by equations. A consequence of this is the existence for each n of a free algebra on n generators—that is, there exists an algebra F_n in $\mathcal{V}$ defined by the *universal* property: if A is an algebra in $\mathcal{V}$ generated by n elements, then A is a subobject of F_n.

What is the analogue of free algebras for hereditary classes? A reasonable definition is the following. A *sequence of universal models* for a hereditary class $\mathcal{T}$ of geometries is a sequence (T_n) of geometries such that

U1. T_n is in $\mathcal{T}$ and is of rank n.

U2. If G is a geometry of rank n in $\mathcal{T}$, then G is a subgeometry of T_n.

In contrast to the situation in universal algebra, sequences of universal models need not exist. Take for example, the hereditary class of all geometries, or the class of all totally unimodular geometries.

DEFINITION. A *variety* $\mathcal{T}$ of combinatorial geometries is a hereditary class of geometries with a sequence of universal models.

There are two important examples of varieties. Let $\mathcal{L}(q)$ be the collection of all geometries coordinatizable over the finite field $GF(q)$. The projective geometries $P_n(q)$—here, n is the rank of $P_n(q)$ as a geometric lattice—are universal models for $\mathcal{L}(q)$. A less well-known example is the collection $\mathfrak{X}(A)$ of voltage-graphic geometries with voltages in a finite group A; the universal models here are the Dowling geometries $Q_n(A)$ (see **[7]**). These voltage-graphic geometries were discovered and studied by Zaslavsky **[14, 15]**. We shall need only the basic properties of Dowling geometries and they will be recounted at the appropriate moment. We should mention that when A is the trivial group, $Q_n(A)$ is the graphic geometry of the complete graph (on $n + 1$ vertices) and the associated variety $\mathfrak{X}(A)$ is the variety of graphic geometries.

One is at this point hard put to think of additional examples. The reason for this is that, apart from degenerate cases, there are no others. This statement, which will be made precise in §9, is our main result. We proceed first to its proof.

3. Modular flats. Henceforth, (T_n) is a sequence of universal models for a variety $\mathcal{T}$ of geometries.

LEMMA 1. *Let G be a geometry in $\mathcal{T}$ and let t_i be a subgeometry of G isomorphic to T_i. Then t_i is a modular flat of G.*

PROOF. First observe that t_i is of rank i and its closure is a subgeometry of T_i by universality. Hence, t_i is a closed set. Now, let u be a modular complement of t_i. The

VARIETIES OF COMBINATORIAL GEOMETRIES 487

function from the interval $[u, \hat{1}]$ to $[\hat{0}, t_i]$ defined by $x \mapsto x \wedge t_i$ is a surjection. Since u is a modular complement, $[u, \hat{1}]$ contains a copy of T_i as a subgeometry (viz. the image of $[0, t_i]$ under the map $x \mapsto x \vee u$); as T_i is universal, $[u, \hat{1}]$ is isomorphic to T_i. Thus, the surjection is in fact a bijection.

In particular, if v strictly contains u, $v \wedge t_i \neq \hat{0}$. Thus, none of the flats lying above u can be a complement of t_i: that is, no two complements of t_i are comparable. By a result of Stanley [**11**, Theorem 1], t_i is modular. □

An immediate consequence of the lemma is that the universal model T_n is supersolvable. Recall that a geometric lattice of rank n is *supersolvable* if there exists a (saturated) chain $\hat{0} = x_0 < x_1 < \cdots < x_{n-1} < x_n = \hat{1}$ with rank $x_i = i$ such that every flat x_i is modular [**12**].

PROPOSITION 2. *The universal model T_n is supersolvable with a chain $\hat{0} = t_0 < t_1 < \cdots < t_n = \hat{1}$ of modular flats such that the subgeometry t_i is isomorphic to T_i.*

PROOF. As $\mathfrak{T}$ is closed under direct sum, $T_{n-1} \oplus T_1$ occurs as a subgeometry of T_n; in particular, there is a subset t_{n-1} in T_n isomorphic to T_{n-1}. This set t_{n-1} is a modular copoint. We can now repeat the argument on the subgeometry t_{n-1}. □

Another consequence is a technical tool we shall use time and again.

LEMMA 3 (THE PROJECTION ARGUMENT). *Let t_{n-1} be a copoint of T_n isomorphic to T_{n-1} and let $u = \{x_1, \ldots, x_m\}$ be a set of points of T_n. Let z be a point of T_n which is not in the union $t_{n-1} \cup \bar{u}$. Then the map $x_i \to \overline{x_i z} \wedge t_{n-1}$ determines an isomorphism of u with a subgeometry of t_{n-1}.*

PROOF. By Lemma 1, $\overline{x_i z} \wedge t_{n-1}$ is a point. □

We abbreviate the above situation by the locution: The configuration u is projected onto the copoint t_{n-1} via contraction by z.

4. Two disconnected examples. The *order* of the sequence (T_n) of universal models is the positive integer $|T_2| - 1$. We first consider the case when q equals one; that is, when T_2 is a two-point line.

PROPOSITION 4. *Let $q = 1$. Then T_n is the Boolean algebra (or free geometry) B_n on n elements.*

PROOF (GREENE [**8**]). We proceed by induction on the index n. By hypothesis, all the copoints in T_n are isomorphic to the Boolean algebra B_{n-1} and hence all n-element sets are independent. Thus, T_n is a truncated Boolean algebra. But if it were nontrivially truncated, there would exist a rank two upper minor containing a three-point line. □

The Boolean algebras B_n are the universal models of the *variety of free geometries*. They are also the only geometries in this variety.

We can now assume the *the order q is greater than one.*

LEMMA 5. *Suppose that T_3 is connected. Then T_n is connected for all n.*

PROOF. It suffices to show that for $n \geqslant 4$, if T_{n-2} and T_{n-1} are connected, then so is T_n. Suppose that T_n is disconnected. As T_{n-1} is connected and T_n contains $T_{n-1} \oplus T_1$ as a subgeometry, we conclude that $T_n \cong T_{n-1} \oplus T_1$. But T_n also contains

$T_{n-2} \oplus T_2$. Let t_{n-2} be the subset in T_n isomorphic to T_{n-2}; then $T_n/t_{n-2} \cong T_2$. Now, t_{n-2}, being connected, must be contained in the summand isomorphic to T_{n-1}. Hence, $T_n/t_{n-2} \cong (T_{n-1}/t_{n-2}) \oplus T_1 \cong T_1 \oplus T_1$. This implies that T_2 is a two-point line, contrary to our standing assumption that $q \geqslant 2$. □

Is there a sequence (M_n) of universal models with M_3 disconnected? If so, we must have $M_3 \cong M_2 \oplus M_1$ where M_2 is a line with $q+1$ points. A moments thought yields: $M_4 \cong M_2 \oplus M_2$. We can now show by induction that

$$M_{2n+1} \cong M_2 \oplus \cdots \oplus M_2 \oplus M_1 \quad \text{and} \quad M_{2n} \cong M_2 \oplus \cdots \oplus M_2$$

(where there are n copies of M_2 in each sum). By taking all subgeometries of the geometries M_n, we obtain a variety of geometries. The reader should have no difficulty in seeing why we call it the *variety of matchstick geometries of order q*.

We can now assume that *the universal models T_n are all connected.*

5. Having enough points. We have arrived at the first branch (of two) in our proof. First, a definition. We say that a geometry *splits* if it is the union of two of its proper flats (where we regard flats as point sets).

From here until §8, we assume that *no T_n splits*. This condition ensures that there are "enough points" to perform certain contractions.

LEMMA 6 (UPPER HOMOGENEITY). *Let p be any point in T_{n+1}. Then there exists a copoint in T_{n+1} isomorphic to T_n and not containing p. In particular, the contraction T_{n+1}/p is isomorphic to T_n.*

PROOF. The result being trivial for $n = 1$, we assume that n is at least 2. Let p be a point in T_{n+1} and consider an embedding i_{2n} of T_{n+1} into T_{2n}. As T_{2n} contains the direct sum $T_n \oplus T_n$ and the image of p can be in at most one of the summands, there exists a flat t_n in T_{2n} isomorphic to T_n not containing the point $i_{2n}(p)$. Now as T_{2n} does not split, there exists a point z such that z is not in the union $(t_n \vee i_{2n}(p)) \cup i_{2n}(T_{n+1})$. By our choice of z, the image of $i_{2n}(T_{n+1})$ in the contraction T_{2n}/z is isomorphic to T_{n+1} and the image t_n' of t_n is a flat in T_{2n}/z isomorphic to T_n and not containing the image of $i_{2n}(p)$. By universality, there is an embedding of T_{2n}/z into T_{2n-1}. Let i_{2n-1} be the composition of the maps

$$T_{n+1} \to T_{2n} \to T_{2n}/z \to T_{2n-1}.$$

In T_{2n-1}, there is a flat (the image of t_n') which is isomorphic to T_n and does not contain $i_{2n-1}(p)$. As $T_{2n-1}, T_{2n-2}, \ldots, T_{n+2}$ do not split, we can repeat this argument to obtain an isomorphism i_{n+1} of T_{n+1} into T_{n+1} such that there is a flat t_n'' of T_{n+1} isomorphic to T_n and not containing $i_{n+1}(p)$. The inverse image $i_{n+1}^{-1}(t_n'')$ is a copoint of T_{n+1} not containing p. □

Now, consider the set of all copoints in T_n isomorphic to T_{n-1}. By the lemma, their intersection is $\hat{0}$. Thus, there exists a set of n such copoints for which the intersection is $\hat{0}$. Let $t^1, \ldots, t^n$ be such a set of copoints. As these copoints are modular, the meet sublattice they generate in T_n is a Boolean algebra. In particular, the intersections

$$p_i = t^1 \wedge \cdots \wedge t^{i-1} \wedge t^{i+1} \wedge \cdots \wedge t^n$$

are points in T_n. The subgeometry of T_n obtained by restricting T_n to the set of points $\bigcup t^i$ is denoted by $\langle p_1,\ldots,p_n\rangle$; it is called the *frame of T_n generated by the joints* $p_1,\ldots,p_n$.

What is the span of $p_1,\ldots,p_k$? This span, being the intersection $t_{k+1}\wedge\cdots\wedge t_n$ of modular copoints, is a modular flat of rank k. It is in fact isomorphic to T_k. (For, by modularity, $[\hat{0},\overline{p_1\cdots p_k}]\cong[u,\hat{1}]$, where u is any complement of $\overline{p_1\cdots p_k}$; but T_n is upper homogeneous.)

We now arrive at the second branch in our proof. Choose a frame in T_3. Two possibilities may occur: the frame is all of T_3; or there may be more points in T_3. The first possibility leads to Dowling geometries while the second leads to projective geometries.

6. Projective geometries. We assume in this section that T_3 *contains more points than a frame*. Ultimately, we shall show that this implies the order q of (T_n) is a prime power and T_n is the projective geometry $P_n(q)$.

We begin with an easy fact.

LEMMA 7. *Let (T_n) be of order q. Then*

$$|T_n|\leqslant 1+q+q^2+\cdots+q^{n-1},$$

with equality if and only if T_n is a projective geometry of order q.

PROOF. For $n\geqslant 1$, let $a_n=|T_n|-|T_{n-1}|$. We prove by induction that $a_n\leqslant q^{n-1}$; this will clearly imply the inequality. For $n=1$ or 2, the assertion holds trivially. Now consider $n\geqslant 3$. Let $t_{n-2}<t_{n-1}$ be a chain in T_n consisting of a coline and a copoint isomorphic to T_{n-2} and T_{n-1} (respectively). There are q other copoints covering t_{n-2} and by induction, these copoints u satisfy: $|u|-|t_{n-2}|\leqslant q^{n-2}$. Hence, $|T_n|-|T_{n-1}|\leqslant q\cdot q^{n-2}$.

Suppose now that equality is attained. As T_n is connected, it suffices to show that T_n is a modular geometry [2, p. 93]; as is easy to show, this is equivalent to showing that every line is modular, which we prove by showing that every line contains at least $q+1$ points. We proceed by induction. In T_1 and T_2, every line has $q+1$ points. Now, let t_{n-2} be a coline isomorphic to T_{n-2} in T_n. By the equality, every copoint lying above t_{n-2} is isomorphic to T_{n-1}. Using the induction hypothesis, every line contained in one of these copoints has $q+1$ points. Finally, consider a line l not contained in any of these copoints. This line l must intersect every copoint covering t_{n-2} and hence must have $q+1$ points. □

Let $\langle p_1p_2p_3\rangle$ be a frame in T_3. By assumption, there is a point—an *exterior point*—in T_3 not in the frame. Consider now a frame $\langle p_1p_2p_3p_4\rangle$ in T_4. Let z be an exterior point of the plane $\overline{\langle p_2p_3p_4\rangle}$ and let p be the intersection of the lines $\overline{p_4z}$ and $\overline{p_2p_3}$. Contracting by the point z, we project the line $\overline{p_1p_4}$ onto the line $\overline{p_1p}$. We conclude that $\overline{p_1p}$ has $q+1$ points, of which $q-1$ are exterior points. Similarly, each contraction (of T_4) by one of these $q-1$ points projects $\overline{p_3p_4}$ onto a line containing p_4 and $q-1$ exterior points in $\overline{p_1p_2p_4}$. This gives at least $(q-1)^2$ exterior points in $\overline{p_1p_2p_4}$ (hence in T_3), which by the lemma makes T_3 a projective plane of order q.

Our final step is to show that T_n is a projective geometry. Again, we proceed by induction, having already proved the assertion for $n \leqslant 3$. Suppose that the assertion is true for $m \leqslant n-1$. Let $\langle p_1 \cdots p_{n+1}\rangle$ be a frame in T_{n+1} and let $t = \overline{p_1 \cdots p_{n-2}}$. As observed earlier, t is isomorphic to T_{n-2} and is thus a projective geometry of rank $n-2$. Let u be any flat covering t and contained in the modular flat $\overline{p_1 \cdots p_n}$. The flat $\overline{p_1 \cdots p_n}$ is isomorphic to T_n. We shall show that it is in fact a projective geometry. Consider the intersection $u \wedge \overline{p_{n-1}p_n}$. As $\overline{p_{n-1}p_n}$ is a modular line, this intersection is a point a. Obviously, u is the closure of $p_1,\ldots,p_{n-2}, a$. Now, in T_{n+1}, consider the line $\overline{ap_{n+1}}$. This line is contained in the projective plane $\overline{p_{n-1}p_n p_{n+1}}$ and hence there exists a point z in $\overline{ap_{n+1}}$ distinct from a and p_{n+1}. Contracting by the point z, we project the projective geometry (of rank $n-1$) $\overline{p_1 \cdots p_{n-2}p_{n+1}}$ onto u (u being contained in $\overline{p_1 \cdots p_n}$). We conclude that u is a projective geometry of rank $n-1$; indeed, every flat contained in $\overline{p_1 \cdots p_n}$ covering t is a projective geometry. Counting up all the points, we obtain the equality

$$|T_n| = |\overline{p_1 \cdots p_n}| = 1 + q + \cdots + q^{n-1}.$$

By the lemma, T_n is a projective geometry of rank n. Finally, we note that, as is well known, projective geometries of order q and rank higher than three exist only if q is a prime power.

7. Dowling geometries. We now tackle the possibility that T_3 *consists only of a frame*. Our first result is that this property propagates upwards.

LEMMA 8. *If T_3 consists only of a frame, then T_n also consists only of a frame.*

PROOF. Let n be the first index for which the lemma is false. Choose a frame $\langle p_1,\ldots,p_n\rangle$ in T_n and let z be a point in T_n not in the frame. Consider the intersection $\overline{p_1 \cdots p_{n-1}} \wedge \overline{p_n z}$. By modularity, this is a point contained in the flat $\overline{p_1 \cdots p_{n-1}}$ (which is isomorphic to T_{n-1}), but it is not contained in the frame $\langle p_1,\ldots,p_{n-1}\rangle$. This is a contradiction. □

Our next task is to determine the structure of T_3. Let $\langle p_1 p_2 p_3\rangle$ be a frame for T_3. There are exactly three modular lines: $l_{12} = \overline{p_1 p_2}$, $l_{23} = \overline{p_2 p_3}$ and $l_{13} = \overline{p_1 p_3}$. Let a be an *interior* point (i.e. a point distinct from the joints) of l_{12}, and let b be any interior point of l_{23}. The line $\overline{ab}$ meets the modular line l_{13} in a unique point c, and consists precisely of a, b and c. Such lines $\overline{ab}$ are called *transversals*. We interpret the transversal $\{a, b, c\}$ as the equation $ab = c$ and form a multiplication table with the interior points of l_{12} labelling the rows, the interior points of l_{23} labelling the columns, and the interior points of l_{13} as the entries. This multiplication table is a Latin square. Our next step is to show that it can be labelled so as to form the multiplication table of a group.

To do so, we move up to T_4; there are four copies of T_3 (called *facets*) in T_4, viz. $t_{ijk} = \overline{p_i p_j p_k}$ where i, j and k are elements from the set $\{1,2,3,4\}$. We shall always choose i, j and k so that $i<j<k$. Again, we denote the modular line $\overline{p_i p_j}$ by l_{ij}.

We now label the interior points in the six modular lines l_{ij}. Choose interior points a_{12}, a_{13} and a_{14} in the lines l_{12}, l_{13} and l_{14} respectively. Label these points ε; ε will turn out to be the identity element of the group. Further, label the interior points in l_{12} (by Greek letters, say). We will make use of the following two procedures for extending the labelling.

(A) Suppose that the points b on l_{ij} and a_{jk} on l_{jk} have been labelled α and ε respectively, and that b' is the point of l_{ik} for which $\{b, a_{jk}, b'\}$ is a transversal. Then label b' by α.

(B) Suppose that the points a_{ij} on l_{ij} and b on l_{ik} have been labelled ε and α respectively, and that b' is the point of l_{jk} for which $\{a_{ij}, b', b\}$ is a transversal. Then label b' by α.

The reader should keep in mind here that multiplication in the quasigroup corresponding to t_{ijk} is given by:

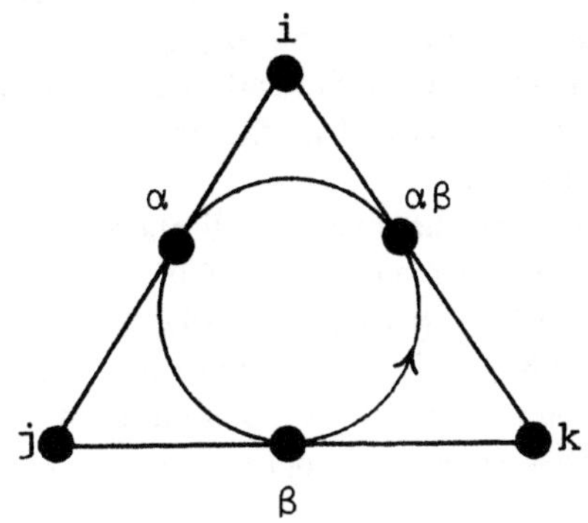

The following steps extend the labelling to all interior points of the line l_{ij}.

(1) Apply (B), first with $(i, j, k) = (1,2,3)$ and $b = a_{13}$, then with $(i, j, k) = (1,2,4)$ and $b = a_{14}$. This gives us points labelled ε on l_{23} and l_{24}.

(2) Apply (A) with $(i, j, k) = (1,2,3)$ and b ranging over all interior points of l_{12} to obtain labels for the points of l_{13}. Repeat, replacing 3 by 4, to obtain labels for l_{14}.

(3) Apply (B) with $(i, j, k) = (1,2,3)$ and b ranging over the interior points of l_{13} to obtain labels for l_{23}. Repeat, replacing 3 by 4, to obtain labels for l_{24}.

(4) Apply (B) with $(i, j, k) = (1,3,4)$ and b ranging over the interior points of l_{14} to obtain labels for l_{34}.

This completes the labelling. Notice that step (4) above was one of four different ways of obtaining a labelling of l_{34}, the other three being application of (A) with $(i, j, k) = (1,3,4)$, and application of (A) or (B) with $(i, j, k) = (2,3,4)$. We will show momentarily that all of these procedures lead to the same labelling. Here, the basic observation is

Lemma 9. *Let* $\{a, b, c\}$, $\{c, d, e\}$ *and* $\{a, f, e\}$ *be transversals on three distinct facets of* T_4. *Then,* $\{b, d, f\}$ *is a transversal on the fourth facet.*

Proof. Without loss of generality, we may assume that $\{a, b, c\}$, $\{c, d, e\}$ and $\{a, f, e\}$ are transversals on the facets t_{123}, t_{134} and t_{124} respectively. In the contraction of T_4 by b, a is projected onto c, $\overline{ae}$ onto $\overline{ce}$ and l_{24} onto l_{34}. Therefore, $f = \overline{ae} \wedge l_{24}$ is projected onto $\overline{ce} \wedge l_{34} = d$. □

In what follows we use α_{ij} to denote the point of l_{ij} which is labelled α.

We may now verify that each of the three alternate procedures mentioned above produces our present labelling of l_{34}. We first observe that since $\{\varepsilon_{12}, \varepsilon_{23}, \varepsilon_{13}\}$, $\{\varepsilon_{12}, \varepsilon_{24}, \varepsilon_{14}\}$ and $\{\varepsilon_{13}, \varepsilon_{34}, \varepsilon_{14}\}$ are transversals, Lemma 9 says that $\{\varepsilon_{23}, \varepsilon_{34}, \varepsilon_{24}\}$ is also. But then, as $\{\alpha_{12}, \varepsilon_{23}, \alpha_{13}\}$ and $\{\alpha_{12}, \varepsilon_{24}, \alpha_{14}\}$ are also transversals, $\{\alpha_{13}, \varepsilon_{34}, \alpha_{14}\}$ must be one as well. That is, we would have obtained the same labelling of l_{34} by applying (A) with $(i, j, k) = (1, 3, 4)$. That the remaining two procedures give the same result is shown similarly: $\{\alpha_{23}, \varepsilon_{34}, \alpha_{24}\}$ is a transversal because $\{\varepsilon_{12}, \alpha_{23}, \alpha_{13}\}$, $\{\alpha_{13}, \varepsilon_{34}, \alpha_{14}\}$ and $\{\varepsilon_{12}, \alpha_{24}, \alpha_{14}\}$ are; and $\{\varepsilon_{23}, \alpha_{34}, \alpha_{24}\}$ is a transversal because $\{\varepsilon_{12}, \varepsilon_{23}, \varepsilon_{13}\}$, $\{\varepsilon_{12}, \alpha_{24}, \alpha_{14}\}$ and $\{\varepsilon_{13}, \alpha_{34}, \alpha_{14}\}$ are.

Now each of the four facets t_{ijk} defines a multiplication as described above. We need to know that these multiplications are the same. But contraction by a_{34} (which equals ε_{34}) fixes α_{12} and projects β_{23} onto β_{24}, $(\alpha\beta)_{13}$ onto $(\alpha\beta)_{14}$, so that t_{123} and t_{124} define the same multiplication. Similarly we can equate the multiplications of t_{124} and t_{134} by contracting a_{23} and of t_{134} and t_{234} by contracting a_{12}. Thus all multiplications are the same.

We are now ready to show that our multiplication table is the multiplication table of a group. The only axiom requiring proof is the associative law. Consider the transversals $\{\alpha_{12}, \beta_{23}, (\alpha\beta)_{13}\}$, $\{(\alpha\beta)_{13}, \gamma_{34}, (\alpha\beta)\gamma_{14}\}$, and $\{\beta_{23}, \gamma_{34}, (\beta\gamma)_{24}\}$. By the lemma, $\{\alpha_{12}, (\beta\gamma)_{24}, (\alpha\beta)\gamma_{14}\}$ is also a transversal. Hence, $(\alpha\beta)\gamma = \alpha(\beta\gamma)$. We denote the abstract group defined by our multiplication table by A. As Dowling has shown (his proof of Theorem 8 in [7] applies here), the group A is determined up to isomorphism by T_3.

What are the dependencies in T_n? The easiest way to describe them is to consider the following graphical representation. (The ideas and terminology here are based on Zaslavsky [**14**, **15**].) We construct the graph $\overline{K}_n$ on the vertex set $\{1, 2, \ldots, n\}$ by inserting between each pair of vertices i and j (with $i < j$) $|A|$ multiple edges labelled by the elements of A and directed from i to j. The graph $\overline{K}_n$ is thus a "flattened" representation of the geometry T_n. We identify the points of T_n with the vertices and edges of $\overline{K}_n$ by: $p_i \mapsto i$ and $\alpha_{ij} \mapsto$ the edge labelled α between the vertices i and j.

Let S be the union of $\{1, 2, \ldots, n\}$ and the edge set of $\overline{K}_n$. A subset $C \subseteq S$ is called a *cycle* if it consists of a single vertex or it is a cycle (which is not necessarily directed) in the graph-theoretic sense: that is, if the elements of C (which are all edges) can be rearranged so that e_1 is an edge between i_0 and i_1, e_2 is an edge between i_1 and $i_2, \ldots$ and e_m is an edge between i_{m-1} and i_0, and $i_0, \ldots, i_m$ are distinct vertices. Fix a cycle C of edges. Consider the *label map* $\lambda: C \to A$ defined on the edges of C by: for e_j labelled α, $\lambda(e_j) = \alpha$ if $i_{j-1} < i_j$ and α^{-1} if $i_{j-1} > i_j$. The cycle C is said to be *balanced* if $\lambda(e_1) \cdots \lambda(e_m) = \varepsilon$ (this condition is clearly independent of our choice of starting point); C is *unbalanced* otherwise. We also insist that a cycle consisting of a single vertex is unbalanced; cycles of this kind are called *unbalanced loops*. Besides cycles, we need three other kinds of subgraphs of $\overline{K}_n$. A *theta graph* is a set of edges $\{e_1, \ldots, e_k, f_1, \ldots, f_l, g_1, \ldots, g_m\}$ such that $\{f_1, \ldots, f_l\}$ is nonempty, and

$$\{e_1, \ldots, e_k, f_1, \ldots, f_l\}, \quad \{e_1, \ldots, e_k, g_1, \ldots, g_m\}, \quad \text{and} \quad \{f_1, \ldots, f_l, g_1, \ldots, g_m\}$$

are cycles. A *handcuff* is a subset of S of the form

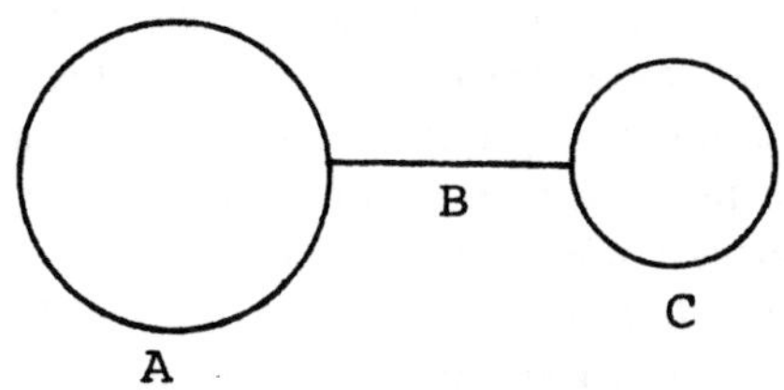

where A and C are cycles (thus A or C may consist of a single vertex) and B is a simple path. In the special case that B is a single vertex, we also call the cycle a *figure-of-eight*. A *bicycle* (or *bicircular graph*) is a theta graph or a handcuff.

There are four kinds of circuits in the plane t_{ijk} (which equals the span of p_i, p_j and p_k), namely $\{\alpha_{ij}, \beta_{jk}, (\alpha\beta)_{ik}\}$, $\{r, \alpha_{rs}, s\}$, $\{r, \alpha_{rs}, \beta_{rs}\}$ and $\{\alpha_{rs}, \beta_{rs}, \gamma_{rs}\}$; here, in the latter three types, we use r and s to stand for any pair of elements from $\{i, j, k\}$. Circuits of the first type are balanced cycles; those of the remaining three types are bicycles containing no balanced cycles. We call these circuits the *atomic circuits* of T_n.

For brevity, we use the term *Z-set* for a set which is either a balanced cycle, or a bicycle none of whose cycles is balanced. Clearly, no Z-set properly contains another. We shall show that the Z-sets are all the circuits of T_n by showing, firstly, that any subset of S not containing a Z-set is independent, and secondly, that any Z-set is dependent.

Suppose that I is a dependent subset of S which contains no Z-set. We may assume that I is a circuit. Decompose I into its connected components (in the graph-theoretic sense). As I contains no Z-set, any connected component of I can contain at most one unbalanced cycle. Let m be the maximum size of an unbalanced cycle in I. Consider first the case $m \geqslant 3$. Let $\{e_1, \ldots, e_m\}$ (arranged so that e_{i-1} and e_i are incident) be an unbalanced cycle in I of size m. Take the circuit I and the atomic circuit $\{e_{m-1}, e_m, f\}$ (where $\{e_{m-1}, e_m, f\}$ is a transversal). By the circuit elimination property (see e.g. [**13**, p. 23]) we can eliminate the edge e_m from $I \cup \{f\}$, thus obtaining a circuit I' containing fewer unbalanced cycles of size m than I. Using this argument repeatedly, we can reduce to the case $m \leqslant 2$. If $m = 2$, let $\{e_1, e_2\}$ be an unbalanced cycle of size two in I. Consider the atomic circuit $\{e_1, e_2, i\}$, where i is one of the endpoints of e_1 (and e_2). By the circuit elimination property, we can obtain a circuit I' containing one less unbalanced cycle of size two than I. Iterating this argument, we may assume that I contains at most one unbalanced loop (which is the only possible unbalanced cycle of length one) in each connected component.

Now let i be an unbalanced loop in I occurring in a connected component of I which has size at least two. Let e be an edge of I incident on i and let j be the other endpoint of e. Consider the atomic circuit $\{i, e, j\}$; by circuit elimination, we can eliminate e from $I \cup \{j\}$ to obtain another circuit I' containing fewer edges than I. Similarly, if e is an edge (with endpoints i and j) in a component of I containing no unbalanced loop, we can eliminate e from the union of $\{i, e, j\}$ and I to obtain a circuit I containing fewer edges than I. All told, these elimination arguments allow us to assume that I contains no edges—that is to say, I is a union of unbalanced loops. But the set $\{1,2,\ldots,n\}$ of unbalanced loops is an independent set. This contradiction proves that any set which contains no Z-set is independent.

The proof that all Z-sets are dependent consists in showing that every Z-set may be obtained from the atomic circuits by iterating circuit elimination. The details, which are easy but tedious, are left to the reader.

We have now shown that the circuits of T_n must be the Z-sets. That the Z-sets do indeed form the circuits of a geometry is proved in [**14** and **15**]. The universal model T_n thus obtained is the *Dowling geometry* $Q_n(A)$ *of rank n based on the finite group A* (see [**7**]). The variety specified by these universal models is called the variety of *voltage-graphic geometries with voltages in A*. (Our description of $Q_n(A)$ differs somewhat from that of [**7**]. For a proof of the equivalence when A is cyclic—which should indicate the proof in the general case—see [**5** or **6**].)

8. The split case: origami geometries. We return now to the first branch of the proof. Assume that some T_m splits and choose n to be the least integer for which T_{n+1} splits. Since T_2 is connected it does not split, and we may assume that n is at least two. We shall in fact show that n must be equal to two.

LEMMA 10. *Suppose $T \in \mathfrak{T}$ contains flats t_m and t_r isomorphic to T_m and T_r respectively and that the rank of their intersection $t_m \wedge t_r$ is k. Then if $r \leq n$, $t_m \wedge t_r$ is isomorphic to T_k.*

PROOF. We induct on r, the (smallest) case $r = k$ being trivial. Let $t_m \wedge t_r = t$.

Let t_{r-1} be a flat in t_r isomorphic to T_{r-1}, and let x be a point of t_r not in $t_{r-1} \cup t$. (Such a point exists since t_r does not split.) If we let t'_m and t'_{r-1} be the images of t_m and t_r under contraction by x, then t'_m and t'_{r-1} are flats of T/x isomorphic to T_m and T_{r-1} respectively. Moreover, $t'_m \wedge t'_{r-1}$ is isomorphic to t (since it is the image of t), and to T_k (by the induction hypothesis). Thus t is isomorphic to T_k. □

We must investigate in some detail the structure of the geometries T_m. A first step in this direction is

LEMMA 11. *For each $m = n+1,\ldots,2n$, there exist flats t and t' in T_m, isomorphic to T_n, such that $t \vee t' = T_m$.*

PROOF. The result is true for $m = 2n$ as T_{2n} contains the direct sum $T_n \oplus T_n$. Moreover, if it is true for $m+1$ and there exists a point x in T_{m+1} but not in $t \cup t'$, then contraction by x gives the result for m.

So if the lemma is false, there must exist an m_0 ($> n$) such that the lemma holds for $m_0+1,\ldots,2n$, and $T_{m_0+1} = t \cup t'$.

By connectivity $m_0 \leq 2n-2$. By Lemma 1, we may regard T_{m_0+1} as a flat of T_{m_0+2}. Since the lemma is true for m_0+2, there is a flat t_n of T_{m_0+2} which is isomorphic to T_n and not contained in T_{m_0+1}. By Lemmas 1 and 10, $t_{n-1} = t_n \wedge T_{m_0+1}$ is isomorphic to T_{n-1}. Since t_{n-1} is contained in $t \cup t'$ and does not split, we may assume that t_{n-1} is contained in t. This proves the lemma for $m = n+1$, since the flat $t \vee t_n$ is of rank $n+1$, and by universality is a subgeometry of T_{n+1}.

Finally, the validity of the lemma for $m = n + 1$ rules out the existence of an $\dot{m}_0$ as described above. For if u, v, w, x are isomorphic to T_n with $u \vee v = T_{n+1}$ and $w \vee x = T_{m_0+1}$ ($m_0 > n$), then we may regard T_{n+1} as a flat of T_{m_0+1}, and it is easily seen (because rank$(u \wedge v) >$ rank$(w \wedge x)$) that u and v cannot both be contained in $w \cup x$. $\square$

We see in particular that T_{n+1} is the union of two flats isomorphic to T_n. In fact we can show that for each $m \geqslant n$, T_{m+1} is the union of a copy of T_m and a copy of T_n, and can obtain a fairly precise description of how these additional copies of T_n must be attached.

LEMMA 12. *Let* $m \geqslant n$.

(a) *There exist flats* t_m *and* t_n *in* T_{m+1} *with* t_m *isomorphic to* T_m, t_n *isomorphic to* T_n, *and* $T_{m+1} = t_m \cup t_n$.

(b) *Let* $t_n \subseteq t_{n+1} \subseteq \cdots \subseteq t_m$ *with* t_i *isomorphic to* T_i, *and for* $n \leqslant r \leqslant m$ *let* $t_r = \bigcup_{i=1}^{r-n+1} t^i$ *with* t^i *isomorphic to* T_n. (*This assumption is justified by* (a).) *If* $k \leqslant n$, *and* t *is a flat of* t_m *isomorphic to* T_k, *then* $t \subseteq t^i$ *for some* $i \in \{1,\ldots,m-n+1\}$. *In particular*, $t^1,\ldots,t^{m-n+1}$ *are the only copies of* T_n *in* t_m.

PROOF. (a) By Lemma 11 there is a collection of copies of T_n which span T_{m+1}. (For $m \geqslant 2n$ this follows from $T_{m+1} \supseteq T_{m-n+1} \oplus T_n$.) Thus, given a copy t_m of T_m in T_{m+1}, there exists a flat t_n of T_{m+1} isomorphic to T_n and not contained in t_m.

By Lemma 10, $t_m \wedge t_n$ is isomorphic to T_{n-1}.

Suppose there exists a point x in T_{m+1} not in $t_m \cup t_n$. If $t = t_m \wedge (t_n \vee x)$, then contraction by x projects t_n onto t, and shows that t is isomorphic to T_n. Now it is easy to show (since T_n does not split) that if $T_{n+1} = t' \cup t''$ with t', t'' isomorphic to T_n, then T_{n+1} contains no further copies of T_n. On the other hand, the flat $t_n \vee x$ is a subgeometry of T_{n+1} (by universality) and contains the copies t_n and t of T_n, plus the point x outside $t_n \cup t$. This is a contradiction and we have proved (a).

(b) Let r be the least index for which $t \subseteq t_r$. By (a), $t = (t \cap t^{r-n+1}) \cup (t \cap t_{r-1})$. Since $t \not\subseteq t_{r-1}$ and t does not split, we must have $t \subseteq t^{r-n+1}$. $\square$

Let $m \geqslant n$. By Lemma 12(b), there are exactly $m - n + 1$ copies of T_n in T_m. We take these to be the vertices of a graph G_m in which vertices t and t' are joined if and only if $t \wedge t'$ is isomorphic to T_{n-1}. We will speak of t both as a vertex of G_m and as a flat of T_m. We remark that G_m is connected by Lemma 12.

LEMMA 13. G_m *is a path. Moreover, if we take* $t^0,\ldots,t^{m-n}$ *to be one of the* (*two*) *natural orderings of its vertices, then* rank$(t^i \wedge t^j) = \max\{n - j + i, 0\}$ *for all* $0 \leqslant i \leqslant j \leqslant m$.

PROOF. Let VG_m denote the set of vertices of G_m. We first remark that for any t, t', $t'' \in VG_m$,

$$(*)\qquad \begin{aligned} n + \text{rank}(t \wedge t') &\leqslant \text{rank}[(t \wedge t'') \vee (t' \wedge t'')] + \text{rank}(t \wedge t' \wedge t'') \\ &= \text{rank}(t \wedge t'') + \text{rank}(t' \wedge t''). \end{aligned}$$

(The equality of the last two expressions follows from Lemmas 10 and 1.)

The *length* of a path between two vertices is the number of edges in the path, and the *distance*, $d(t, t')$, between two vertices t and t' is the length of a shortest path joining them. An easy consequence of (*) is:

(**) If P is a path of length d joining t and t', then $\operatorname{rank}(t \wedge t') \geqslant n - d$. Moreover, if t'' is a vertex in P such that the segment of P from t to t'' has length c, and if $\operatorname{rank}(t \wedge t'') > n - c$, then $\operatorname{rank}(t \wedge t') > n - d$.

We first prove the lemma for $m = 2n$. As T_{2n} contains $T_n \oplus T_n$, there exist t and t' in VG_{2n} with $t \wedge t' = \hat{0}$. By (**), $d(t, t') \geqslant n$. But G_{2n} has only $n + 1$ vertices, so it must be a path from t to t'. Moreover, if this path has vertex sequence $t = t^0, t^1, \ldots, t^n = t'$, then (**) implies $\operatorname{rank}(t^i \wedge t^j) = n - j + i$ for $0 \leqslant i \leqslant j \leqslant 2n$. This proves the lemma for $m = 2n$.

For $m < 2n$, we regard T_m as a flat of T_{2n}. Then G_m is a connected subgraph of G_{2n} and the result follows.

For $m > 2n$, we proceed by induction. If $t^1, \ldots, t^{m-n}$ are the vertices of a connected subgraph of G_m, then their span (in T_m) is isomorphic to T_{m-1}. (Induction on distance shows that the rank of the span is at most $m - 1$; on the other hand, Lemma 12(b) shows that their union cannot be contained in a proper subgeometry of T_{m-1}.) Thus (by induction) every connected proper subgraph of G_m is a path, so that G_m is either a path or a cycle. But if G_m is a cycle, say $t^0, \ldots, t^{m-n}$, then our inductive hypothesis applied to the paths $t^1, \ldots, t^{m-n}$ and t^{m-n}, t^0, t^1 gives $0 = \operatorname{rank}(t^1 \wedge t^{m-n}) = n - 2$, a contradiction. It follows that G_m is a path, say $t^0, \ldots, t^{m-n}$. Since each of the (two) paths of length $m - 1$ in G_m is the graph of a flat of T_m isomorphic to T_{m-1}, we also obtain the equality

$$\operatorname{rank}(t^i \wedge t^j) = \max\{n - j + i, 0\}$$

by induction for all pairs (i, j) other than $(0, m)$. To see the equality in the outstanding case, let $t_{m-1} \subseteq T_m$ be the flat spanned by $t^1, \ldots, t^{m-n}$. Then $t^0 \wedge t_{m-1} = t^0 \wedge t^1$, so that $t^0 \wedge t^{m-n} \subseteq t^1 \wedge t^{m-n} = \hat{0}$. □

We are now in a position to prove that n equals two. Before proceeding, we observe that for $n = 2$, we do obtain a variety. Define the geometries $O_m(q)$ recursively by: $O_2(q)$ is a line with $q + 1$ points; $O_m(q)$ is the union of a copy t_{m-1} of $O_{m-1}(q)$ and a copy t_2 of the line $O_2(q)$ taken in such a way that the intersection $t_{m-1} \wedge t_2$ is a point and t_{m-1} contains a copy of $O_{m-2}(q)$ disjoint from t_2. It is easy to deduce from Lemmas 12 and 13 that this sequence is the only sequence of universal models for which $|T_2| = q + 1$ and T_3 splits.

Another way to describe $O_n(q)$ is to take the Boolean algebra on the point set $\{1, 2, \ldots, n\}$. On each of the lines $\overline{12}, \overline{23}, \ldots, \overline{i(i+1)}, \ldots, \overline{(n-1)n}$, add $q - 1$ points in general position. The resulting geometry is the geometry $O_n(q)$ defined above. The geometries $O_n(q)$ are called the *full origami geometries of order q*. Their subgeometries form a variety called the *variety of origami geometries of order q*. Note that the variety of free geometries is just the variety of origami geometries of order one.

It remains to show that n cannot be greater than two. First we fix some notation. Let $m = |T_{n-1}| + 2n - 1$ and let $T_m = t^0 \cup \cdots \cup t^{m-n}$ (notation as in Lemma 13). Further, for $1 \leqslant i \leqslant m - n$, let $t^i_{n-1} = t^{i-1} \cap t^i$; for $1 \leqslant i \leqslant m - n - 1$, let $t^i_{n-2} = t^{i-1} \cap t^i \cap t^{i+1}$. Note that t^i_{n-1} is isomorphic to T_{n-1} and t^i_{n-2} is isomorphic to T_{n-2}.

Let us call a point x of t^i_{n-1} an *exceptionable point* if

(e) there is a (possibly empty) set of points (called a set of *foci* for x) $\{z_1,\ldots,z_k\}$ in $T_m \setminus \bigcup_{j=i}^{m-n} t^j$ such that there are n (distinct) copies of T_n containing the image of x in the contraction $T_m/\{z_1,\ldots,z_k\}$.

Note that, in any geometry T in the variety $\mathfrak{T}$, no point can be on more than n distinct copies of T_n. This follows from the fact that, by Lemma 13, any set of n copies of T_n having nonempty intersection must be of the form $\{t^i,\ldots,t^{i+n-1}\}$. For the same reason, a point x is on n copies of T_n in T_m itself if and only if it is the unique point in the intersection of the flats $t^i,\ldots,t^{i+n-1}$ for some i between 0 and $m-2n+1$. We call such exceptionable points x *n-points*.

Assume that $n \geqslant 3$. We shall show that

(f) for $1 \leqslant i \leqslant m-2n+1$, the number of exceptionable points on t^{i+1}_{n-1} is strictly greater than the number of exceptionable points on t^i_{n-1}.

Of course, (f) leads to a contradiction: since t^1_{n-1} contains an exceptionable point (the intersection of $t^0,\ldots,t^{n-1}$), t^{m-2n+2}_{n-1} contains at least $m-2n+2$ exceptionable points. But by our choice of m, t^{m-2n+2}_{n-1} contains only $m-2n+1$ $(=|T_{n-1}|)$ points of any kind.

Let, then, $1 \leqslant i \leqslant m-2n+1$. Observe that there is a unique n-point, x_{i+1} say, which is in t^{i+1}_{n-1} but not in t^i_{n-2}. This point is the intersection of $t^i,\ldots,t^{i+n-1}$. We will have proved (f) if we can construct an injection from the set of exceptionable points in $t^i_{n-1} \setminus t^i_{n-2}$ to the set of exceptionable points which are not n-points in $t^{i+1}_{n-1} \setminus (t^i_{n-2} \cup \{x_{i+1}\})$. To this end, choose a point z in $t^i \setminus (t^i_{n-1} \cup t^{i+1}_{n-1})$. Such a point exists since T_n does not split. Consider the mapping $x \mapsto x := (x \vee z) \wedge t^{i+1}_{n-1}$; that is, x is mapped onto the intersection x' of the line $\overline{xz}$ and the modular flat t^{i+1}_{n-1}. This map is clearly an injection from $t^i_{n-1} \setminus t^i_{n-2}$ into $t^{i+1}_{n-1} \setminus t^i_{n-2}$. Let x be an exceptionable point in $t^i_{n-1} \setminus t^i_{n-2}$ and suppose that x' is on r copies of T_n: these must be the flats $t^i,\ldots,t^{i+r-1}$. Suppose that $\{z_1,\ldots,z_k\}$ is a set of foci for x. Then, in the contraction $T_m/\{z_1,\ldots,z_k, z\}$, the image of x' (which is also the image of x) is on $n+r-2$ copies of T_n. (For the image of x in $T_m/\{z_1,\ldots,z_k\}$ is on n copies of T_n and the image of x' in $T_m/\{z_1,\ldots,z_k\}$ is still on r copies of T_n—remember that the foci are points in $T_m \setminus \bigcup_{j=i}^{m-n} t^j$—and contraction by z destroys one copy each from the two collections of copies of T_n.) Since x' is in the intersection $t^i \wedge t^{i+1}$, $r \geqslant 2$. Hence, as $n+r-2 \leqslant n$, $r=2$. This implies that x' is indeed an exceptionable point (with a set of foci $\{z_1,\ldots,z_k, z\}$) and that x' is not an n-point (in particular, $x' \neq x_{i+1}$). Thus, $x \mapsto x'$ is an injection satisfying the required properties and we have proved (f). This completes the proof that $n=2$.

9. The main theorem. We have now proved the following theorem.

THEOREM 14. *Let $\mathfrak{T}$ be a variety of geometries with a sequence of universal models (T_n). Then $\mathfrak{T}$ is one of the following collections:*

A. *the variety of free geometries,*

B. *the variety of matchstick geometries of order q* (§4),

C. *the variety of origami geometries of order q* (§8),

D. *the variety of geometries coordinatizable over the finite field $GF(q)$,*

E. *the variety of voltage-graphic geometries with voltages in a fixed finite group A* (§7).

The corresponding universal models T_n are

A. *the Boolean algebra B_n,*

B. *the full matchstick geometry $M_n(q)$ of order q,*

C. *the full origami geometry $O_n(q)$ of order q,*

D. *the projective geometry $P_n(q)$ over the finite field $GF(q)$,*

E. *the Dowling geometry $Q_n(A)$ based on the finite group A.*

One can easily deduce further results from the theorem. For example, the only varieties of binary geometries are the free geometries, the matchstick and origami geometries of order two, the graphic geometries and all binary geometries.

A more interesting result which seems difficult to prove independently is

COROLLARY 15. *The only varieties which are closed under orthogonal duality are the varieties of free geometries and the varieties of geometries coordinatizable over a fixed finite field.*

The proof consists of checking that the other varieties are not closed under orthogonal duality. This is obvious for the matchstick and origami geometries (with $q > 1$). For voltage-graphic geometries, it is easy to show by a counting argument that the orthogonal dual of $Q_4(A)$ cannot be an A-labelled voltage-graphic geometry.

Our theorem says that in some sense the only nondegenerate examples of well-structured hereditary classes are the geometries coordinatizable over a finite field and voltage-graphic geometries. While the study of projective geometry over a finite field is a classical subject, the study of voltage-graphic geometries has just been initiated by Zaslavsky in [**14** and **15**]. Our present work would have been unimaginably more difficult had voltage-graphic geometries not been discovered.

Voltage-graphic geometries, like the coordinatizable geometries, can be described in an "economical" fashion. More precisely, to describe a voltage-graphic geometry, one need only specify a multigraph whose edges are labelled with the group elements, just as to describe a coordinatizable geometry, one need only specify a finite set of n-tuples of field elements. These descriptions are in general more compact (require less storage space in a computer) and easier to manipulate than purely set-theoretic descriptions such as a listing of all the circuits or bases. Thus, an important practical consequence of our theorem is that geometries belonging to a variety have compact descriptions.

What happens if we drop the restriction that our geometries be finite? Two unpleasant phenomena arise. The first is that (as we have defined it) a sequence of universal models need not be unique. For example, if k is a field which is isomorphic to a subfield of itself, the projective geometry $P_n(k)$ is embeddable in the affine geometry $A_n(k)$. The second, even harder to deal with, is that there are other varieties of geometries. Apart from the existence of varieties of geometries related to the origami geometries, we also encounter varieties of algebraic geometries as in [**9**]; these are geometries in which the dependence relation is given by algebraic dependence over a field. Over fields of positive characteristic, it is known [**13**, p. 185] that these varieties are different from the varieties of coordinatizable geometries. Moreover, the natural universal models are not even supersolvable.

We conclude with a question:

Can varieties of finite combinatorial geometries be characterised by a finite list of excluded minors?

That is to say, for any variety $\mathfrak{T}$ of geometries, does there exist a finite number of geometries $E_1, \ldots, E_n$ such that G is in the variety $\mathfrak{T}$ if and only if none of the geometries $E_1, \ldots, E_n$ occurs as a minor of G?

ACKNOWLEDGEMENT. The authors would like to thank the referee for his careful reading of the manuscript and many helpful suggestions.

REFERENCES

1. G. Birkhoff, *On the structure of abstract algebras*, Proc. Cambridge Philos. Soc. **31** (1935), 433–454.

2. ______, *Lattice theory*, 3rd ed., Amer. Math. Soc. Colloq. Publ., vol. 25, Amer. Math. Soc., Providence, R.I., 1967.

3. P. M. Cohn, *Universal algebra*, Harper and Row, New York, 1965.

4. H. H. Crapo and G.-C. Rota, *On the foundations of combinatorial theory: Combinatorial geometries* (preliminary edition), M.I.T. Press, Cambridge, 1970.

5. P. Doubilet, G.-C. Rota and R. Stanley, *On the foundations of combinatorial theory*. VI: *The idea of generating function*, Sixth Berkeley Sympos. on Math. Statist. and Prob., Vol. II: Probability Theory, Univ. California, Berkeley, Calif., 1972, pp. 267–318.

6. T. A. Dowling, *A q-analog of the partition lattice*, A Survey of Combinatorial Theory (J. N. Srivastava, editor), North-Holland, Amsterdam, 1973, pp. 101–115.

7. ______, *A class of geometric lattices based on finite groups*, J. Combin. Theory Ser. B **14** (1973), 61–86.

8. C. Greene, *Lectures in combinatorial geometries*, Notes from the NSF Seminar in Combinatorial Theory, Bowdoin College, 1971, unpublished.

9. S. Mac Lane, *A lattice formulation for transcendence degrees and p-bases*, Duke Math. J. **4** (1938), 455–468.

10. D. Kelly and G.-C. Rota, *Some problems in combinatorial geometry*, A Survey of Combinatorial Theory (J. N. Srivastava, editor), North-Holland, Amsterdam, 1973, pp. 309–312.

11. R. P. Stanley, *Modular elements of geometric lattices*, Algebra Universalis **1** (1971), 214–217.

12. ______, *Supersolvable lattices*, Algebra Universalis **2** (1972), 197–217.

13. D. J. A. Welsh, *Matroid theory*, Academic Press, London, 1976.

14. T. Zaslavsky, *Biased graphs*, preprint, 1977.

15. ______, *Signed graphs*, Discrete Appl. Math. **4** (1982), 47–74.

DEPARTMENT OF MATHEMATICS, MASSACHUSETTS INSTITUTE OF TECHNOLOGY, CAMBRIDGE, MASSACHUSETTS 02139

DEPARTMENT OF MATHEMATICS, NORTH TEXAS STATE UNIVERSITY, DENTON, TEXAS 76203

Subject index

GPSR Compliance

The European Union's (EU) General Product Safety Regulation (GPSR) is a set of rules that requires consumer products to be safe and our obligations to ensure this.

If you have any concerns about our products, you can contact us on ProductSafety@springernature.com

In case Publisher is established outside the EU, the EU authorized representative is:

Springer Nature Customer Service Center GmbH
Europaplatz 3
69115 Heidelberg, Germany

Batch number: 10577969

Printed by Printforce, the Netherlands